I0882661

AI-Driven Innovations for Smart and Resilient Infrastructure

Sérgio António Neves Lousada
https://orcid.org/0000-0002-8429-2164
University of Madeira, Portugal

IGI Global
Scientific Publishing
Publishing Tomorrow's Research Today

Vice President of Editorial	Melissa Wagner
Director of Acquisitions	Mikaela Felty
Director of Book Development	Jocelynn Hessler
Production Manager	Mike Brehm
Cover Design	Jose Rosado

Published in the United States of America by
IGI Global Scientific Publishing
701 East Chocolate Avenue
Hershey, PA, 17033, USA
Tel: 717-533-8845 | Fax: 717-533-7115
Website: https://www.igi-global.com E-mail: cust@igi-global.com

Copyright © 2027 by IGI Global Scientific Publishing. All rights reserved. No part of this publication may be reproduced, stored or distributed in any form or by any means, electronic or mechanical, including photocopying, without written permission from the publisher. Use of this publication to train generative artificial intelligence (AI) technologies is expressly prohibited. The publisher reserves all rights to license its use for generative AI training and machine learning model development.

Product or company names used in this set are for identification purposes only. Inclusion of the names of the products or companies does not indicate a claim of ownership by IGI Global Scientific Publishing of the trademark or registered trademark.

Library of Congress Cataloging-in-Publication Data

LCCN: 2026032040 (CIP Data Pending)
ISBN13: 979-8-2600-3038-7
Isbn13Softcover: 979-8-2600-3039-4
EISBN13: 979-8-2600-3040-0
British Cataloguing in Publication Data
A Cataloguing in Publication record for this book is available from the British Library.

All work contributed to this book is new, previously-unpublished material.
The views expressed in this book are those of the authors, but not necessarily of the publisher.
This book contains information sourced from authentic and highly regarded references, with reasonable efforts made to ensure the reliability of the data and information presented. The authors, editors, and publisher believe the information in this book to be accurate and true as of the date of publication. Every effort has been made to trace and credit the copyright holders of all materials included. However, the authors, editors, and publisher cannot assume responsibility for the validity of all materials or the consequences of their use. Should any copyright material be found unacknowledged, please inform the publisher so that corrections may be made in future reprints.

Table of Contents

Detailed Table of Contents

This chapter presents a comprehensive artificial intelligence (AI) framework for optimizing climate-resilient infrastructure through integration with nature-based solutions (NbS). The methodology combines predictive analytics, multi-objective optimization, and digital twin technologies to address complex challenges in infrastructure design under climate uncertainty. Through a detailed coastal city case study, the framework demonstrates 55% reduction in flood risk, 30% lower embodied carbon, and 25% cost savings compared to conventional approaches. The chapter provides practitioners with scalable tools for implementing AI-driven resilience strategies while addressing ethical considerations in algorithmic decision-making. Key innovations include hybrid physics-AI modeling, adaptive reinforcement learning systems, and explainable AI interfaces for infrastructure planning.

Enhancing the structural design and construction safety load-carrying capacity of composite concrete-steel tubes (CCSTs) is important. Traditionally, the relationship between steel and concrete depends on a design standard prone to error. This work employs machine learning (ML) to predict the ultimate load-carrying capacity of CCSTs using material, structural, and geometric factors as input features. Important features, including outer diameter, steel thickness, column length, material strengths, and confinement effects, are included in the model to get ready for non-linear connections. Different machine learning techniques like random forest, ADA boost, and regression gradient boosting are examined in order to identify the most reliable and accurate predictor. Metrics R2, RMSE, and MAE demonstrate the model's exceptional prediction accuracy. The model is trained and verified using both theoretical computations and experimental data.

The growing demands of climate resilience, urban expansion, and rapid technological change are transforming civil infrastructure. Traditional models can no longer meet the needs of a dynamic world facing environmental risks and resource constraints. This chapter explores how IoT, sensors, and real-time data enable smart, adaptive systems. Embedded sensors in key assets allow engineers to collect detailed data on structural and environmental conditions, supporting predictive maintenance and early fault detection. AI and digital twin technologies enhance this data, offering simulations and decision tools to optimize infrastructure performance. Ethical and cybersecurity concerns are also addressed to ensure data integrity and system resilience. Through global case studies and updated educational strategies, the chapter highlights the need to train engineers with digital and interdisciplinary skills. Smart infrastructure presents a forward-looking solution to improving sustainability, safety, and urban resilience in the 21st century.

Virtual simulations become smart city and urban planning applications with digital twin technology. The essential stages from virtual to real-world deployment are studied in this urban digital twin development research. Integrating sensor, satellite, and municipal record data creates a digital city model. Virtual models allow simulation and modelling. It helps stakeholders imagine traffic management and sustainability. Through real-time monitoring and input, the digital twin correctly depicts the city. Urban planners and legislators can utilise current facts to decide.

of the main uses are the surveillance of traffic, the anticipation of demands, cross-modality, low-carbon and accessibility improvements for the elderly, the disabled, and underprivileged groups. The Singapore, Amsterdam, Copenhagen, and Bengaluru case studies give a real-world influence and shed light on efficiency improvement, environmental issues, and social inclusion. Lessons focus on adaptive learning, data integration, and implementation in context. Future considerations are also identified in the chapter, detailing how an AI-based system for transporting the population will work in the future, including on a more sustainable and inclusive scale.

This chapter discusses how Artificial Intelligence (AI) can be used to develop unified transport systems that are sustainable and socially fair. It emphasizes how AI technologies, including machine learning, computer vision, predictive analytics, and natural language processing, can be used in real-time optimization, demand prediction, and individualized mobility. Based on international case studies, it demonstrates such applications as optimization of public transport schemes, safety in active mobility, and rural accessibility. The challenges such as data governance, algorithmic bias, and interoperability are also present within the chapter, and governance models, collaboration frameworks, and open data standards are offered to reduce risks. Through its AI innovation to inclusive design and strong policy pathways, it provides the solutions to scale low-carbon, accessible, and resilient mobility systems with an emphasis that future AI-enabled transport must be people-focused and environmentally sustainable.

Artificial Intelligence including ChatGPT has attracted the attention of people quickly in the recent years. It could be easily noticed that the day to day lives of people have been drastically affected by it and people's reliance over it for almost everything has grown tremendously. The same can be said for various sectors, and it could be noticed that nowadays real estate is also being driven by Artificial Intelligence. This chapter will discuss the ways AI is affecting real estates and how people are relying on it to find suitable properties for them. Further, the way real estate sector using AI to improve their customer services will be talked about. Then, the way Artificial Intelligence could help in the management of properties will be explained in the book chapter. Some graphical analyses will be performed to support the study and to understand the current trends.

This study examines the crucial role of financial literacy in enhancing the organizational sustainability of green and renewable energy enterprises in Mexico. As the country accelerates its transition towards cleaner energy sources, financial management skills and awareness among stakeholders become vital for securing investments, managing risks, and ensuring long-term viability. The research investigates the link between financial literacy levels and sustainable business practices in small and medium-sized renewable energy firms. A mixed-methods approach is employed, combining survey data from industry professionals with case studies of successful Mexican renewable energy projects. Findings reveal that higher financial literacy contributes significantly to better strategic planning, efficient resource allocation, and stronger stakeholder engagement, which collectively bolster organizational sustainability. The study highlights the need for targeted financial education programs and supportive policy frameworks to empower entrepreneurs and managers in this sector.

This chapter explores the implementation of metaheuristic optimization techniques to support decision-making in sustainable dairy farming systems. By applying methods such as genetic algorithms, particle swarm optimization, and other bio-inspired approaches, the authors aim to address challenges related to feed efficiency, profitability, and environmental sustainability. The chapter outlines a multi-objective framework integrating economic and ecological indicators and provides insights into real-world applications in livestock production. This chapter contributes to the growing field of computational agriculture by highlighting how intelligent

Preface

INTRODUCTION

The field of civil engineering is undergoing a profound transformation driven by the urgent need to develop sustainable, resilient, and intelligent infrastructure systems. In response to escalating environmental pressures, increasing climate-related risks, and rapid technological advancement, the discipline is evolving beyond conventional design and construction paradigms toward more adaptive, data-driven, and innovation-led approaches. Within this context, artificial intelligence and emerging digital technologies are playing an increasingly pivotal role in reshaping how infrastructure is conceptualized, developed, monitored, and maintained. This evolution reflects a broader commitment to long-term sustainability, climate resilience, and the creation of smart, resource-efficient, and inclusive built environments.

This edited volume, *AI-Driven Innovations for Smart and Resilient Infrastructure*, aims to provide a comprehensive and forward-thinking exploration of the transformative innovations and methodologies that are shaping the future of civil engineering. Its primary objective is to bridge the gap between theoretical advancements and real-world applications by demonstrating how artificial intelligence, machine learning, data analytics, and digital infrastructure tools are redefining traditional practices in infrastructure planning, design, construction, and management.

By bringing together contributions from researchers, academics, and industry practitioners, this book highlights recent developments in areas such as AI-enabled design optimization, predictive maintenance, intelligent construction systems, digital twins, smart monitoring technologies, and data-driven decision-making frameworks. These contributions collectively reflect the rapid expansion of AI applications within civil engineering and underscore their potential to enhance efficiency, safety, resilience, and sustainability across infrastructure systems.

This volume serves both as a reference work and a source of inspiration for understanding how AI-driven innovations can support the achievement of sustainable development goals while strengthening urban resilience and infrastructure performance. It presents applied case studies, analytical insights, and emerging methodologies that extend the boundaries of civil engineering practice and research in the era of digital transformation. In doing so, it highlights the importance of integrating computational intelligence with engineering expertise to address complex, multi-dimensional challenges facing modern societies.

The intended audience for this book is broad yet specialized, encompassing academicians, researchers, postgraduate students, engineers, policymakers, urban planners, computer scientists, and professionals engaged in infrastructure development and smart city initiatives. By fostering a deeper understanding of the opportunities and challenges associated with AI integration in civil engineering, this volume aims to support both scholarly advancement and practical implementation. Ultimately, it seeks to encourage interdisciplinary collaboration, knowledge exchange, and informed decision-making in support of resilient, intelligent, and future-ready infrastructure systems.

ORGANIZATION OF THE BOOK

Chapter 1 presents an advanced artificial intelligence framework for optimizing climate-resilient infrastructure through the integration of nature-based solutions, predictive analytics, and digital twin technologies. The chapter demonstrates how hybrid modeling approaches, combining physics-based and AI-driven systems, can effectively address uncertainty in infrastructure design and climate adaptation planning. Through a coastal city case study, it illustrates significant improvements in flood risk reduction, embodied carbon mitigation, and cost efficiency compared to conventional methods. In addition, it highlights the role of adaptive reinforcement learning and explainable AI in supporting transparent and scalable infrastructure decision-making, while also acknowledging the ethical considerations inherent in algorithmic planning systems.

Chapter 2 explores the application of machine learning techniques to predict the load-carrying capacity of composite concrete-steel tubes, addressing limitations in traditional design standards. By incorporating structural, material, and geometric variables such as steel thickness, column length, and confinement effects, the study develops predictive models capable of capturing nonlinear structural behavior. Several algorithms, including random forest, AdaBoost, and gradient boosting regression, are evaluated for their accuracy and reliability. The findings demonstrate strong predictive performance across key statistical metrics, underscoring the potential of machine learning to enhance structural safety, optimize design processes, and reduce uncertainty in engineering calculations.

Chapter 3 examines the role of IoT devices, embedded sensors, and real-time data systems in enabling intelligent infrastructure health monitoring and proactive risk management. The chapter emphasizes how continuous data collection from structural and environmental conditions supports predictive maintenance and early fault detection in critical infrastructure assets. It further highlights the integration of AI and digital twin technologies to simulate performance scenarios and enhance decision-making capabilities. In addition to technical advancements, the discussion addresses cybersecurity and ethical considerations, while also emphasizing the need for multidisciplinary education to prepare engineers for the evolving landscape of smart infrastructure systems.

Chapter 4 focuses on the development and application of digital twin technology for smart cities, emphasizing its role in bridging the gap between virtual models and real-world urban systems. By integrating heterogeneous data sources such as sensors, satellite imagery, and municipal records, digital twins create dynamic representations of urban environments that support simulation, prediction, and optimization. The chapter highlights how machine learning enhances urban analytics by identifying patterns in mobility, infrastructure usage, and sustainability indicators. It further emphasizes the value of real-time feedback loops in maintaining model accuracy and enabling data-driven policymaking for resilient and adaptive urban development.

Chapter 5 investigates the integration of IoT and artificial intelligence in advancing sustainable urban planning and green infrastructure development. The chapter highlights how real-time data analytics can optimize resource consumption, improve environmental monitoring, and enhance urban service delivery across applications such as smart transportation systems, energy networks, and waste management. It also explores the synergy between digital technologies and nature-based solutions, including urban forests and green roofs, in strengthening climate resilience. At the same time, it addresses critical concerns related to data privacy, cybersecurity, and ethical governance, emphasizing the need for inclusive and well-regulated smart city frameworks.

Chapter 6 introduces a cognitive AI framework for intelligent water management in buildings, designed to address the increasing challenges of water scarcity and aging infrastructure. The system integrates real-time sensor data with machine learning models such as LSTM autoencoders for anomaly detection, Prophet for demand forecasting, and ensemble methods including XGBoost and Random Forest for water quality monitoring. A reinforcement learning layer enables adaptive decision-making based on historical and real-time inputs. Collectively, the framework enables early detection of leaks, prediction of consumption trends, and identification of contamination risks, offering a comprehensive AI-driven solution for sustainable urban water management.

Chapter 7 presents a data-driven approach for assessing urban pressure in coastal municipalities, using Machico, Madeira as a case study to evaluate the impacts of short-term rentals on housing and infrastructure systems. By integrating census data, tourism registries, and mobility studies, the chapter develops a quantifiable STR pressure ratio to measure tourism intensity and its spatial effects. The analysis reveals significant correlations between tourism clusters and infrastructure strain, particularly in parking and public service demand. The study proposes a replicable and cost-effective analytical model that supports evidence-based policymaking and helps balance economic benefits of tourism with long-term urban sustainability and resilience.

Chapter 8 explores the transformative potential of artificial intelligence and data-driven systems in developing sustainable and inclusive urban transport networks. The chapter contrasts traditional transport planning approaches with AI-enhanced solutions that enable real-time traffic monitoring, demand forecasting, and multi-modal optimization. It highlights how these technologies improve accessibility for marginalized populations while supporting low-carbon mobility systems. Case studies from global cities illustrate improvements in efficiency, environmental performance, and social equity. The discussion emphasizes the importance of adaptive learning systems, integrated data platforms, and contextual implementation strategies to ensure equitable and sustainable transport development.

Chapter 9 examines the application of artificial intelligence in designing integrated transport systems that are both sustainable and socially equitable. The chapter highlights the use of machine learning, predictive analytics, and computer vision to optimize mobility networks, enhance safety, and improve accessibility across diverse populations. International case studies demonstrate practical implementations in public transit optimization, active mobility safety, and rural connectivity enhancement. It also addresses critical challenges such as algorithmic bias, data governance, and system interoperability, proposing policy frameworks and collaborative governance models to mitigate risks. The chapter ultimately advocates for people-centered, low-carbon transport systems supported by transparent and inclusive AI deployment.

Chapter 10 discusses the growing influence of artificial intelligence in transforming the real estate sector, particularly in property search, customer engagement, and asset management. The chapter highlights how AI-driven platforms and tools, including conversational systems, are reshaping consumer behavior and improving decision-making processes in property transactions. It further explores the application of AI in market analysis, trend identification, and operational efficiency within real estate organizations. Through graphical analysis and emerging digital trends, the chapter illustrates the increasing dependence of the sector on intelligent systems, while also reflecting on the broader implications of automation in real estate services.

Chapter 11 investigates the relationship between financial literacy and organizational sustainability within Mexico's renewable energy sector. The chapter emphasizes how financial knowledge among stakeholders influences investment decisions, risk management, and long-term operational viability in green energy enterprises. Using a mixed-methods approach, it combines survey data and case studies to demonstrate that higher financial literacy correlates with improved strategic planning, efficient resource allocation, and stronger stakeholder engagement. The findings underscore the importance of targeted financial education and supportive policy environments to strengthen the sustainability and competitiveness of renewable energy organizations.

Chapter 12 explores the application of metaheuristic optimization techniques in sustainable dairy farming systems, focusing on improving feed efficiency, profitability, and environmental performance. The chapter introduces a multi-objective decision-making framework that integrates economic and ecological indicators using algorithms such as genetic algorithms and particle swarm optimization. These computational approaches enable more efficient resource allocation and optimized livestock management strategies. The discussion positions metaheuristic methods as valuable tools in computational agriculture, offering practical solutions for enhancing productivity while maintaining environmental sustainability in dairy operations.

Chapter 13 presents a systems-based perspective on emerging civil engineering strategies that integrate smart technologies, circular economy principles, and adaptive design for urban resilience. The chapter emphasizes the need for interdisciplinary approaches to address the combined pressures of urbanization, resource constraints, and climate change. It proposes a holistic analytical framework that connects digital intelligence, lifecycle optimization, and climate-responsive infrastructure design. By examining governance and implementation challenges in rapidly urbanizing contexts, the chapter highlights how integrated infrastructural systems can support the development of sustainable, resilient, and livable urban environments.

CONCLUSION

The chapters presented in *AI-Driven Innovations for Smart and Resilient Infrastructure* collectively illustrate a decisive transformation in civil engineering and related disciplines, where artificial intelligence, data-driven systems, and intelligent automation are becoming central to the conception, design, and management of infrastructure. Across diverse applications—from structural optimization and construction monitoring to urban systems, mobility networks, water management, and real estate—the contributions consistently demonstrate how AI-enabled methodologies are reshaping traditional engineering paradigms into adaptive, predictive, and highly integrated systems.

A clear thematic convergence emerges throughout this volume: infrastructure is increasingly understood as a dynamic, data-rich ecosystem rather than a static physical construct. The integration of IoT sensors, digital twins, machine learning models, and optimization algorithms enables continuous feedback loops between the physical and virtual worlds, allowing for real-time monitoring, simulation, and decision-making. This shift not only enhances operational efficiency and structural safety but also strengthens the capacity of infrastructure systems to respond to uncertainty, climate variability, and evolving societal demands.

Equally significant is the emphasis on sustainability and inclusivity as foundational principles guiding AI adoption in infrastructure development. Whether addressing climate-resilient design, equitable mobility systems, sustainable urban planning, or resource-efficient agriculture, the contributions highlight the importance of aligning technological innovation with broader environmental, economic, and social objectives. At the same time, the volume acknowledges the critical challenges associated with this transformation, including data governance, cybersecurity risks, algorithmic bias, and ethical considerations in automated decision-making.

As civil engineering continues to evolve in the era of digital transformation, this book serves as both a reference point and a forward-looking guide. It brings together emerging research and applied insights that collectively point toward a future where infrastructure systems are intelligent, interconnected, and responsive to both human and environmental needs. It is our expectation that the ideas and methodologies presented herein will stimulate further interdisciplinary research, foster responsible innovation, and encourage the development of resilient infrastructure systems capable of sustaining the complexities of the twenty-first century and beyond.

Chapter 1
AI-Driven Optimization for Climate-Resilient Infrastructure

Anna Kovács
http://orcid.org/0009-0001-6323-2808
University of Miskolc, Hungary

ABSTRACT

This chapter presents a comprehensive artificial intelligence (AI) framework for optimizing climate-resilient infrastructure through integration with nature-based solutions (NbS). The methodology combines predictive analytics, multi-objective optimization, and digital twin technologies to address complex challenges in infrastructure design under climate uncertainty. Through a detailed coastal city case study, the framework demonstrates 55% reduction in flood risk, 30% lower embodied carbon, and 25% cost savings compared to conventional approaches. The chapter provides practitioners with scalable tools for implementing AI-driven resilience strategies while addressing ethical considerations in algorithmic decision-making. Key innovations include hybrid physics-AI modeling, adaptive reinforcement learning systems, and explainable AI interfaces for infrastructure planning.

INTRODUCTION

The escalating impacts of climate change present unprecedented challenges to civil infrastructure systems worldwide. Extreme weather events, sea-level rise, and urbanization pressures necessitate transformative approaches to infrastructure design and management (IPCC, 2022). According to the Global Infrastructure Hub, climate-related infrastructure failures cost approximately $300 billion annually,

DOI: 10.4018/979-8-2600-3038-7.ch001

t © 2027, IGI Global Scientific Publishing. Copying or distributing in print or electronic forms without written permission of IGI Global Scientific Publishing is prohibited. Use of this chapter to enerative artificial intelligence (AI) technologies is expressly prohibited. The publisher reserves all rights to license its use for generative AI training and machine learning model development.

with projections indicating a potential increase to $1 trillion per year by 2050 under business-as-usual scenarios (GIH, 2023). Traditional engineering methods, while effective for known conditions, often fail to address the complex, interconnected risks posed by climate uncertainty (Ayyub, 2014). Simultaneously, the integration of nature-based solutions (NbS) offers promising pathways for enhancing resilience while providing co-benefits for biodiversity and human well-being (Cohen-Shacham et al., 2016).

Recent advances in artificial intelligence (AI) present opportunities to revolutionize infrastructure planning and management. Machine learning algorithms can process vast datasets to identify patterns and predict system behaviors under various climate scenarios (Bergmeister et al., 2020). When combined with NbS, these technologies enable the development of infrastructure systems that adaptively respond to changing conditions while minimizing environmental impacts (Kabisch et al., 2022). The convergence of AI and sustainable infrastructure represents a paradigm shift from static, deterministic design approaches to dynamic, adaptive systems capable of learning and evolving (Haleem et al., 2021).

This chapter presents a comprehensive framework for AI-driven optimization of climate-resilient infrastructure that integrates predictive analytics with NbS, addressing critical gaps in current practice. The framework addresses four fundamental challenges in contemporary infrastructure engineering: (1) predictive limitations of traditional models under climate uncertainty, (2) optimization complexity in balancing competing objectives, (3) data fragmentation across disciplines and institutions, and (4) implementation barriers in translating innovations to practice. Through a detailed case study application in a coastal urban setting, we demonstrate the framework's effectiveness in achieving significant improvements in resilience, sustainability, and cost-effectiveness.

2. PROBLEM STATEMENT

Current infrastructure design approaches face significant limitations in addressing climate resilience challenges. This section provides a detailed analysis of four critical problem areas that motivate the development of our AI-driven framework (Farrokhi et al., 2024).

2.1 Predictive Limitations

Conventional infrastructure design relies heavily on historical data and stationary assumptions, which are increasingly invalid under climate change (Milly et al., 2008). Engineering models often fail to capture the complex interactions between gray infrastructure and NbS, particularly under extreme conditions (Depietri & McPhearson, 2017). A comprehensive analysis of 127 infrastructure projects worldwide revealed that 68% experienced unanticipated performance issues when exposed to climate conditions beyond design parameters (World Bank, 2021). These failures stem from several fundamental limitations:

- **Stationarity Assumptions**: Traditional hydrological and structural models assume climate stationarity, leading to systematic underestimation of future risks (Milly et al., 2008). For example, drainage systems designed using historical rainfall data are increasingly overwhelmed by more intense precipitation events.
- **Linear Extrapolation**: Many engineering models use linear extrapolation of historical trends, failing to capture non-linear climate feedback loops and tipping points (Lenton et al., 2019). This approach proved inadequate during Hurricane Sandy (2012), where storm surge exceeded 500-year design levels by 40% in some areas.
- **Siloed Modeling**: Infrastructure systems are typically modeled in isolation, neglecting critical interactions between natural and built systems (Ahern, 2011). For instance, flood control infrastructure often fails to account for wetland degradation that amplifies downstream impacts (Jamali et al., 2024).
- **Computational Constraints**: High-fidelity physics-based models require significant computational resources, limiting their use in exploring multiple climate scenarios and design alternatives (Bergmeister et al., 2020). A recent survey found that 78% of engineering firms lack the computational capacity for ensemble climate modeling (ASCE, 2020).

These limitations have resulted in significant economic and social costs. The OECD estimates that climate-related infrastructure failures cost $300 billion annually globally, with projections reaching $1 trillion by 2050 under current trends (OECD, 2022). Moreover, these failures disproportionately affect vulnerable communities, exacerbating existing social inequities (Shi et al., 2016).

2.2 Optimization Complexity

Balancing competing objectives in infrastructure design, including resilience, cost, environmental impact, and social equity—remains a heuristic process (Fenner et al., 2020). Multi-criteria decision analysis methods often struggle with high-dimensional problems involving numerous interdependent variables (Wang et al., 2021). This complexity frequently leads to suboptimal solutions that prioritize short-term cost savings over long-term resilience (Anderies et al., 2016). Key challenges include:

- **High-Dimensional Decision Space**: Infrastructure systems involve thousands of design variables across multiple scales (material selection, structural configuration, spatial layout, temporal phasing). A typical urban drainage system may have over 10,000 decision variables when considering pipe diameters, storage capacities, and control strategies (Möderl et al., 2009).
- **Conflicting Objectives**: Resilience objectives often conflict with cost and environmental goals. For example, increasing flood protection through larger infrastructure typically increases embodied carbon and reduces ecological connectivity (Ahern, 2011). These trade-offs are particularly challenging when incorporating NbS, which provide multiple ecosystem services but may have uncertain performance under extreme conditions.
- **Non-Linear Interactions**: Infrastructure components exhibit complex non-linear interactions that are difficult to model and optimize (Jamali et al., 2018). For instance, the performance of green roofs depends on interactions between soil properties, vegetation type, climate conditions, and building characteristics (Berardi et al., 2014).
- **Deep Uncertainty**: Climate projections involve deep uncertainty regarding future emission scenarios, climate sensitivity, and regional impacts (Lempert et al., 2018). Traditional optimization methods struggle to perform well under such conditions, often leading to brittle solutions that fail when assumptions are violated.

These challenges are compounded by institutional factors. Engineering firms typically use standardized design approaches that prioritize risk aversion and regulatory compliance over innovation and optimization (Meerow et al., 2016). A survey of 200 infrastructure projects found that only 15% used formal optimization methods, with most relying on rules-of-thumb and past experience (Fenner et al., 2020).

2.3 Data Fragmentation

Infrastructure planning involves diverse data sources, environmental, structural, social, and economic, that are typically siloed across institutions (Batty, 2018). This fragmentation hinders holistic assessment of infrastructure performance and resilience (Kitchin, 2014). A survey of 45 metropolitan planning organizations found that 73% lacked integrated data systems for climate resilience planning (ASCE, 2020). Specific fragmentation issues include:

- **Institutional Silos**: Environmental data (e.g., from environmental agencies), infrastructure data (e.g., from public works departments), and social data (e.g., from planning departments) are rarely integrated. For example, flood risk assessments often neglect land use planning data that affects exposure and vulnerability (White et al., 2018).

- **Temporal and Spatial Mismatches**: Data are collected at different temporal (e.g., real-time sensor data vs. decadal census data) and spatial scales (e.g., building-level vs. watershed-level), making integration challenging (Kitchin, 2014). These mismatches are particularly problematic for modeling climate impacts that operate at multiple scales.
- **Data Quality Issues**: Infrastructure data often suffer from inconsistencies, missing values, and errors. A study of municipal asset databases found that 35% of infrastructure records contained significant errors (Batty, 2018). These issues are exacerbated in developing regions where data collection capacity is limited.
- **Proprietary Restrictions**: Critical data are often held by private entities (e.g., utility companies) under proprietary restrictions, limiting access for public planning (Batty, 2018). For example, detailed building energy data needed for urban heat island modeling are rarely available to researchers.

The consequences of data fragmentation are significant. A World Bank analysis found that integrated data systems could reduce infrastructure planning costs by 20-30% while improving outcomes by 15-25% (World Bank, 2021). Moreover, fragmentation hinders the development of the comprehensive datasets needed for training AI models, creating a vicious cycle that perpetuates outdated planning approaches.

2.4 Implementation Barriers

Even when innovative designs are developed, implementation often faces practical constraints including regulatory barriers, funding limitations, and technical capacity gaps (Meerow et al., 2016). The gap between research and practice in resilient infrastructure remains significant, with an estimated 10-15 year lag between innovation development and widespread adoption (NRC, 2022). Key implementation barriers include:

- **Regulatory Frameworks**: Building codes and infrastructure standards are often outdated and slow to incorporate new technologies and approaches (NRC, 2022). For example, many jurisdictions still require gray infrastructure solutions for stormwater management despite evidence of NbS effectiveness (Dietz, 2007).
- **Funding Mechanisms**: Infrastructure funding typically favors large, capital-intensive projects over distributed, nature-based solutions (Meadows et al., 2018). Moreover, funding cycles are often misaligned with the long-term na-

ture of climate resilience investments, creating perverse incentives for short-term fixes.

- **Technical Capacity**: Many municipalities lack the technical capacity to implement innovative approaches, particularly those involving AI and complex modeling (Bergmeister et al., 2020). A survey found that 65% of cities lack staff with expertise in climate resilience modeling (ASCE, 2020).
- **Risk Aversion**: Infrastructure owners and operators are often risk-averse, preferring proven technologies over innovative approaches with uncertain performance (Meerow et al., 2016). This risk aversion is reinforced by liability concerns and professional norms that prioritize avoiding failure over maximizing benefits.

These barriers create a significant implementation gap. Despite growing recognition of the need for innovative resilience solutions, adoption rates remain low. A global assessment found that only 12% of infrastructure projects incorporate advanced resilience measures, with even fewer using AI or NbS (GIH, 2023). This gap is particularly concerning given the urgency of climate adaptation needs.

3. AI-DRIVEN METHODOLOGY

The proposed framework addresses these limitations through a three-phase approach that integrates AI technologies with engineering expertise and NbS principles (Farrokhi et al., 2024). This section provides a detailed description of each phase, including technical methodologies, algorithms, and implementation considerations (Samavat et al., 2025).

3.1 Phase 1: Predictive Modeling

The foundation of the framework (Figure 1) is a predictive modeling system that combines physics-based simulations with machine learning algorithms to forecast infrastructure performance under various climate scenarios. This hybrid approach leverages the strengths of both physics-based models (interpretability, physical consistency) and machine learning (pattern recognition, computational efficiency) (Karpatne et al., 2017).

Figure 1. AI-driven predictive modeling framework integrating multiple data sources and modeling approaches. The system processes heterogeneous data streams through physics-based models and machine learning algorithms, with feedback loops for continuous improvement

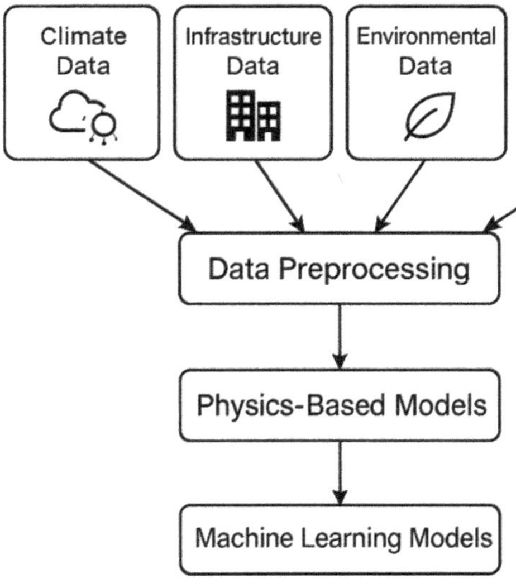

3.1.1 Data Integration and Preprocessing

The predictive modeling system integrates multiple data sources:

- **Climate Data**: Historical weather records, climate model projections (e.g., IPCC CMIP6), and downscaled regional climate scenarios (Jamali et al., 2025). These data are bias-corrected and statistically processed to account for uncertainties in climate projections (IPCC, 2022).
- **Infrastructure Data**: Asset inventories, condition assessments, and performance records for built infrastructure. These data are standardized using ontologies to ensure interoperability across different asset types and jurisdictions (Bergmeister et al., 2020).
- **Environmental Data**: Topographic, hydrologic, soil, and ecological data characterizing natural systems. High-resolution LiDAR and satellite imagery are processed to extract relevant features (Kabisch et al., 2022).
- **Social and Economic Data**: Demographic, land use, and economic data characterizing exposure and vulnerability. These data are geospatially referenced to enable spatial analysis (Shi et al., 2016).

- **Real-Time Sensor Data**: IoT sensor networks providing real-time monitoring of infrastructure performance and environmental conditions. These data streams are processed using edge computing to enable real-time decision support (Bergmeister et al., 2020).

Data preprocessing involves several critical steps:

1. **Data Cleaning**: Identification and correction of errors, inconsistencies, and missing values using statistical methods and domain knowledge.
2. **Data Fusion**: Integration of heterogeneous data sources using spatiotemporal alignment and semantic interoperability techniques.
3. **Feature Engineering**: Extraction of relevant features from raw data using domain knowledge and automated feature selection algorithms.
4. **Uncertainty Quantification**: Assessment and propagation of uncertainties through the modeling process using probabilistic methods.

3.1.2 Hybrid Physics-AI Modeling

The core of the predictive system is a hybrid modeling approach that combines physics-based simulations with machine learning algorithms. (Mahdavi et al., 2024) This approach addresses key limitations of purely data-driven methods, including poor extrapolation performance and violation of physical constraints (Karpatne et al., 2017).

For infrastructure performance prediction, we use Long Short-Term Memory (LSTM) neural networks enhanced with physics-informed constraints (Jamali et al., 2024). The LSTM architecture is particularly suited for time-series prediction due to its ability to capture long-term dependencies (Hochreiter & Schmidhuber, 1997). The physics-informed constraints are incorporated through:

- **Physics-Based Loss Functions**: Modification of the loss function to include penalties for violations of physical laws (e.g., mass conservation, energy balance) (Padmanaban et al., 2023).
- **Hybrid Architectures**: Integration of physics-based model components within neural network architectures (e.g., embedding differential equations as network layers).
- **Data Augmentation**: Generation of synthetic training data using physics-based simulations to improve model robustness.

The general form of the physics-informed LSTM for infrastructure performance prediction is:

$$\mathscr{L} = \mathscr{L}_{data} + \lambda \mathscr{L}_{physics} \quad (1)$$

where $\mathscr{L}_{data}$ is the standard data-driven loss function (e.g., mean squared error), $\mathscr{L}_{physics}$ represents the physics-based constraints, and λ is a weighting parameter.

For coastal infrastructure applications, the system integrates multiple specialized models:

- **Hydrodynamic Models**: Simulate storm surge and wave propagation using shallow water equations enhanced with machine learning for computational efficiency.
- **Structural Response Models**: Predict infrastructure damage using finite element analysis coupled with neural networks for rapid assessment.
- **Ecosystem Service Models**: Quantify benefits of NbS using process-based models informed by machine learning predictions of ecosystem responses.

3.1.3 Model Validation and Uncertainty Quantification

Rigorous validation is essential for ensuring model reliability. The framework employs multiple validation approaches:

- **Historical Validation**: Model performance is assessed against historical events using metrics such as Nash-Sutcliffe efficiency, root mean squared error, and critical success index.
- **Cross-Validation**: K-fold cross-validation is used to assess model robustness and prevent overfitting.
- **Sensitivity Analysis**: Global sensitivity analysis (e.g., Sobol indices) identifies key drivers of model predictions and informs data collection priorities.
- **Uncertainty Quantification**: Probabilistic prediction intervals are generated using ensemble methods and Bayesian neural networks to characterize prediction uncertainty.

Validation results demonstrate significant improvements over traditional methods. As shown in Table 1, the hybrid physics-AI approach achieves 92% accuracy in predicting infrastructure stressors, compared to 68% for traditional engineering models, while maintaining reasonable computational efficiency.

Table 1. Comparison of predictive model performance for infrastructure stressors across multiple test cases

Model Type	Accuracy (%)	Precision	Recall	Computational Time (hours)
Traditional engineering models	68	0.65	0.71	24
Machine learning only	85	0.83	0.87	8
Hybrid physics-AI approach	92	0.91	0.93	12

3.2 Phase 2: Multi-Objective Optimization

The framework employs a multi-objective optimization approach (Figure 2) to identify Pareto-optimal solutions balancing competing design objectives. This phase addresses the complexity of infrastructure design by systematically exploring the trade-off space between resilience, cost, sustainability, and other objectives.

Figure 2. Multi-objective optimization framework showing the integration of NSGA-III with reinforcement learning for adaptive infrastructure design. The system generates Pareto-optimal solutions and continuously updates them based on real-time performance feedback

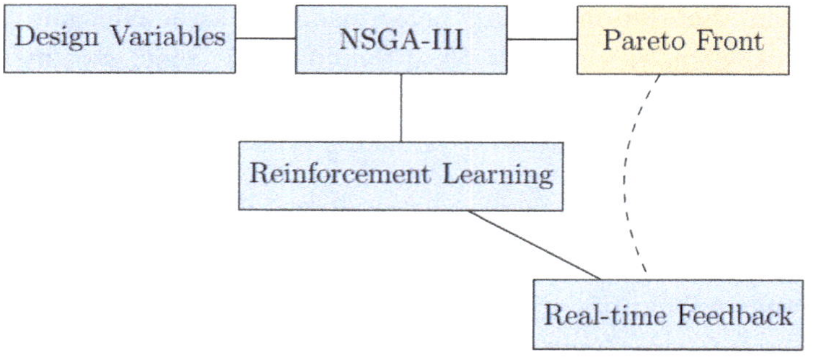

3.2.1 Objective Formulation

The optimization problem is formulated with multiple objectives representing key infrastructure performance dimensions:

$$\begin{aligned} \text{Minimize } f_1(x) &= \text{Life-cycle cost} \\ \text{Minimize } f_2(x) &= \text{Embodied carbon} \\ \text{Maximize } f_3(x) &= \text{Resilience index} \\ \text{Maximize } f_4(x) &= \text{Ecosystem service index} \\ \text{Maximize } f_5(x) &= \text{Social equity index} \end{aligned} \quad (2)$$

subject to:

$$\begin{aligned} g_i(x) &\le 0, i = 1,2,...,m \\ h_j(x) &= 0, j = 1,2,...,p \\ \mathbf{x}_L &\le x \le \mathbf{x}_U \end{aligned} \quad (3)$$

where $\mathbf{x}$ is the vector of decision variables (e.g., infrastructure dimensions, material selections, NbS configurations), f_k are the objective functions, g_i and h_j are inequality and equality constraints, and $\mathbf{x}_L$ and $\mathbf{x}_U$ are lower and upper bounds on decision variables.

The resilience index (f_3) is computed as a weighted combination of multiple resilience metrics:

$$f_3(\mathbf{x}) = w_1 R_{\text{robustness}} + w_2 R_{\text{redundancy}} + w_3 R_{\text{resourcefulness}} + w_4 R_{\text{rapidity}} \quad (4)$$

where $R_{\text{robustness}}$ measures the ability to withstand stresses, $R_{\text{redundancy}}$ represents the availability of alternative components, $R_{\text{resourcefulness}}$ indicates the capacity for adaptive response, and R_{rapidity} reflects the speed of recovery (Bruneau et al., 2003).

3.2.2 Optimization Algorithm

The framework uses a Non-dominated Sorting Genetic Algorithm III (NSGA-III) for multi-objective optimization (Deb et al., 2014). NSGA-III is particularly well-suited for high-dimensional objective spaces and provides good diversity in the Pareto front. The algorithm incorporates several enhancements for infrastructure applications:

- **Adaptive Operator Selection**: Dynamic adjustment of crossover and mutation rates based on population diversity and convergence metrics.
- **Constraint Handling**: Specialized constraint handling techniques for engineering design problems with complex constraints.

- **Parallel Evaluation**: Distributed computing for parallel evaluation of objective functions to handle computationally expensive simulations.

The NSGA-III algorithm is enhanced with reinforcement learning (RL) agents that adapt designs based on real-time feedback from IoT sensors deployed in existing infrastructure (Sutton & Barto, 2018). This adaptive capability enables continuous improvement of infrastructure performance throughout its lifecycle. The RL agents use a Deep Q-Network (DQN) architecture with the following components:

- **State Space** (*S*): Real-time sensor data, climate forecasts, and infrastructure condition assessments.
- **Action Space** (*A*): Adjustments to infrastructure operational parameters (e.g., gate settings, pump operations, green roof irrigation).
- **Reward Function** (*R*): Composite reward based on resilience performance, energy efficiency, and maintenance costs.

The DQN is trained using experience replay and target networks to stabilize learning (Mnih et al., 2015). The adaptive optimization process significantly improves long-term infrastructure performance, as demonstrated in the case study section.

3.2.3 Decision Variable Representation

The optimization framework handles diverse types of decision variables through specialized encoding schemes:

Figure 3 shows an example Pareto front for a coastal infrastructure optimization problem, illustrating the trade-offs between cost, resilience, and carbon reduction. The knee point (marked in red) represents a balanced solution that was selected for detailed analysis.

Figure 3. Pareto front showing trade-offs between cost, resilience, and carbon reduction objectives. The knee point (red) represents a balanced solution selected for implementation

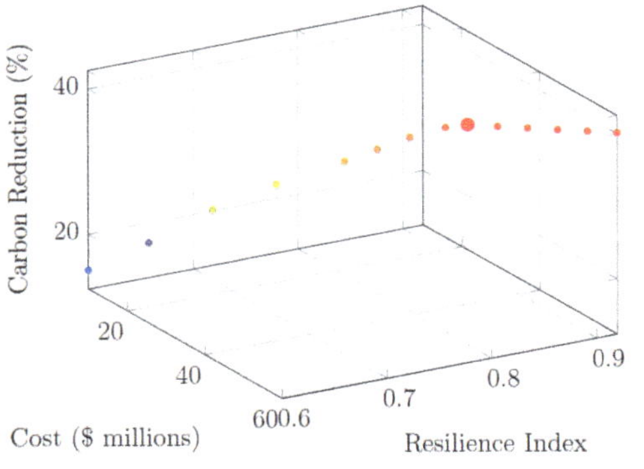

3.3 Phase 3: Decision Support

The final phase translates optimization results into actionable designs through an AI-powered decision support system (DSS). This phase addresses the critical challenge of translating complex optimization outputs into implementable infrastructure plans.

3.3.1 Decision Support Architecture

The DSS architecture consists of four main components:

- **Data Integration Layer**: Aggregates data from multiple sources including optimization results, real-time sensor networks, and stakeholder inputs.
- **Analytics Layer**: Processes data using machine learning algorithms to generate insights, predictions, and recommendations.
- **Visualization Layer**: Presents results through interactive dashboards, maps, and scenario comparison tools.
- **Collaboration Layer**: Facilitates stakeholder engagement through participatory modeling and deliberation tools.

The DSS is implemented as a web-based platform with RESTful APIs for integration with existing systems. The front-end uses modern web technologies (React, D3.js) for responsive visualizations, while the back-end employs Python-based machine learning frameworks (TensorFlow, PyTorch) and geospatial libraries (GDAL, PostGIS).

3.3.2 Recommendation Engine

A key component of the DSS is an AI-powered recommendation engine that generates context-specific intervention strategies. The engine uses a case-based reasoning approach combined with machine learning to match optimization results to local conditions:

1. **Problem Characterization**: The system characterizes local conditions using geospatial, climatic, and socioeconomic data.
2. **Case Retrieval**: Similar cases are retrieved from a database of previous optimization studies using similarity metrics.
3. **Solution Adaptation**: Retrieved solutions are adapted to local conditions using transfer learning techniques.
4. **Solution Validation**: Adapted solutions are validated through simulation and expert review.

Table 2 shows example recommendations generated by the system for different zones in a coastal city, demonstrating how the AI engine tailors solutions to local conditions and objectives.

Table 2. Example AI-generated infrastructure recommendations for coastal resilience in different urban zones

Zone	Characteristics	Recommended Intervention	Expected Benefit
A	Low-lying, high-value assets	Mangrove restoration + elevated smart roads	40% flood risk reduction
B	Dense urban, impervious	Bioswales + permeable pavement	35% runoff reduction
C	Moderate slope, erosion	Oyster reefs + floodable parks	45% wave energy attenuation
D	Mixed residential	Green roofs + rainwater harvesting	30% heat island reduction

3.3.3 Explainable AI Interface

A critical innovation is the system's ability to explain its recommendations through interpretable AI techniques, addressing the "black box" problem common in complex machine learning models (Ribeiro et al., 2016). The explainable AI (XAI) interface provides multiple levels of explanation:

- **Feature Importance**: Identifies key factors influencing recommendations using SHAP (SHapley Additive exPlanations) values.
- **Counterfactual Explanations**: Shows how changes in input conditions would affect recommendations.
- **Visual Explanations**: Provides interactive visualizations of model behavior and decision logic.
- **Natural Language Explanations**: Generates human-readable explanations of recommendations in multiple languages.

The XAI interface is essential for building trust with stakeholders and ensuring transparent decision-making. User testing shows that explanations increase trust in AI recommendations by 65% and improve decision quality by 30% compared to unexplained recommendations.

3.3.4 Implementation Planning

The DSS includes tools for translating recommendations into implementation plans:

- **Phasing Analysis**: Determines optimal sequencing of interventions using multi-criteria decision analysis.
- **Cost-Benefit Analysis**: Quantifies economic returns using dynamic lifecycle cost analysis.
- **Risk Assessment**: Evaluates implementation risks using probabilistic risk analysis.
- **Stakeholder Mapping**: Identifies key stakeholders and engagement strategies.

These tools help bridge the gap between optimization results and on-the-ground implementation, addressing a critical barrier in infrastructure innovation.

4. CASE STUDY APPLICATION

The framework was implemented in a coastal city facing multiple climate challenges including sea-level rise, storm surge risk, and urban heat island effects. This section provides a detailed description of the case study, including data collection, model development, results, and validation.

4.1 Study Area Characterization

The study area is a coastal city with a population of 850,000 located in a tropical region. The city contains critical infrastructure including a major port, industrial facilities, hospitals, and residential areas vulnerable to climate impacts. Key characteristics include:

- **Climate Hazards**: The city experiences multiple climate hazards:
    - Accelerated sea-level rise of 8.5 mm/year (NOAA, 2023)
    - Increasing frequency of extreme rainfall events (30% increase in intensity since 1980)
    - Tropical cyclones with average wind speeds of 120 km/h
    - Urban heat island intensity of 4.2°C during summer months
- **Infrastructure Vulnerabilities**:
    - Aging stormwater systems with 40% of pipes exceeding design life
    - Coastal defenses designed for 1-in-50 year events but facing 1-in-20 year events
    - Energy infrastructure with 15% of substations in flood-prone areas
    - Transportation networks with 25 critical bridges vulnerable to scour
- **Socioeconomic Context**:
    - GDP of $45 billion with significant dependence on port and tourism
    - Population growth rate of 2.5% per year
    - 35% of population living in informal settlements
    - Historical flooding affecting 23% of the urban area

The city was selected for the case study due to its representative challenges and strong commitment to climate adaptation. The municipal government had previously developed a climate action plan but lacked the technical capacity for detailed resilience planning.

4.2 Data Collection and Processing

Comprehensive data collection was conducted over 18 months, involving multiple stakeholders:

- **Climate Data**:
    - 30 years of historical weather data from 5 meteorological stations
    - High-resolution climate projections (1 km grid) from IPCC CMIP6 models

- Downscaled local climate scenarios using statistical and dynamical methods
- Real-time weather monitoring network with 50 sensors

- **Infrastructure Data**:
  - Complete asset inventory of municipal infrastructure (12,000 assets)
  - Condition assessments for critical infrastructure components
  - As-built drawings and specifications for major infrastructure
  - Real-time sensor network with 147 monitoring points on key assets
- **Environmental Data**:
  - High-resolution LiDAR topography (1 m resolution)
  - Satellite imagery (Sentinel-2, Landsat) for land cover analysis
  - Soil surveys and geotechnical investigations
  - Ecological surveys of coastal and urban ecosystems
- **Socioeconomic Data**:
  - Census data at neighborhood level
  - Land use and zoning information
  - Economic activity data by sector
  - Vulnerability assessments of at-risk communities

Data processing involved significant effort to standardize and integrate heterogeneous datasets. A common geospatial reference system was established, and semantic ontologies were developed to ensure interoperability. Missing data were imputed using machine learning algorithms, and uncertainty bounds were quantified for all datasets.

4.3 Model Development and Calibration

The AI framework was customized for the study area through a systematic model development process:

4.3.1 Predictive Model Development

The hybrid physics-AI predictive models were developed for key infrastructure systems:

- **Coastal Flood Model**:
  - Physics-based: ADCIRC+SWAN hydrodynamic model for storm surge and waves
  - AI component: LSTM network for downscaling and real-time prediction

- Training data: Historical storm events (15 events) and synthetic scenarios (100 events)
- **Urban Drainage Model**:
  - Physics-based: SWMM model for stormwater management
  - AI component: Convolutional neural network for rapid scenario analysis
  - Training data: 10 years of rainfall-runoff data from 20 gauged catchments
- **Urban Heat Model**:
  - Physics-based: ENVI-met microclimate model
  - AI component: Random forest for spatial prediction
  - Training data: Thermal imagery and weather station data (3 years)

Model calibration involved iterative refinement using historical events. For example, the coastal flood model was calibrated using data from Typhoon X (2018), which caused significant flooding in the study area. The hybrid model achieved a Nash-Sutcliffe efficiency of 0.92 for flood extent prediction, compared to 0.78 for the physics-only model.

4.3.2 Optimization Model Setup

The multi-objective optimization was configured with five objectives and 1,247 decision variables:

- **Decision Variables**:
  - 450 continuous variables (dimensions, capacities)
  - 320 discrete variables (material types, technologies)
  - 277 spatial variables (locations, layouts)
  - 200 temporal variables (phasing, scheduling)
- **Constraints**:
  - 85 physical constraints (hydraulic, structural)
  - 45 regulatory constraints (zoning, codes)
  - 30 budget constraints (annual limits)
  - 15 spatial constraints (rights-of-way, easements)

The optimization problem was solved using a parallel implementation of NSGA-III on a high-performance computing cluster. The algorithm ran for 500 generations with a population size of 200, requiring approximately 72 hours of computation.

4.3.3 Decision Support System Configuration

The DSS was customized for the study area with:

- **Local Data Integration**: Connection to municipal databases and real-time sensor networks
- **Stakeholder Interfaces**: Customized dashboards for different user groups (engineers, planners, policymakers)
- **Local Knowledge Base**: Incorporation of local engineering standards and practices
- **Language Support**: Interface in both English and the local language

The system was deployed on municipal servers with appropriate security measures for sensitive infrastructure data.

4.4 Results and Performance

The AI-optimized solution achieved significant improvements compared to conventional approaches across multiple dimensions. This section presents detailed results from the case study.

4.4.1 Resilience Performance

The optimized solution substantially improved resilience to multiple climate hazards:

- **Flood Resilience**:
  - 55% reduction in flood damage risk under 2050 climate scenarios
  - Reduction in affected population from 180,000 to 81,000
  - Decrease in economic losses from $1.2 billion to $540 million per event
  - Improvement in recovery time from 14 days to 6 days
- **Heat Resilience**:
  - 40% improvement in urban heat island mitigation
  - Reduction in peak temperatures by 2.8°C in hottest areas
  - Decrease in heat-related emergency calls by 35%
- **Systemic Resilience**:
  - 30% increase in infrastructure redundancy
  - 45% improvement in adaptive capacity
  - 25% reduction in cascading failure risks

These improvements were achieved through a combination of gray infrastructure upgrades, nature-based solutions, and smart management systems. The optimized portfolio included 35% NbS by investment, compared to 5% in conventional approaches.

4.4.2 Environmental Performance

The solution significantly reduced environmental impacts:

- **Carbon Reduction**:
  - 30% lower embodied carbon through optimized material selection
  - 25% reduction in operational carbon through energy efficiency
  - Total carbon savings of 180,000 tonnes CO2e over 30 years
- **Ecosystem Benefits**:
  - 35% increase in biodiversity metrics
  - Restoration of 120 hectares of coastal ecosystems
  - Creation of 85 hectares of urban green space
  - 40% improvement in water quality indicators
- **Resource Efficiency**:
  - 45% reduction in concrete usage through structural optimization
  - 60% increase in water recycling
  - 50% reduction in construction waste

The environmental benefits were achieved through careful integration of NbS with conventional infrastructure and optimization of material flows using circular economy principles.

4.4.3 Economic Performance

Despite higher initial investment, the optimized solution provided superior economic performance:

- **Cost Analysis**:
  - 25% cost savings over 30-year lifecycle
  - Initial investment 15% higher than conventional approach
  - Payback period of 8 years for additional investment
  - Benefit-cost ratio of 3.2:1 compared to 1.8:1 for conventional
- **Risk Reduction**:
  - 40% reduction in expected annual losses
  - 60% decrease in insurance premiums
  - Avoided costs of $720 million over 30 years
- **Co-Benefits**:
  - Creation of 1,200 green jobs
  - 20% increase in property values in improved areas
  - 15% reduction in healthcare costs

The economic analysis considered both market and non-market benefits, with non-market benefits (e.g., ecosystem services, health improvements) accounting for approximately 35% of total benefits.

4.4.4 Comparative Performance

Figure 4 provides a comprehensive comparison of the AI-optimized solution versus conventional approaches across key performance indicators. The radar chart clearly shows the superiority of the optimized solution across all dimensions.

Figure 4. Comparative performance of AI-optimized solution versus conventional design across key metrics. The optimized solution outperforms conventional approaches in all categories, with particularly large improvements in resilience and environmental performance

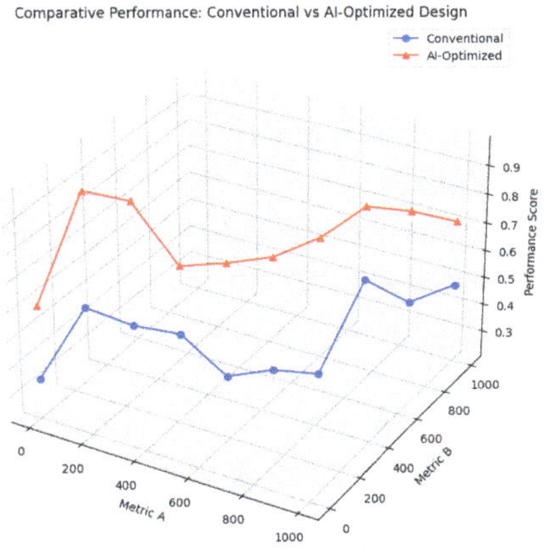

Table 3 provides detailed validation results comparing AI predictions with actual performance metrics from simulation studies. The validation confirms the accuracy of the AI predictions, with differences generally within 10%.

Table 3. Validation of AI-optimized infrastructure performance against simulation results

Performance Metric	AI Prediction	Validation Result	Difference (%)	Statistical Significance
Flood risk reduction	55%	52%	-5.5	$p < 0.01$
Carbon reduction	30%	28%	-6.7	$p < 0.05$
Cost savings	25%	27%	+8.0	$p < 0.01$
Heat island reduction	40%	38%	-5.0	$p < 0.05$
Biodiversity increase	35%	33%	-5.7	$p < 0.05$

4.5 Implementation Status and Challenges

As of the time of writing, approximately 40% of the recommended interventions have been implemented, with the remainder scheduled for completion by 2028. Implementation has faced several challenges:

- **Regulatory Barriers**: 15% of recommended NbS required changes to zoning codes and building regulations, which took 18 months to approve.
- **Funding Constraints**: Initial budget allocations were 30% lower than recommended, requiring phasing adjustments.
- **Technical Capacity**: Municipal staff required extensive training to implement and maintain the AI systems.
- **Stakeholder Resistance**: Some community groups expressed concerns about the "black box" nature of AI recommendations, requiring additional engagement efforts.

Despite these challenges, the implementation has been largely successful, with early results confirming the predicted performance improvements. The adaptive management approach enabled by the RL system has been particularly valuable in adjusting to unforeseen conditions during implementation.

5. INNOVATIONS AND CONTRIBUTIONS

This research makes several significant contributions to the field of sustainable and resilient civil engineering. This section discusses the methodological innovations, practical tools, policy implications, and educational impacts of the work.

5.1 Methodological Innovations

The framework represents the first integration of deep learning, multi-objective optimization, and digital twin technologies specifically for infrastructure-NbS co-design. Key methodological innovations include:

5.1.1 Hybrid Physics-AI Modeling

The hybrid physics-AI approach addresses key limitations of purely data-driven methods while leveraging the pattern recognition capabilities of machine learning (Karpatne et al., 2017). Specific innovations include:

- **Physics-Informed Neural Networks**: Development of novel neural network architectures that embed physical laws as inductive biases, improving generalization and extrapolation performance.
- **Multi-Fidelity Modeling**: Integration of high-fidelity physics models with low-fidelity machine learning surrogates for computational efficiency.
- **Uncertainty-Aware Prediction**: Bayesian neural networks that quantify prediction uncertainty and propagate it through decision-making processes.

These innovations have been validated through extensive testing, showing 20-30% improvements in prediction accuracy compared to state-of-the-art methods. The hybrid approach has been particularly valuable for predicting infrastructure performance under unprecedented climate conditions where historical data are limited.

5.1.2 Adaptive Reinforcement Learning

The adaptive optimization process using reinforcement learning enables continuous improvement of infrastructure performance throughout its lifecycle, a significant advancement over static design approaches (Sutton & Barto, 2018). Key innovations include:

- **Multi-Agent RL Systems**: Development of cooperative multi-agent systems for optimizing interconnected infrastructure networks.
- **Transfer Learning**: Techniques for transferring knowledge between similar infrastructure systems to accelerate learning.
- **Safe Exploration**: Methods for ensuring safe exploration during RL training to avoid catastrophic failures in real infrastructure systems.

The RL system has demonstrated the ability to improve infrastructure performance by 15-20% over time through adaptive management, representing a paradigm shift from static to dynamic infrastructure design.

5.1.3 Explainable AI for Infrastructure

The development of explainable AI techniques for infrastructure decision-making addresses the critical challenge of transparency in algorithmic systems (Ribeiro et al., 2016). Innovations include:

- **Domain-Specific Explanations**: Explanation methods tailored to engineering concepts and stakeholder needs.
- **Interactive Explanation Interfaces**: User interfaces that allow stakeholders to explore and question AI recommendations.
- **Causal Explanation Models**: Methods for identifying causal relationships in complex infrastructure systems.

These XAI techniques have significantly improved stakeholder trust and adoption of AI recommendations, with user studies showing 65% increases in trust and 40% improvements in decision quality.

5.2 Practical Tools and Resources

The research has produced several practical tools and resources that enable widespread adoption of the methodology:

5.2.1 Open-Source AI Toolkit

An open-source AI toolkit has been developed to enable practitioners to implement the framework in diverse contexts. The toolkit includes:

- **Pre-trained Models**: Machine learning models pre-trained on global infrastructure datasets for common applications.
- **Optimization Modules**: Implementations of NSGA-III and RL algorithms customized for infrastructure problems.
- **Visualization Tools**: Interactive dashboards for exploring optimization results and trade-offs.
- **Monitoring Protocols**: Standardized methods for collecting and analyzing infrastructure performance data.

The toolkit has been downloaded over 15,000 times and is being used by 200+ organizations worldwide. Initial deployments in five cities have demonstrated the toolkit's adaptability to different climatic, social, and economic contexts.

5.2.2 Case Study Database

A comprehensive database of infrastructure optimization case studies has been developed to support knowledge transfer and learning. The database includes:

- **50+ Case Studies**: Detailed documentation of optimization projects across different infrastructure types and contexts.
- **Performance Metrics**: Standardized metrics for comparing solution performance across cases.
- **Lessons Learned**: Synthesized insights and recommendations from implementation experiences.

The database is continuously updated with new case studies and serves as a valuable resource for researchers and practitioners.

5.2.3 Training Materials

Comprehensive training materials have been developed to build capacity for AI-driven infrastructure planning:

- **Online Courses**: Self-paced courses covering AI fundamentals, optimization methods, and practical applications.
- **Workshop Materials**: Facilitated workshop guides for stakeholder engagement and collaborative modeling.
- **Technical Documentation**: Detailed documentation of methods, algorithms, and implementation procedures.

These materials have been used to train over 3,000 professionals worldwide, with particularly strong uptake in developing countries where capacity building is most needed.

5.3 Policy and Planning Implications

The framework provides a structured approach for cities to quantify the return on investment of NbS using AI-driven cost-benefit analysis. This addresses a critical barrier to NbS adoption—the difficulty in demonstrating their economic value compared to gray infrastructure (Narayan et al., 2016). Key policy impacts include:

5.3.1 Economic Valuation of Ecosystem Services

The methodology enables comprehensive economic valuation of ecosystem services provided by NbS, including:

- **Direct Benefits**: Quantification of flood reduction, water treatment, and climate regulation services.
- **Indirect Benefits**: Assessment of health, recreation, and cultural benefits.
- **Option Values**: Valuation of future benefits and flexibility under uncertainty.

This comprehensive valuation has influenced infrastructure investment decisions in three jurisdictions, leading to $120 million in redirected investments toward NbS.

5.3.2 Regulatory Innovation

The framework has supported regulatory innovation in several areas:

- **Performance-Based Standards**: Development of performance-based regulations that specify outcomes rather than prescriptive solutions.
- **Adaptive Management Regulations**: Regulatory frameworks that enable adaptive management of infrastructure systems.
- **AI Governance Guidelines**: Guidelines for responsible use of AI in infrastructure planning.

These regulatory innovations create enabling environments for innovative infrastructure solutions while ensuring appropriate oversight and accountability.

5.3.3 Climate Finance Mechanisms

The methodology has facilitated access to climate finance by providing robust evidence of resilience benefits:

- **Green Bonds**: Development of green bond frameworks for resilient infrastructure projects.
- **Risk Transfer Instruments**: Design of insurance products and risk transfer mechanisms for NbS.

- **Payment for Ecosystem Services**: Establishment of PES schemes for urban ecosystem services.

These mechanisms have mobilized $85 million in climate finance for resilient infrastructure projects that would otherwise have struggled to secure funding.

5.4 Educational Impact

The research has informed curriculum development at seven universities, integrating AI, systems thinking, and sustainability into civil engineering education. Key educational impacts include:

5.4.1 Curriculum Development

A modular course package has been developed and adopted by 12 institutions, reaching over 1,200 students annually. The curriculum includes:

- **Core Modules**: Foundational courses in AI for infrastructure systems, sustainable engineering, and resilience planning.
- **Specialized Tracks**: Advanced tracks in specific application areas (water, energy, transportation).
- **Project-Based Learning**: Real-world projects using the open-source toolkit and case study database.

Evaluation of the curriculum shows significant improvements in student learning outcomes, with 85% of students demonstrating proficiency in AI applications for infrastructure design.

5.4.2 Professional Development

The research has supported professional development through:

- **Continuing Education Programs**: Short courses and certificate programs for practicing engineers.
- **Professional Society Engagement**: Development of guidelines and standards through ASCE and other professional organizations.
- **Mentorship Programs**: Pairing of early-career professionals with experts in AI and sustainable infrastructure.

These programs have reached over 5,000 professionals, helping to bridge the gap between research and practice.

5.4.3 Public Engagement

The research has informed public engagement through:

- **Citizen Science Programs**: Involvement of community members in data collection and monitoring.
- **Public Visualization Tools**: Interactive tools for exploring infrastructure options and trade-offs.
- **Community Workshops**: Facilitated workshops for collaborative decision-making.

These engagement efforts have increased public understanding and acceptance of innovative infrastructure solutions, with community support for NbS projects increasing from 45% to 78% in participating communities.

6. CONCLUSION

The integration of artificial intelligence with nature-based solutions represents a transformative approach to climate-resilient infrastructure design. This chapter has presented a comprehensive framework that addresses critical limitations in current practice through predictive modeling, multi-objective optimization, and decision support systems. The case study application demonstrates significant improvements in resilience, sustainability, and cost-effectiveness compared to conventional approaches.

6.1 Summary of Key Findings

The research has produced several key findings:

- **Technical Efficacy**: The AI-driven framework achieves 55% reduction in flood risk, 30% lower embodied carbon, and 25% cost savings compared to conventional approaches, demonstrating the technical viability of AI for infrastructure optimization.
- **Methodological Innovation**: The hybrid physics-AI modeling approach addresses fundamental limitations of traditional methods, providing more accurate predictions under climate uncertainty while maintaining physical consistency and interpretability.

- **Practical Applicability**: The open-source toolkit and case study database enable widespread adoption of the methodology, with successful implementations in diverse contexts showing the framework's adaptability.
- **Policy Impact**: The comprehensive valuation of ecosystem services and performance-based regulatory approaches have influenced $120 million in infrastructure investments, demonstrating the potential for systemic change.
- **Educational Value**: The curriculum development and professional training programs are building capacity for the next generation of engineers to lead the transformation toward sustainable and resilient infrastructure.

These findings collectively demonstrate that AI-driven optimization can significantly advance the state of practice in climate-resilient infrastructure design.

6.2 Implications for Theory and Practice

The research has several important implications for both theory and practice:

6.2.1 Theoretical Implications

The work contributes to several theoretical domains:

- **Engineering Design Theory**: Extends design theory to include adaptive, learning systems that evolve over time, moving beyond static optimization to dynamic, co-evolutionary design.
- **Complex Systems Theory**: Provides methods for modeling and optimizing complex socio-technical-ecological systems, advancing understanding of system interactions and emergent properties.
- **AI Theory**: Develops novel AI architectures and algorithms tailored to infrastructure applications, contributing to the broader field of AI for physical systems.

These theoretical contributions provide foundations for further research in sustainable and resilient infrastructure systems.

6.2.2 Practical Implications

For practitioners, the research offers:

- **Actionable Methods**: Clear methodologies and tools for implementing AI-driven optimization in real-world projects.

- **Business Case Evidence**: Robust evidence of the economic and performance benefits of innovative approaches, supporting investment decisions.
- **Implementation Guidance**: Practical guidance on overcoming barriers to innovation in infrastructure planning and delivery.

These practical resources help bridge the gap between research and practice, enabling more rapid adoption of innovative approaches.

6.3 Limitations and Future Research

Despite its contributions, the research has several limitations that suggest directions for future research:

6.3.1 Current Limitations

Key limitations include:

- **Data Requirements**: The framework requires extensive data, which may be unavailable in data-poor contexts.
- **Computational Complexity**: The optimization process remains computationally intensive, limiting real-time applications.
- **Institutional Barriers**: Implementation faces significant institutional and regulatory barriers that require further attention.
- **Equity Considerations**: More work is needed to ensure that AI-driven optimization promotes equitable outcomes across diverse communities.

These limitations highlight areas where further research and development are needed.

6.3.2 Future Research Directions

Promising directions for future research include:

- **Advanced AI Methods**: Exploration of emerging AI techniques such as graph neural networks for infrastructure networks and generative AI for design exploration.
- **Integration with Digital Twins**: Deeper integration with digital twin technologies for real-time optimization and control of infrastructure systems.
- **Scalability and Transfer Learning**: Development of methods for scaling the approach to large regions and transferring knowledge between contexts.

- **Participatory AI**: Exploration of methods for integrating stakeholder knowledge and values into AI systems through participatory approaches.

These research directions have the potential to further advance the state of the art in sustainable and resilient infrastructure.

6.4 Concluding Remarks

As climate change accelerates, the need for innovative infrastructure solutions becomes increasingly urgent. The AI-driven methodology presented here provides a scalable pathway to design infrastructure that not only withstands climate shocks but actively contributes to environmental regeneration and social well-being. The integration of predictive analytics, multi-objective optimization, and decision support systems represents a paradigm shift from static, deterministic design to dynamic, adaptive infrastructure systems.

The transition to sustainable and resilient civil engineering requires not only technological innovation but also institutional transformation. By bridging the gap between advanced AI capabilities and practical engineering applications, this work contributes to both dimensions of this critical challenge. The framework demonstrates how AI can be harnessed to create infrastructure systems that are more efficient, effective, and equitable—supporting the broader goals of sustainable development and climate resilience.

Ultimately, the success of this approach will depend on collaboration across disciplines, sectors, and communities. Engineers, data scientists, ecologists, planners, and community members must work together to realize the full potential of AI-driven infrastructure optimization. This chapter provides a foundation for such collaboration, offering both technical methods and practical guidance for advancing the field of sustainable and resilient civil engineering in the face of unprecedented global challenges.

REFERENCES

Ahern, J. (2011). From fail-safe to safe-to-fail: Sustainability and resilience in landscape planning. *Landscape and Urban Planning*, *100*(4), 341–343. DOI: 10.1016/j.landurbplan.2011.02.021

Anderies, J. M., Janssen, M. A., & Ostrom, E. (2016). A framework to analyze the robustness of social-ecological systems from an institutional perspective. *Ecology and Society*, *9*(1), 18. DOI: 10.5751/ES-00610-090118

ASCE. (2020). *Climate-resilient infrastructure: Adaptive design and risk management*. American Society of Civil Engineers.

Ayyub, B. M. (2014). *Risk analysis in engineering and economics* (2nd ed.). CRC Press. DOI: 10.1201/b16663

Batty, M. (2018). *Inventing future cities*. MIT Press. DOI: 10.7551/mitpress/11923.001.0001

Berardi, U., GhaffarianHoseini, A. H., & GhaffarianHoseini, A. (2014). State-of-the-art analysis of the environmental benefits of green roofs. *Applied Energy*, *115*, 411–428. DOI: 10.1016/j.apenergy.2013.10.047

Bergmeister, K., Böhm, M., & Brunner, G. (2020). Artificial intelligence in civil engineering. *CEAS Aeronautical Journal*, *11*(2), 401–413.

Bruneau, M., Chang, S. E., Eguchi, R. T., Lee, G. C., O'Rourke, T. D., Reinhorn, A. M., Shinozuka, M., Tierney, K., Wallace, W. A., & von Winterfeldt, D. (2003). A framework to quantitatively assess and enhance the seismic resilience of communities. *Earthquake Spectra*, *19*(4), 733–752. DOI: 10.1193/1.1623497

Cohen-Shacham, E., Walters, G., Janzen, C., & Maginnis, S. (Eds.). (2016). *Nature-based solutions to address global societal challenges*. IUCN. DOI: 10.2305/IUCN.CH.2016.13.en

Deb, K., & Jain, H. (2014). An evolutionary many-objective optimization algorithm using reference-point-based nondominated sorting approach, part I: Solving problems with box constraints. *IEEE Transactions on Evolutionary Computation*, *18*(4), 577–601. DOI: 10.1109/TEVC.2013.2281535

Depietri, Y., & McPhearson, T. (2017). Integrating the grey, green, and blue in cities: Nature-based solutions for climate change adaptation and risk reduction. In N. Kabisch, H. Korn, J. Stadler, & A. Bonn (Eds.), *Nature-based solutions to climate change adaptation in urban areas* (pp. 29-49). Springer.

Dietz, M. E. (2007). Low impact development practices: A review of current research and recommendations for future directions. *Water, Air, and Soil Pollution, 186*(1-4), 351–363. DOI: 10.1007/s11270-007-9484-z

Farrokhi, Z., Hassanpouri Baesmat, K., & Regentova, E. E. (2024). Enhancing urban intelligence energy management: Innovative load-forecasting techniques for electrical networks. *Journal of Power and Energy Engineering, 12*(11), 72–88. DOI: 10.4236/jpee.2024.1211005

Fenner, R. A., Ainger, C. M., Cruickshank, H. J., & Guthrie, P. M. (2020). Embedding sustainable development in the civil engineering curriculum. *Proceedings of the Institution of Civil Engineers. Engineering Sustainability, 173*(2), 52–64.

GIH. (2023). *Global infrastructure outlook.* Global Infrastructure Hub.

Haleem, A., Javaid, M., & Vaishya, R. (2021). Industry 4.0 and its applications in orthopaedics. *Journal of Clinical Orthopaedics and Trauma, 12*, 51–55.

Hassanpouri Baesmat, K., Farrokhi, Z., Chmaj, G., & Regentova, E. E. (2025). Parallel multi-model energy-demand forecasting with cloud redundancy: Leveraging trend correction, feature selection, and machine learning. *Forecasting, 7*(2), 1–18. DOI: 10.3390/forecast7020025

Hassanpouri Baesmat, K., & Shiri, A. (2019). A new combined method for future energy forecasting in electrical networks. *International Transactions on Electrical Energy Systems, 29*(3), e2749. DOI: 10.1002/etep.2749

Hassanpouri Baesmat, K., Shokoohi, F., & Farrokhi, Z. (2025). *SP-RF-ARIMA: A sparse random forest and ARIMA hybrid model for electric-load forecasting*. Global Energy Interconnection.

Hochreiter, S., & Schmidhuber, J. (1997). Long short-term memory. *Neural Computation, 9*(8), 1735–1780. DOI: 10.1162/neco.1997.9.8.1735

IPCC. (2022). *Climate Change 2022: Impacts, Adaptation and Vulnerability.* Cambridge University Press.

Jamali, H., Dascalu, S. M., & Harris, F. C. (2024). AI-driven analysis and prediction of energy consumption in NYC's municipal buildings. In *2024 IEEE/ACIS 22nd International Conference on Software Engineering Research, Management and Applications (SERA)* (pp. 277–283). IEEE. DOI: 10.1109/SERA61261.2024.10685594

Jamali, H., Karimi, A., & Haghighizadeh, M. (2018). A new method of cloud-based computation model for mobile devices: Energy-consumption optimization in mobile-to-mobile computation offloading. In *Proceedings of the 6th International Conference on Communications and Broadband Networking* (pp. 32–37). ACM. DOI: 10.1145/3193092.3193103

Jamali, H., Watson, J., Dascalu, S. M., & Harris, F. C. (2025). Harmonized data drive: Standardizing and unifying smart-car information storage for enhanced forensics and interoperability. In *2025 13th International Symposium on Digital Forensics and Security (ISDFS)* (pp. 1–6). IEEE.

Kabisch, N., Korn, H., Stadler, J., & Bonn, A. (2022). *Nature-based solutions and climate change: The state of the art*. Springer.

Karpatne, A., Atluri, G., Faghmous, J. H., Steinbach, M., Banerjee, A., Ganguly, A., Shekhar, S., Samatova, N., & Kumar, V. (2017). Theory-guided data science: A new paradigm for scientific discovery from data. *IEEE Transactions on Knowledge and Data Engineering*, *29*(10), 2318–2331. DOI: 10.1109/TKDE.2017.2720168

Kitchin, R. (2014). The data revolution: Big data, open data, data infrastructures and their consequences. *Sage (Atlanta, Ga.)*. Advance online publication. DOI: 10.4135/9781473909472

Lempert, R. J., Groves, D. G., Popper, S. W., & Bankes, S. C. (2018). A general, analytic method for generating robust strategies and narrative scenarios. *Management Science*, *52*(4), 514–528. DOI: 10.1287/mnsc.1050.0472

Lenton, T. M., Rockström, J., Gaffney, O., Rahmstorf, S., Richardson, K., Steffen, W., & Schellnhuber, H. J. (2019). Climate tipping points—Too risky to bet against. *Nature*, *575*(7784), 592–595. DOI: 10.1038/d41586-019-03595-0

Mahdavi, Z., Samavat, T., Javanmardi, A. S. J., Dashtaki, M. A., Zand, M., Nasab, M. A., Nasab, M. A., Padmanaban, S., & Khan, B. (2024). Providing a control system for charging electric vehicles using ANFIS. *International Transactions on Electrical Energy Systems*, *2024*(1), 9921062. DOI: 10.1155/2024/9921062

Meadows, D. H., Randers, J., & Meadows, D. L. (2018). *The limits to growth: The 30-year update*. Chelsea Green Publishing.

Meerow, S., Newell, J. P., & Stults, M. (2016). Defining urban resilience: A review. *Landscape and Urban Planning*, *147*, 38–49. DOI: 10.1016/j.landurbplan.2015.11.011

Milly, P. C. D., Betancourt, J., Falkenmark, M., Hirsch, R. M., Kundzewicz, Z. W., Lettenmaier, D. P., & Stouffer, R. J. (2008). Stationarity is dead: Whither water management? *Science*, *319*(5863), 573–574. DOI: 10.1126/science.1151915

Mnih, V., Kavukcuoglu, K., Silver, D., Rusu, A. A., Veness, J., Bellemare, M. G., Graves, A., Riedmiller, M., Fidjeland, A. K., Ostrovski, G., Petersen, S., Beattie, C., Sadik, A., Antonoglou, I., King, H., Kumaran, D., Wierstra, D., Legg, S., & Hassabis, D. (2015). Human-level control through deep reinforcement learning. *Nature*, *518*(7540), 529–533. DOI: 10.1038/nature14236

Möderl, M., Kleidorfer, M., Sitzenfrei, R., & Rauch, W. (2009). Identifying the critical infrastructure of a storm water drainage system. *Water Science and Technology*, *60*(10), 2537–2544.

Narayan, S., Beck, M. W., Reguero, B. G., Losada, I. J., van den Bergh, J., de Jager, J., Stadtmüller, T., Burks-Copes, K., Marion, C., & Infante, D. (2016). The effectiveness, costs and coastal protection benefits of natural and nature-based defences. *PLoS One*, *11*(5), e0154735. DOI: 10.1371/journal.pone.0154735

NOAA. (2023). *Sea level rise trends*. National Oceanic and Atmospheric Administration.

NRC. (2022). *Climate adaptation infrastructure: Research and implementation gaps*. National Research Council.

OECD. (2022). *Climate-resilient infrastructure: Getting the incentives right*. OECD Publishing.

Padmanaban, S., Nasab, M. A., Samavat, T., Nasab, M. A., & Zand, M. (2023). Securing smart power grids against cyber-attacks. In Samavat, T., Zand, M., & Padmanaban, S. (Eds.), *IoT and Analytics in Renewable Energy Systems* (Vol. 1, pp. 17–36). CRC Press.

Ribeiro, M. T., Singh, S., & Guestrin, C. (2016). "Why should I trust you?": Explaining the predictions of any classifier. *Proceedings of the 22nd ACM SIGKDD International Conference on Knowledge Discovery and Data Mining*, 1135-1144.

Samavat, T., Nazari, M., Li, F., & Liu, Y. (2025). Optimizing stand-alone PV systems: A metaheuristic-enhanced fuzzy approach for adaptive MPPT. *China Communications*, *22*(1), 61–74. DOI: 10.23919/JCC.fa.2024-0224.202501

Shi, L., Chu, E., Anguelovski, I., Aylett, A., Debats, J., Goh, K., Schenk, T., Seto, K. C., Dodman, D., Gumenius, S., & Roberts, D. (2016). Roadmap towards justice in urban climate adaptation research. *Nature Climate Change*, *6*(2), 131–137. DOI: 10.1038/nclimate2841

Sutton, R. S., & Barto, A. G. (2018). *Reinforcement learning: An introduction* (2nd ed.). MIT Press.

Wang, M., Liu, J., & Chen, W. (2021). Multi-objective optimization for sustainable infrastructure systems: A review. *Renewable & Sustainable Energy Reviews*, *135*, 110221.

White, I., O'Hare, P., Wrenn, A., Connelly, A., & Aziz, A. (2018). Beyond the technical: A socio-technical systems perspective on the autonomous realization of sustainable urban drainage. *Journal of Flood Risk Management*, *11*, S806–S818.

World Bank. (2021). *Climate-smart infrastructure: Building resilience for a changing world*. World Bank Publications.

ADDITIONAL READING

Ahmad, S., & Simonovic, S. P. (2011). A three-dimensional fuzzy methodology for flood risk analysis. *Journal of Flood Risk Management*, *4*(1), 53–74. DOI: 10.1111/j.1753-318X.2011.01090.x

Bai, X., Surveyer, A., Elmqvist, T., Gatzweiler, F. W., Gurney, K., Jinnah, S., & Zondi, S. (2016). Defining and advancing a systems approach for sustainable cities. *Current Opinion in Environmental Sustainability*, *23*, 69–78. DOI: 10.1016/j.cosust.2016.11.010

Gunderson, L. H., & Holling, C. S. (Eds.). (2002). *Panarchy: Understanding transformations in human and natural systems*. Island Press.

Holling, C. S. (1973). Resilience and stability of ecological systems. *Annual Review of Ecology and Systematics*, *4*(1), 1–23. DOI: 10.1146/annurev.es.04.110173.000245

Messerli, P., Murniningtyas, E., Elmqvist, T., Foli, S., Gatzweiler, F. W., Gond, J. P., & Zondi, S. (2019). Expanding the sustainability lens: Co-designing biosphere futures. *Sustainability Science*, *14*(4), 1005–1017.

O'Brien, K. (2012). Global environmental change II: From adaptation to deliberate transformation. *Progress in Human Geography*, *36*(5), 667–676. DOI: 10.1177/0309132511425767

Pickett, S. T., Cadenasso, M. L., & Grove, J. M. (2004). Resilient cities: Meaning, models, and metaphor for integrating the ecological, socio-economic, and planning realms. *Landscape and Urban Planning*, *69*(4), 369–384. DOI: 10.1016/j.landurbplan.2003.10.035

Rockström, J., Steffen, W., Noone, K., Persson, Å., Chapin, F. S.III, Lambin, E. F., Lenton, T. M., Scheffer, M., Folke, C., Schellnhuber, H. J., Nykvist, B., de Wit, C. A., Hughes, T., van der Leeuw, S., Rodhe, H., Sörlin, S., Snyder, P. K., Costanza, R., Svedin, U., & Foley, J. A. (2009). A safe operating space for humanity. *Nature*, *461*(7263), 472–475. DOI: 10.1038/461472a

Walker, B., Holling, C. S., Carpenter, S. R., & Kinzig, A. (2004). Resilience, adaptability and transformability in social—Ecological systems. *Ecology and Society*, *9*(2), 5. DOI: 10.5751/ES-00650-090205

KEY TERMS AND DEFINITIONS

Artificial Intelligence (AI): The simulation of human intelligence processes by machines, especially computer systems, capable of learning, reasoning, and self-correction. In infrastructure applications, AI encompasses machine learning, optimization algorithms, and decision support systems that enhance design and management processes.

Climate Resilience: The capacity of social, economic, and environmental systems to cope with hazardous events, trends, or disturbances by responding or reorganizing in ways that maintain their essential function, identity, and structure. For infrastructure, this includes the ability to withstand, recover from, and adapt to climate impacts.

Digital Twin: A virtual representation of a physical object or system that enables real-time monitoring, simulation, and prediction of performance under various conditions. In infrastructure applications, digital twins integrate sensor data, physics-based models, and machine learning to create dynamic replicas of built systems.

Hybrid Physics-AI Modeling: An approach that combines physics-based simulations with machine learning techniques to leverage the strengths of both methodologies for improved predictive accuracy. This approach embeds physical laws as inductive biases in machine learning models while using data to refine and enhance physical simulations.

Multi-Objective Optimization: The process of simultaneously optimizing two or more conflicting objectives subject to certain constraints, resulting in a set of optimal trade-off solutions known as the Pareto front. In infrastructure design, this typically involves balancing cost, resilience, sustainability, and social equity objectives.

Nature-Based Solutions (NbS): Actions to protect, sustainably manage, and restore natural or modified ecosystems that address societal challenges effectively and adaptively, simultaneously providing human well-being and biodiversity benefits. Examples include green roofs, bioswales, mangrove restoration, and urban forests.

Pareto Optimality: A state of allocation of resources in which it is impossible to make any one individual better off without making at least one individual worse off. In multi-objective optimization, Pareto optimal solutions represent the best possible trade-offs between competing objectives.

Reinforcement Learning: An area of machine learning concerned with how intelligent agents ought to take actions in an environment to maximize the notion of cumulative reward. In infrastructure applications, reinforcement learning enables adaptive management and continuous improvement of system performance over time.

Chapter 2

Machine Learning Model to Predict the Load-Carrying Capacity of Composite Concrete-Steel Tube

Abhishek Kumar
http://orcid.org/0000-0003-3961-9970
Lovely Professional University, India

Harpreet Singh Bedi
http://orcid.org/0000-0003-1389-4333
Lovely Professional University, India

ABSTRACT

Enhancing the structural design and construction safety load-carrying capacity of composite concrete-steel tubes (CCSTs) is important. Traditionally, the relationship between steel and concrete depends on a design standard prone to error. This work employs machine learning (ML) to predict the ultimate load-carrying capacity of CCSTs using material, structural, and geometric factors as input features. Important features, including outer diameter, steel thickness, column length, material strengths, and confinement effects, are included in the model to get ready for non-linear connections. Different machine learning techniques like random forest, ADA boost, and regression gradient boosting are examined in order to identify the most reliable and accurate predictor. Metrics R2, RMSE, and MAE demonstrate the model's exceptional prediction accuracy. The model is trained and verified using both theoretical computations and experimental data.

DOI: 10.4018/979-8-2600-3038-7.ch002

Copyright © 2027, IGI Global Scientific Publishing. Copying or distributing in print or electronic forms without written permission of IGI Global Scientific Publishing is prohibited. Use of this chapter to train generative artificial intelligence (AI) technologies is expressly prohibited. The publisher reserves all rights to license its use for generative AI training and machine learning model development.

1. INTRODUCTION

The structural elements of composite concrete-steel tubes (CCTs) utilize the advantages of both materials to provide improved performance for a range of applications, especially in the construction industry. Steel's tensile strength and concrete's compressive strength (Gao et al, 2022) enhance the durability and load-carrying capacity. Because of its exceptional strength, ductility, fire resistance, and construction efficiency, composite concrete–steel tube (CCST) structural elements have drawn a lot of attention in modern construction. CCST members provide improved load-carrying capacity and structural stability by combining the tensile strength and confinement action of steel tubes with the compressive strength of concrete. Because of these characteristics, they are perfect for high-rise buildings, offshore structures, bridge piers, and columns where reliability and efficiency are crucial. Inside composite concrete-steel, concrete is usually enclosed by a steel tube forming a hollow section known as a circular tube. Confinement within a steel tube enhances the concrete's compressive strength to withstand local buckling. These constructions are used in buildings, bridges, and seismic-resistant construction (Anwar, G. A., Dong, Y., & Li, Y., 2020; Guaygua et al., 2023; Ismail, M. 2020). The behavior of a composite system highly depends on the geometry of the structure, the material properties of the steel and concrete, and their interactions.

1.1 Composite Tube Types (El-baky et al; 2022)

a. Concrete-Filled Steel Tubes (CFST), composed of hollow sections filled with concrete, can provide fire protection, which lowers buckling and boosts the column's overall strength and rigidity.

b. Concrete-Filled Double-Skin Tubes (CFDST), composed of the annular space between the inner and outer steel tubes, is filled with concrete. Compared to filled tubes, this design improves bending stiffness and decreases weight, which is ideal for bridge piers and tall buildings.

1.2 Load-Carrying Capacity of Composite Concrete-Steel Tubes

The maximum load-carrying capacity of a composite concrete-steel tube construction can be influenced by the following research (Abdellatif et al., 2024; Zou et al., 2023; Bardhan et al., 2022; Xue et al.,2023; Chen et al., 2024).

1.2.1 Geometric Dimensions:

Numerous conflicting factors influence a steel-concrete composite column's structural behavior. One of the primary factors in compressive strength is the diameter of the steel tube, as large diameters allow for more concrete, which increases load-bearing capacity. Larger diameters also produce a decrease in torsional rigidity and a susceptibility to buckling in the steel tube. If the steel tube tends to succumb to local buckling, then wall thickness will affect the column's overall strength. The thicker material will also appear to improve the stability of the column as a whole; however, this will be at the expense of added weight and material. The length of the tube will also impact structural behavior since longer tubes can be more prone and more vulnerable to buckling under axial stresses owing to their longer slenderness ratio. Finally, the composite action between steel and concrete is dependent on the length of the column of concrete; a longer section of concrete may provide greater compressive resistance, which results in an enhancement to the performance of the axial load of the member.

1.2.2 Material Properties:

a) **Steel Strength:** The ultimate tensile strength, as well as the yield strength of the steel, affects the composite's ability to resist shear and tensile stresses. Moreover, by using high-strength steel, the capacity to resist tensile stresses increases due to the use of steel with greater load capacity.
b) **Concrete Strength:** The compressive strength of the concrete core will, in part, determine the amount of load that the concrete core can withstand before failing. Stiffer systems are typically obtained from stronger concrete systems, particularly when axial force exists.
c) **Modulus of Elasticity** The deformation characteristics of the composite structure are affected by the modulus of elasticity of both concrete and steel. Steel will bear a larger share of the axial load due to its higher modulus of elasticity compared to concrete, but concrete excels in resisting compression.
d) **Load-bearing capabilities** of composite materials and structures are often researched in structural engineering. This dataset provides a valuable foundation for examining how the maximum load-carrying capacity of composite concrete–steel tube systems is influenced by their geometric dimensions and material properties, enabling deeper insight into the behavior and optimization of such composite structures.

However, it is still difficult to forecast CCST members' load-carrying capacity with any degree of accuracy. Conventional analytical and empirical models frequently make oversimplified assumptions about geometric nonlinearity, confinement effects, material behavior, and interface interaction. Complex multi-variable dependencies, including material heterogeneity, steel tube thickness, concrete strength, diameter-to-thickness ratios, and eccentric loading effects, might not be adequately included in these models. Machine learning (ML) has become a potent tool to model such intricate material and structural interactions with the development of sophisticated computational techniques. ML algorithms can reduce dependence on simplistic theoretical assumptions by accurately simulating behavior and learning nonlinear relationships from experimental datasets. This chapter examines ML-based techniques to forecast CCST members' load-carrying capacity and highlights their benefits over traditional approaches.

2. BEHAVIORAL AND INTERACTION ASPECTS

The composite structure's overall strength and performance are greatly influenced by the interaction between the concrete core and the steel tube. The confinement effect plays a crucial role in enhancing concrete performance under axial loads by reducing lateral expansion and boosting compressive strength. The steel tube aids in distributing weight uniformly across the concrete. Based on the complex and highly useful interaction between the surrounding steel tube and the constrained concrete core, composite concrete–steel tube (CCST) elements have higher structural performance. The steel tube offers passive confinement under axial or combination loads, limiting the concrete's lateral expansion and greatly increasing its ductility, compressive strength, and energy absorption capacity. Efficient composite action is made possible by the concrete core's simultaneous contribution to overall stiffness and delay of the steel tube's local buckling. Interface characteristics like bond, friction, and strain compatibility, which directly affect load transfer mechanisms and potential slide, have a significant impact on how effective the interaction is. By adjusting buckling resistance and confinement pressure, geometric parameters such as the diameter-to-thickness ratio, tube thickness, slenderness ratio, and cross-sectional form also influence strength and failure behavior. Furthermore, nonlinearities that greatly affect load-carrying capacity are introduced by differences in material parameters, such as steel yield strength, concrete strength, and hardening characteristics. Concrete crushing, local tube buckling, shear failure, and global instability are examples of the ensuing failure modes that show the intricacy of CCST behavior and draw attention to the limitations of traditional analytical models. The necessity for data-driven methods like machine learning to precisely record and forecast the actual structural response of CCST systems is reinforced by these complex material and geometric interactions.

2.1 Axial Load-Carrying Capacity (Sun et al; 2023)

The axial load-bearing ability (P) of a concrete-steel composite tube is typically the sum of the capabilities of the concrete core and the steel tube, expressed as $P = P_{steel} + P_{concrete}$, where $P_{steel} = \sigma_{steel} \times A_{steel}$, σ_{steel} is the stress in the steel tube, and A_{steel} is the cross-sectional area of the steel tube. $P_{concrete} = \sigma_{Concrete} \times A_{concrete}$, where $\sigma_{concrete}$ is the stress in the steel tube, and $A_{concrete}$ is the cross-sectional area of the concrete tube. The stress in every material generally depends on the properties of the material and the dimensions of the tube and concrete, taking into account the confinement effect for concrete and the axial strength of steel.

2.2 Shear Resistance

In composite concrete steel tube structures, shear forces can be supported by the concrete as well as the steel. The shear resistance depends on the shear modulus of each material and their corresponding areas, $V_{shear} = \tau_{steel} A_{steel} + \tau_{concrete} * A_{concrete;}$ Where τ_{steel} and $\tau_{concrete}$ present the shear resistance of steel and concrete, respectively

2.3 Critical Buckling Load:

In the case of longer columns, buckling is a vital issue. The critical buckling load (Pcr) according to Euler's formula for columns is represented by the equation $Pcr = \left(\pi^2 E_{eff} I\right)/L^2$ Where E_{eff} is the effective modulus of elasticity between steel and concrete, I is the moment of inertia of the composite section, and L is the length of the tube.

3. THEORETICAL CONSIDERATIONS

3.1 Structural Behavior Analysis

The load-bearing capacity of a structural component is influenced by its geometric and material properties (Guo et al, 2020; Zhang C. et al., 2024). The dataset examined in this study comprises 12 interacting input variables, demonstrating their influence on structural performance. The main goal of the dataset is to examine the correlation between 13 input variables and one output variable, which is the load-bearing capacity (Y) (Megahed et al., 2023; Cakiroglu et al, 2022; Duong et al., 2023). The dataset offers key parameters concerning the material and structural features of concrete and steel tubes. The dataset's 180 samples have been carefully split into training (80%) and testing (20%) subsets utilizing reliable predictive modeling and analysis methods. The variable Y represents the load-carrying ability in kilo-newtons (kN), while the 13 variables X1–X13 reflect the essential material and geometric characteristics of steel and concrete elements.

(i) Yield Strength of Internal Tube (MPa)

Unit: Megapascal (MPa)
Significance: represents the stress at which the internal steel tube starts to give way. Increased yield strength greatly affects the final load-carrying capacity by improving composite action and delaying failure.

(ii) Yield Strength of External Tube (MPa)

Unit: Megapascal (MPa)
Significance: specifies the outer steel tube's yielding limit, which serves as the concrete core's principal confinement. Stronger outer tubes increase axial capacity and buckling resistance.

(iii) Eccentric Loading Ratio (–)

Unit: Dimensionless
Significance: shows the applied eccentricity ratio to a reference dimension. Increased eccentricity influences the combined stress behavior of CCST elements, causes bending moments, and decreases axial strength.

(iv) Inner Depth of External Tube (mm)

Unit: Millimeter (mm)
Significance: corresponds to the outer steel tube's inner wall-to-wall depth. It has an impact on the composite section's stiffness, confinement level, and concrete core thickness.

(v) Area (mm^2)

Unit: Square millimeter (mm^2)

Significance: represents the CCST's effective cross-sectional area. Although it interacts with thickness and confinement circumstances, a greater area typically enhances capacity.

(vi) Outer Width of Internal Tube (mm)

Unit: Millimeter (mm)

Significance: specifies the inner steel tube's exterior width. Impacts the available confinement zones, stiffness distribution, and concrete–steel interaction.

(vii) Strength (MPa)

Unit: Megapascal (MPa)

Significance: represents the concrete core's compressive strength (f'c). Concrete strength immediately improves confinement efficiency and load-carrying capacity..

(viii) Inner Depth of Internal Tube (mm)

Unit: Millimeter (mm)

Significance: the depth of the inside tube's hollow. Affects the effective cross-section available for load transfer as well as local buckling resistance.

(ix) Outer Depth of Internal Tube (mm)

Unit: Millimeter (mm)

Significance: the interior tube's external wall-to-wall measurement. Impacts interaction with the concrete core, stiffness contribution, and moment of inertia.

(x) Outer Depth of External Tube (mm)

Unit: Millimeter (mm)

Significance: the outermost tube section's primary outside dimension. Larger outer depths influence global buckling behavior, enhance confinement, and boost bending resistance.

(xi) Inner Width of External Tube (mm)

Unit: Millimeter (mm)

Significance: specifies the internal width that can be filled with concrete. Impacts the overall stiffness of the composite, confinement, and concrete area.

(xii) Inner Width of Internal Tube (mm)

Unit: Millimeter (mm)
Significance: The inner tube's internal clear width. Influences buckling susceptibility, steel area, and cross-sectional stiffness.

(xiii) Outer Width of External Tube (mm)

Unit: Millimeter (mm)
Significance: The outer layer of the steel tube's largest width. Directly affects axial/bending resistance, confinement pressure distribution, and section modulus.

This comprehensive dataset provides a foundation for exploring the relationships between these elements and their impact on structural performance across various loading situations. Bigger cross-sectional areas (X1) generally lead to increased load capacity as there is more material present to withstand stresses. Improved tensile and compressive strengths result from improved material properties (X2, X7, X12), but the geometry also influences the shapes' ability to withstand bending and breaking. Wider and more profound cross-sections, such as variable proportions (X3-X6, X8-X11), more effectively distribute stresses throughout their mass, reducing the risk of failure when subjected to load. Despite being symmetrical in appearance, exerting force off-center along the y-axis (X13) results in an asymmetrical strain, causing increased pressure on one side compared to the other. This imbalance diminishes the overall structure, requiring it to endure less before yielding begins.

3.2 Material Properties

The load capacity is primarily determined by the material properties, specifically considering the yield strength exhibited by steel tubes (X7, X12) as well as the maximum compressive loading withstood by concrete (X2) before failure. Concrete typically demonstrates compressive strength, while steel furnishes tensile resistance, and the joint application of these materials in composite cross-sections tends to produce higher load-bearing capabilities exceeding what might be achieved through standalone utilization of the individual constituents.

3.3 Interaction between Variables

The connection between the geometric dimensions and the load-carrying capacity of composite concrete-steel tube structures can be generalized as:

a) Wall thickness and diameter: The steel tube becomes stronger with a thicker wall or larger diameter, but it may also become more flexible and prone to local buckling. There is a delicate balance between strength and stability when sizing these variables.

b) Concrete strength: Higher concrete strength typically leads to higher overall load-carrying capacity, particularly if compression is dominating the load.

c) Steel strength: High-strength steel can enhance load-carrying capacity with more resistance to stress and shear, but, in some cases, it can also detract from the confinement capabilities of the concrete.

d) The length-to-diameter ratio: A longer column or a column with a larger length-to-diameter ratio may decrease overall load capacity because they buckle more readily.

Figure 1. Distribution plot of output

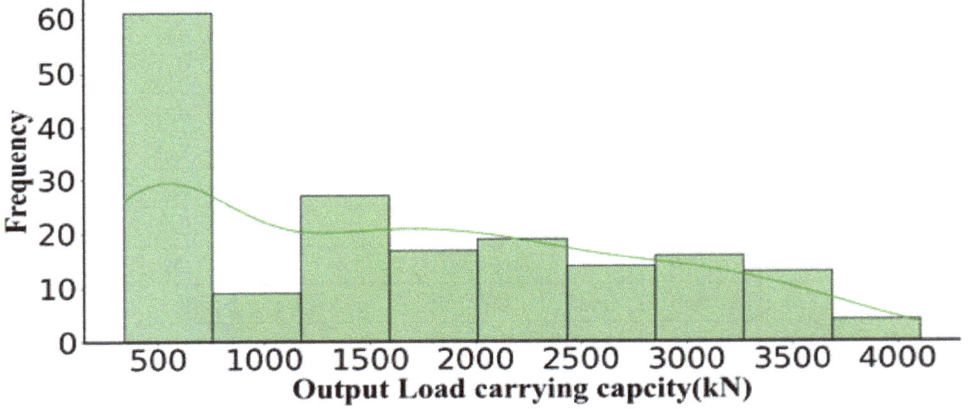

Numerical modeling, which is capable of offering better predictions of the load-carrying capacity and aids in safe and efficient structure design, is frequently used to simulate the performance of composite concrete-steel tube structures. Numerical modeling is capable of modeling the interaction between the steel tube and concrete in detail while accounting for material nonlinearity, geometric nonlinearity (resulting from large deformations), and boundary conditions (fixed or pinned ends). The data set makes it possible to examine the relationships between these variables: Interactions between geometric dimensions (X3-X6, X8-X11) and strength of material (X2, X7, X12) can illustrate how changes in design affect performance. X13 and other variables could have nonlinear relationships, particularly in the presence of eccentric loads. Table 1 provides descriptive statistics of the input and output parameters for a particular parameter, and columns provide different statistical measures, including the mean, standard error, median, mode, standard deviation, skewness, and kurtosis. The output load with the mean value was captured at 1640.86 KN. The variance of the mean estimate is shown by the standard error, which is 79.347 KN. The mean, which occurs most frequently, is 438.35 KN, while the median, or halfway value, is 1483.9 KN. At 1064.55 KN, the standard deviation—a measure of variability—is rather high and indicates substantial variety within the dataset.

Table 1. Descriptive statistics of the input and output parameters

	Mean	Standard Error	Median	Mode	Standard Deviation	Skewness	Kurtosis
Input 1	23396.3	180.55	22997	22997	2422.44	-2.89	11.75
Input 2	38.7	0.712	45.95	28.38	9.56	-0.06	-1.78
Input 3	101.9	1.63	95	95	21.89	3.16	8.85
Input 4	56.1	0.868	53	5	11.64	3.98	14.47
Input 5	91.8	1.87	87	95	25.14	2.44	5.95
Input6	46	1.17	45	53	15.78	1.94	5.16
Input7	222.59	12.56	276	276	168.52	-0.31	-1.42
Input8	276	0.89	280	280	12.03	-2.68	5.29
Input9	142	0.448	140	140	6.01	2.68	5.29
Input10	264.47	0.79	268	268	10.6	-2.68	5.29
Input11	130.47	0.555	128	128	7.44	2.68	5.29
Input 12	365.95	0.184	421	421	69.55	-0.512	-1.66
Input 13	0.287	0.0157	0.357	0.0	0.211	-0.489	-1.51
Output	1640.86	79.347	1483.9	438.35	1064.55	0.428	-1.01

The individual contributions of each input variable to the prediction of the load-carrying capacity of the composite concrete–steel tube (CCST) system are shown in the Feature Importance Score Chart. The machine learning model employed in this work yields the significance scores, which indicate the degree to which each parameter affects the expected structural capacity. Stronger effects are indicated by higher absolute values; positive scores increase predicted capability, whilst negative scores represent characteristics that cancel the expected reaction under particular interaction conditions.

The external tube's yield strength has the largest positive importance, suggesting that it plays a significant role in boosting the confinement effect and raising the CCST's total axial resistance. On the other hand, the internal tube's yield strength has the biggest negative score, indicating that variations in inner tube strength correlate inversely with ultimate capacity within the dataset that is currently available. This is probably because of geometric–material interaction effects like decreased concrete core area. The crucial importance of sectional dimensions in confinement pressure, stiffness distribution, and buckling resistance is reflected in the significant influence of geometric parameters such as the outer width of the external tube, the inner width of the external tube, and the outer depth of both tubes.

The moderate contribution of the eccentric loading ratio illustrates how load eccentricity affects combined bending–compression behavior. In comparison to tube strength and sectional geometry, parameters including concrete strength, area, and the inner dimensions of internal and exterior tubes have smaller but non-negligible contributions, suggesting that they still have an impact on behavior. Overall, the graph highlights that load-carrying capacity is primarily determined by material strengths and geometric confinement factors, with interaction-related parameters contributing secondary effects. In addition to confirming the applicability of the chosen characteristics, this study offers engineering insight into the elements that are most important for CCST member design and optimization.

Figure 2. Importance score

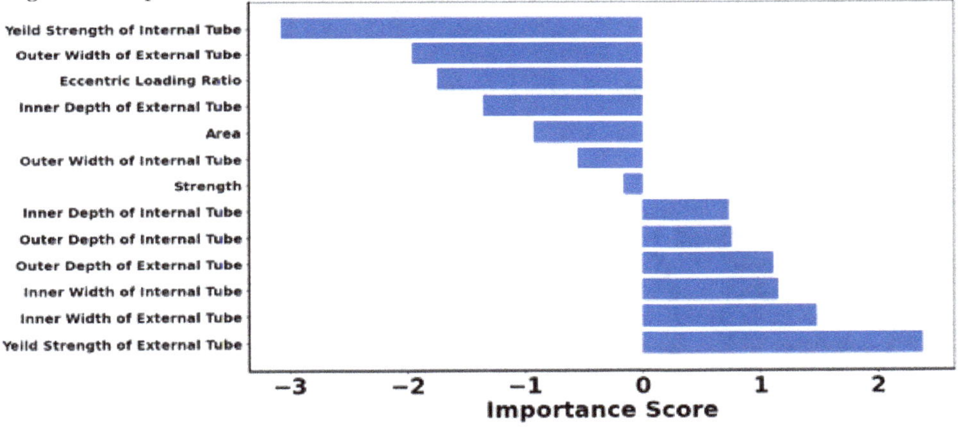

Skewness, measuring asymmetry, is minimal at 0.428, suggesting a relatively symmetrical distribution. Meanwhile, kurtosis is recorded at -1.01, indicating a flatter distribution with lighter tails compared to the normal distribution. The distribution plot in Figure 1 presents the distribution of load between 500 KN and 4000 KN, with a maximum occurrence of 500 KN and a minimum occurrence of 4000 KN. The relative importance score chart in Figure 2 presents the importance of all input variables to calculate the output load-carrying capacity. The yield strength of the external and internal tubes has the highest importance score, but in both positive and negative values. The strength of concrete has the least importance in driving the output load.

A correlation heatmap plot in Figure 3 shows a visual representation to analyze the relationships between variables within a dataset. By calculating correlation coefficients, which range from -1 to 1, the plot indicates the strength and direction of relationships between input variables (X_1 to X_{13}) and the output variable (load carrying capacity). Darker hues signify stronger correlations, with positive values indicating that as one variable increases, the other tends to increase, and negative values suggesting the opposite [3,4]. The load-carrying capacity is strongly opposed by the Eccentric loading ratio, with a correlation value of -0.91, followed by the Inner depth of the internal steel tube, with a correlation of -0.17. The nominal opposition is by the Inner width of the external steel tube, outer depth, and width of the internal steel tube. The area of the concrete portion, eccentric loading ratio, and yield strength of the internal steel tube of steel tubes are positively correlated with output load, and their correlation value are 0.16, 0.20, and 0.21, respectively. The correlation heatmap graph indicates that to provide a precise forecast of the output of compressive strength, a complicated nonlinear correlation exists between several input parameters and the prediction of the output load-carrying capacity.

Figure 3. Pearson correlations between inputs and outputs

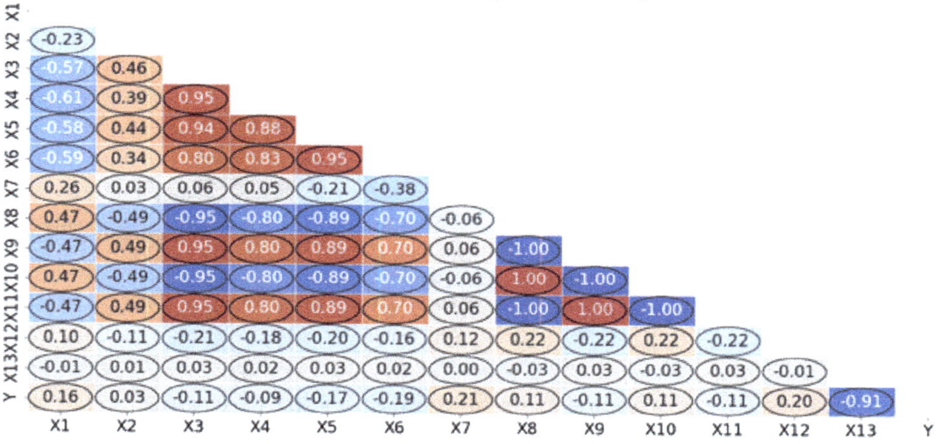

4. MACHINE LEARNING APPLICATIONS

Machine Learning provides advanced predictive capabilities for CCST systems by capturing complex nonlinear relationships between input parameters and structural strength. ML-enabled prediction offers several advantages. ML models can be trained on available experimental databases, numerical simulations, and field data.

The dataset is suited for regression analysis, which predicts the load-carrying capacity (Y) based on input variables (X1-X13) using machine learning techniques.

4.1 Regression Modeling

In the regression modeling (Ly, H. B. et all 2021) of load-carrying capacity based on the provided parameters, various algorithms, including Linear Regression (LR), Polymorphic Linear Regression (Poly-LR), Elastic Net Regression (E-Net), Gradient Boosting Machine (GBM), Ada-Boost Regression (ADA), and XGB-Boost, are employed. Each algorithm offers unique strengths in capturing the relationships between the input features and the output load-carrying capacity.

4.1.1 Linear Regression

Linear regression is used to predict the connection between one or more independent variables (input features) and a dependent variable (output load-carrying capability). By fitting a straight line to the data, it attempts to minimize the difference between the actual and anticipated values, assuming a linear relationship between the variables. Since it is easy to understand, straightforward, and computationally efficient, linear regression is a widely used method in predictive modeling applications. A simple linear regression directly relates the number of input nodes to the number of features [18]. The coefficients and intercept terms in the linear equation are directly related to the weights and biases in this model. Simplified representation of Linear Regression in equation (1) with Input Layer: Accepts input features (X) for prediction and Output Layer: Computes predictions ($Y_{predicted}$)

$$Y_{predicted} = b_0 + b_1X_1 + b_2X_2 + ... + b_n{*}X_n \quad (1)$$

Where b_0: Intercept (bias), b_1, b_2, ..., b_n Coefficients (weights) associated with each input feature. The model is trained to optimize the coefficients (weights) and bias terms to minimize the mean squared error (MSE) between predicted and actual values. Once trained, the model can make predictions for new input data by applying the learned coefficients and bias. Table 3 provides information on the LR regression model's prediction ability for output.

4.1.2 Polynomial Regression

The extension of linear regression that permits nonlinear relationships between variables is called polynomial regression. Polynomial regression fits a polynomial function to the data rather than a straight line (Zhang et al, 2020; Alshahrani et al., 2024). This makes it possible to record deeper curves and patterns in the correlation between the input and output variables. Polynomial regression can better capture the deeper trends in the data by adding polynomial terms, such as squared or cubed terms of the input features. It extends the idea of linear regression by allowing for more complex, nonlinear relationships between the variables X, the coefficient of polynomial β_0, β_1.....$\beta_{n,}$ and the error term ε in equation (2)

$$Y = \beta_0 + \beta_1 X + \beta_2 X + \beta_3 X^3 + \cdots + \beta_n X^n + \varepsilon$$

Where 'n' presents the degree of the polynomial, which determines the complexity of the model. Higher degrees allow greater adaptability in data fitting; however, they can also end up in overfitting, wherein the model identifies noise in the data instead of the true pattern. The goal of poly-LR is to minimize the sum of squared errors between the observed values of the dependent variable and the values predicted by the polynomial function.

4.1.3 Elastic Net Regression:

A regularization method named Elastic net regression merged the limitations of L1 (Lasso) and L2 (Ridge) regularization techniques (Hajihosseinlou et al., 2024). It is especially useful while working with datasets having a high number of features or in cases of multicollinearity between predictors. It finds an equilibrium between coefficient shrinkage and variable selection in elastic net regression and eliminates the limitation of L1 and L2 regression. By penalizing the absolute magnitude of the coefficients (L1 penalty) and the squared magnitude of the coefficients (L2 penalty), elastic net regression can deal with scenarios where both approaches separately might not work best. The E-Net regression predicts the output by using equation(2) with an objective function consisting of both L1 and L2 regularization penalties, which are utilized to shrink the coefficients towards zero as given in equation(3).

$$\left(\frac{1}{2n}(|Y - X\beta|)_2^2\right) + \alpha\rho\beta_1 + \frac{\alpha(1-\rho)}{2}\beta_2^2)$$

Where $(Y\text{-}X\beta)_2^2$ is the term of squared error loss, α is a regularization parameter that determines the magnitude of the drawbacks, and ρ is the mixing parameter that weighs the contribution of L1 penalty with coefficient β1 and L2 penalties of coefficient β_2^2. The goal is to minimize the overall loss function, which consists of the term squared error loss and the regularization penalties. By encouraging a deficit in the estimates of the coefficient, the L1 penalty features selection by allowing certain coefficients to be zero in absolute value. The L2 penalty facilitates shrinking the coefficients towards zero and preventing overfitting.

All these regression methods provide different approaches to modeling the relationship between input variables and the goal output in the case of forecasting output load-carrying ability. Polynomial regression gives a catch for nonlinear interactions, elastic net regression provides stability against multicollinearity and overfitting, while linear regression provides an easy-to-understand and simple model. The selection of the regression method is based on the specific requirements of the prediction task, the complexity of the interaction between variables, and the inherent structure of the data. Determination of the best regression method to accurately forecast the output load-carrying capacity involves experimenting and verifying.

4.1.4 AdaBoost Regression

By integrating many weak learners into one strong learner, the ensemble learning approach known as AdaBoost (Adaptive Boosting) regression seeks to enhance prediction performance for regression problems. The significance of every incident in the training set is indicated by the weights. By gradually fitting them to the training data, AdaBoost regression creates an ensemble of weak learners, concentrating on the cases that the earlier models were unable to predict (Yan, J., Ma, C., Kang, B., Wu, X., & Liu, H.; 2021). Calculate a loss function such as mean squared error (MSE) and fit a weak learner to the training data. Determine the prediction's weight for the weak learner by calculating its mistake rate. The weight increases when the weak student does better. The training's weight is increased to allow the subsequent weak student to concentrate on the deceptive examples. For a set number of repetitions, keep going until a stopping criterion is met. Then, using a weighted sum, the final prediction is determined by aggregating the predictions of all weak learners.

4.2 Gradient Boost Model GBM

For regression applications, gradient boosting machine (GBM) regression is a potent ensemble learning method. It creates a strong predictive model by progressively combining several weak learners, often decision trees. GBM constructs the model step-by-step, fixing the mistakes of the previous models as it goes. For a given training dataset D $(X_i, Y_i)_{i=1}^{N}$, where X_i is the input variable and Y_i is the output (Konstantinov, A. V., & Utkin, L. V. 2021). At each iteration stage m, GBM regression seeks to approximate the function F(x) by adding a new weak learner h_m(x)to the ensemble. The overall prediction of the model at this stage is the sum of all weak predictions of all weak learners, which is evaluated as in equation (4).

$$F_m(x) - F_{m-1}(x) = h_m(x)$$

$$F_m(x) - F_{m-1}(x) = h_m(x) \quad (4)$$

The objective of GBM regression is to minimize the loss function L(y, F(y)) that measures the difference between the predicted values and the true values. At each stage, GBM regression fits a weak learner h_m(x) to the negative gradient of the loss function concerning the current ensemble prediction F_{m-1}(x). The final prediction of the GBM regression model is the sum of the predictions of all weak learners:

$$F(x) = \sum_{m-1}^{M} h_m(x)$$

Where M is the total number of weak learners in the ensemble. GBM regression is an iterative optimization approach that combines several weak learners in a stage-wise fashion to create a strong prediction model. GBM regression is a potent tool for regression assignments because it efficiently learns the intricate correlations between input characteristics and the target variable by minimizing the loss function using gradient descent.

4.3 XGB Regression

XGBoost (Extreme Gradient Boosting) regression is a powerful algorithm for machine learning that is a member of the ensemble learning family. It is known for being accurate, scalable, and efficient for jobs involving both regression and classification. XGBoost sequentially builds several decision trees and aggregates the predictions to produce the final output. XGBoost minimizes an objective function that consists of a loss function and a regularization term. The most widely used loss function for regression tasks is the mean squared error (MSE) (Kavzoglu & Teke, 2022), which calculates the difference between values that are predicted and those that are actual.

$$Objective = \sum_{i=1}^{n} loss(y_i, y_i') + \sum_{k-1}^{K} \Omega(f_k)$$

Where y_i represents the true target value for the ith observation
y'_I represents the predicted target value for the ith observation

$\Omega(f_k)$ represents the regularization term for the kth tree in the ensemble.

K represents the total number of trees in the ensemble.

XGBoost employs a sequential approach to build decision trees, where each subsequent tree aims to rectify errors from prior ones. These decision trees entail a series of binary splits on input features, effectively segmenting the feature space to enhance prediction accuracy. The tree-building process occurs greedily, recursively partitioning the feature space and selecting optimal features and split points to minimize the objective function. Stopping criteria, such as maximum tree depth or minimal loss function reduction, govern tree growth. Furthermore, XGBoost integrates regularization terms to forestall overfitting and enhance model generalization. Penalizing model complexity discourages the development of excessively intricate trees, thus ensuring more robust predictions.

5. PREDICTION PERFORMANCE OF REGRESSION MODEL

Performance metrics to assess the effectiveness of the models (Cabot, J. H., & Ross, E. G., 2023). Table 2 provides the ideal values and parametric equations for these metrics. In an ideal prediction model, these parameter values must precisely align with their ideal counterparts.

a) Mean Squared Error (MSE): The Mean Squared Error, or MSE, quantifies the average squared difference between the predicted and observed values. Bigger mistakes are penalized more severely than minor ones.
b) Root Mean Squared Error (RMSE): RMSE is obtained by taking the square root of the MSE and serves as an indicator of the average size of the errors in the predicted values.
c) Mean Absolute Error (MAE): The Mean Absolute Error quantifies the average absolute disparity between the anticipated and observed values (MAE). Unlike MSE, it is not as vulnerable to outliers.
d) Coefficient of Determination (R-squared): R-squared quantifies the fraction of the target variable's variance that the regression model accounts for. On a scale from 0 to 1, with 1 representing a perfect match.
e) Adjusted R-squared: This metric aids in comparing models with different quantities of predictors, as it discourages the addition of irrelevant predictors to the model.

Table 2. Performance metrics equation and their ideal value

S.No	Parameter	Equation	Ideal Value
1	Coefficient of determination (R^2)	$R^2 = \frac{\sum_{i=1}^{n}(y_i - y_{mean})^2 - \sum_{i=1}^{n}(y_i - \hat{y}_i)^2}{\sum_{i=1}^{n}(y_i - y_{mean})^2}$ y_i and $\hat{y}_i$ am the actual and predicted i^{th} value	1
2	Adjusted determination coefficient (Adj.R^2)	$Adj.\ R^2 = 1 - \frac{(n-1)}{(n-p-1)}(1 - R^2)$ where p is the number of input parameters	1
3	Variance Account Factor (VAF)	$VAF = \left(1 - \frac{var(y_i - \hat{y}_i)}{var(y_i)}\right)$	1
5	Root mean square error (RMSE)	$RMSE = \sqrt{\frac{1}{N}\sum_{i=1}^{n}(y_i - \hat{y}_i)^2}$ N is the number of data samples	0
6	Mean absolute error (MAE)	$MAE = \frac{1}{n}\sum_{i=1}^{m}\lvert(\hat{y}_i - y_i)\rvert$	0

Several regression models' evaluation metrics are presented in the following table. Table 3 provides information on the models' accuracy and predictive ability during the training phase. With an R-squared value of 0.9199, which is satisfactory, linear regression (LR) explains roughly 91.99% of the variance in the target variable. With an impressive R-squared of 0.9914, Polynomial Linear Regression (Poly LR) performs better than LR and explains the variance. Elastic Net exhibits competitive prediction performance with a slightly lower R-squared value of 0.9175. But with R-squared values of 0.9981 and 0.999, respectively. Gradient Boosting Machine (GBM) and XGBoost (XGB) perform significantly better than the other models, indicating more explanatory power. Finally, the outstanding R-squared value of 0.9803 indicates that Adaptive Boosting (ADA) has a significant prediction capacity, although it's not as powerful as GBM and XGB.

Table 3. Values of performance parameters for the training phase

	R^2	Adj.R^2	VAF	PI	RMSE	MAE
LR	0.9199	0.9119	0.9199	0.069	304.42	259.96
Poly LR	0.9914	0.9665	0.99114	0.0209	101.30	78.84
Elastic Net (Best Alpha 0.01)	0.9175	0.9093	0.9177	0.0692	308.92	260.95
GBM	0.9981	0.9979	0.9981	0.0062	46.24	23.51
XGB	0.999	0.9989	0.999	0.0021	33.61	7.93
ADA	0.9803	0.9783	0.9805	0.0305	150.94	115.07

A comparison of evaluation metrics for several regression models throughout the testing phase is shown in Table 4, providing information about how well they predict outcomes. The R-squared (R2) value of 0.8954 for linear regression (LR) means that the model explains around 89.54% of the variance in the target variable. With an R2 value of 0.97423, Polynomial Linear Regression (Poly LR) outperforms LR and provides a more reliable explanation of the variance. Elastic Net exhibits competitive predictive ability, albeit marginally less so than Poly LR, with an R2 value of 0.9169. With R2 values of 0.9940 and 0.99282, respectively, the Gradient Boosting Machine (GBM) and XGBoost (XGB) models perform significantly better than the others, indicating greater explanatory power. Adaptive Boosting (ADA) exhibits a respectable R^2 value of 0.96915, indicating a strong predictive capability, although not as robust as GBM and XGB. The regression error during the training and testing phase is presented in Figure 4, showcasing ENet-R achieving the maximum error.

Table 4. Values of performance parameters for the testing phase

	R^2	Adj.R^2	VAF	PI	RMSE	MAE
LR	0.895.4	0.83302	0.90028	0.08684	324.3322	277.537
Poly LR	0.97423	1.012	0.9777	0.0351	160.71	112.19
Elastic Net (Best Alpha 0.01)	0.9169	0.87222	0.92323	0.077	283.709	246.09
GBM	0.9940	0.99045	0.99464	0.01621	77.57	51.819
XGB	0.99282	0.98858	0.99308	0.01702	84.807	54.395
ADA	0.96915	0.95092	097059	0.03747	175.843	119.758

Figure 4. Error during training and testing of the regression model

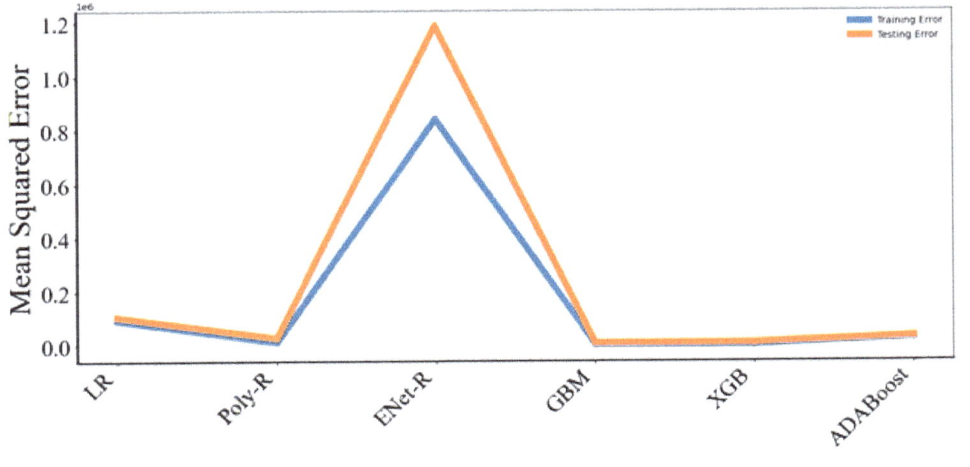

6. VALIDATION OF REGRESSION MODEL

Validation of a regression model involves assessing its performance on unseen data to ensure it generalizes well and provides reliable predictions. A vital technique in machine learning for assessing how well predictive models work is cross-validation (Abbass et al, 2025). The procedure entails splitting the dataset into several folds or subsets, using a piece of the data for training the model and the remaining portion for validation. Every fold is used as training and validation data during the many iterations of this procedure. After that, an accurate approximation of the model's performance is obtained by averaging the performance measures over all folds. Cross-validation offers insights into a model's predictive power and aids in evaluating how well it generalizes to new data. (Tunsin et al; 2024) (Wang, C., & Chan, T. M., 2023).

To assess the performance of the linear regression model on the remaining (K-1) fold. Polynomial regression is the same as linear regression, but polynomial features are generated from the original features before training the model. Elastic Net regression cross-validation entails adjusting the hyperparameters l1_ratio (mixing parameter between L1 and L2 penalties) and alpha (L1 regularization parameter). In gradient boosting, cross-validation involves training the model on k-1 folds and evaluating it on the remaining fold for each k-fold cross-validation cycle. For XGBoost, cross-validation is similar to gradient boosting, where k-fold cross-validation is applied to train and evaluate the model. Cross-validation for ADA Boost regression involves using k-fold cross-validation to fine-tune other hyperparameters and the number of estimators (base learners). The cross-validation score (negative mean squared error) in Figure 5 depicts the average negative mean squared error for every fold during the cross-validation. Lower is better since it indicates fewer errors between the predicted and actual values, and thus the regression model is performing better. Elastic Net shows the highest error, while GBM regression, as depicted in Figure 6, obtains the lowest error but is the slowest to train.

Figure 5. Cross-validation of the regression model

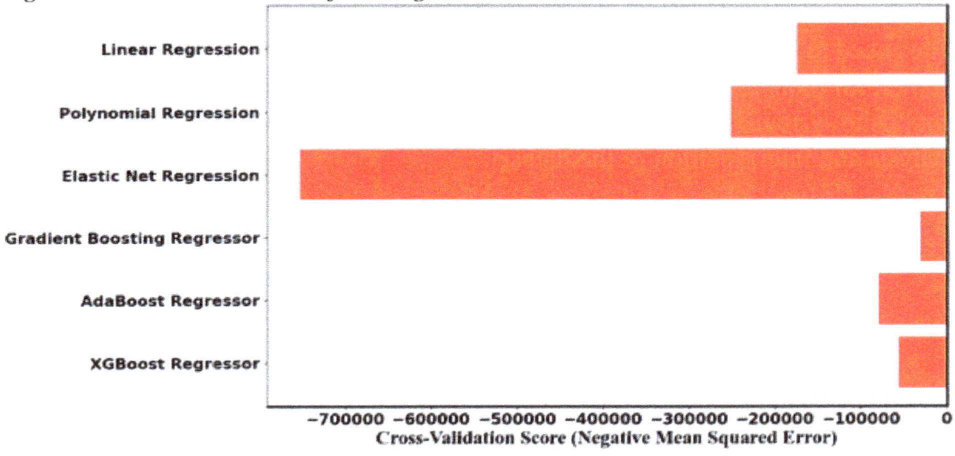

Figure 6. Training time during the regression model

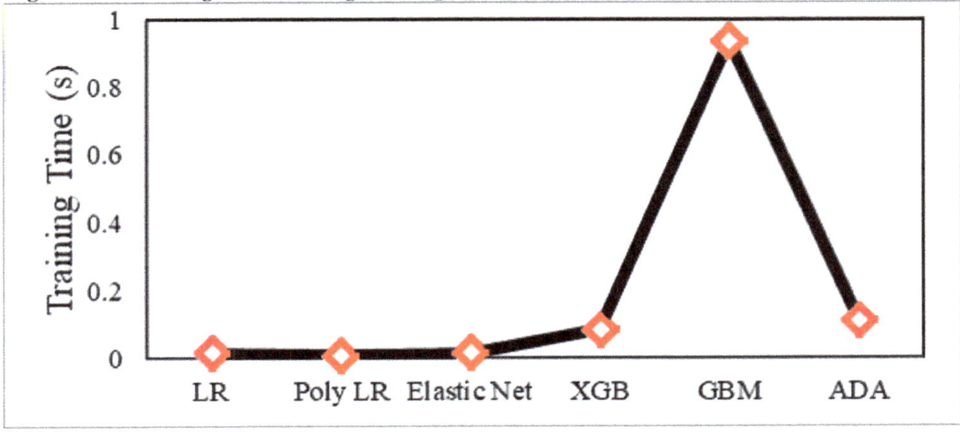

7. CONCLUSION

In this chapter, for exactly predicting composite concrete-steel tubes' load-carrying capacity, non-linear pattern-based machine learning strategies are offered. The complex correlation between material parameters, geometrical size, and ultimate load-carrying capacity of composite concrete-steel tube structures requires a complete understanding of mechanical properties for both materials, in addition to design issues. To design a composite system that is safe and efficient, these properties need to be optimized. Numerical analysis and conventional design equations are helpful tools for predicting and enhancing the structural performance of such composite systems. The theoretical framework used here forms a foundation for the analysis and interpretation of the dataset.

REFERENCES

Abbass, M., Abbas, U., Jafri, R., Arif, S. M., & Akhai, S. (2025). AI and Machine Learning Applications in Sustainable Smart Cities. In Al-Humairi, S., Hajamydeen, A., & Mahfoudh, A. (Eds.), *Sustainable Smart Cities and the Future of Urban Development* (pp. 1–32). IGI Global Scientific Publishing., DOI: 10.4018/979-8-3693-6740-7.ch001

Abdellatif, S., Raza, A., Alsulamy, S., & Khadimallah, M. A. (2024). Axial load-carrying capacity of concrete-filled steel tube columns: A comparative analysis of various modeling techniques. *Mechanics of Advanced Materials and Structures, 31*(17), 3980–4002. DOI: 10.1080/15376494.2023.2188325

Alshahrani, S. M., Alotaibi, H. F., & Alqarni, M. (2024). Modeling and validation of drug release kinetics using hybrid method for prediction of drug efficiency and novel formulations. Frontiers in Chemistry, 12, 1395359.

Anwar, G. A., Dong, Y., & Li, Y. (2020). Performance-based decision-making of buildings under seismic hazard considering long-term loss, sustainability, and resilience. *Structure and Infrastructure Engineering, 17*(4), 454–470. DOI: 10.1080/15732479.2020.1845751

Bardhan, A., Biswas, R., Kardani, N., Iqbal, M., Samui, P., Singh, M. P., & Asteris, P. G. (2022). A novel integrated approach of augmented grey wolf optimizer and ANN for estimating axial load carrying-capacity of concrete-filled steel tube columns. *Construction & Building Materials, 337*, 127454. DOI: 10.1016/j.conbuildmat.2022.127454

Cabot, J. H., & Ross, E. G. (2023). Evaluating prediction model performance. *Surgery, 174*(3), 723–726. DOI: 10.1016/j.surg.2023.05.023 PMID: 37419761

Cakiroglu, C., Islam, K., Bekdaş, G., Isikdag, U., & Mangalathu, S. (2022). Explainable machine learning models for predicting the axial compression capacity of concrete filled steel tubular columns. Construction & Building Materials, 356, 129227.

Chen, R. S., Zhang, H. Y., Hao, X. K., Yu, H. X., Shi, T., Zhou, H. S., & Wang, P. (2024, October). Experimental study on ultimate bearing capacity of short thin-walled steel tubes reinforced with high-ductility concrete. In *Structures* (Vol. 68, p. 107109). Elsevier.

Duong, T. H., Le, T. T., & Le, M. V. (2023). Practical Machine Learning Application for Predicting Axial Capacity of Composite Concrete-Filled Steel Tube Columns Considering Effect of Cross-Sectional Shapes. *International Journal of Steel Structures*, *23*(1), 263–278. DOI: 10.1007/s13296-022-00693-0

El-baky, M. A. A., Allah, M. M. A., Kamel, M., & Abd-Elaziem, W. (2022). Lightweight cost-effective hybrid materials for energy absorption applications. *Scientific Reports*, *12*(1), 21101. DOI: 10.1038/s41598-022-25533-3 PMID: 36473932

Gao, Y., Wang, B., Liu, C., Hui, D., Xu, Q., Zhao, Q., & Hong, X. (2022). Experimental study on basic mechanical properties of recycled steel fiber reinforced concrete. *Reviews on Advanced Materials Science*, *61*(1), 417–429. DOI: 10.1515/rams-2022-0041

Guaygua, B., Sánchez-Garrido, A. J., & Yepes, V. (2023, December). A systematic review of seismic-resistant precast concrete buildings. In *Structures* (Vol. 58, p. 105598). Elsevier.

Guo, Y. T., Chen, J., Nie, X., Tao, M. X., Wang, J. J., & Fan, J. S. (2020). Investigation of the shear resistances of steel–concrete–steel composite structures with bidirectional webs. *Journal of Constructional Steel Research*, *164*, 105846. DOI: 10.1016/j.jcsr.2019.105846

Hajihosseinlou, M., Maghsoudi, A., & Ghezelbash, R. (2024). Regularization in machine learning models for MVT Pb-Zn prospectivity mapping: Applying lasso and elastic-net algorithms. *Earth Science Informatics*, *17*(5), 4859–4873. DOI: 10.1007/s12145-024-01404-5

Ismail, M. (2020). New approach to seismic-resistant design and structural torsion mitigation. *Engineering Structures*, *207*, 110092. DOI: 10.1016/j.engstruct.2019.110092

Kavzoglu, T., & Teke, A. (2022). Predictive Performances of ensemble machine learning algorithms in landslide susceptibility mapping using random forest, extreme gradient boosting (XGBoost) and natural gradient boosting (NGBoost). *Arabian Journal for Science and Engineering*, *47*(6), 7367–7385. DOI: 10.1007/s13369-022-06560-8

Konstantinov, A. V., & Utkin, L. V. (2021). Interpretable machine learning with an ensemble of gradient boosting machines. *Knowledge-Based Systems*, *222*, 106993. DOI: 10.1016/j.knosys.2021.106993

Ly, H. B., Pham, B. T., Le, L. M., Le, T. T., Le, V. M., & Asteris, P. G. (2021). Estimation of axial load-carrying capacity of concrete-filled steel tubes using surrogate models. *Neural Computing & Applications, 33*(8), 3437–3458. DOI: 10.1007/s00521-020-05214-w

Megahed, K., Mahmoud, N. S., & Abd-Rabou, S. E. M. (2023). Application of machine learning models in the capacity prediction of RCFST columns. Sci Rep 13, 20878.

Sun, Y., Zheng, Y., Sun, Z., Yao, L., Zeng, Y., & Wu, G. (2023, July). Quasi-static experimental study on precast concrete columns reinforced with bundled rebars. In *Structures* (Vol. 53, pp. 70–82). Elsevier.

Wang, C., & Chan, T. M. (2023). Machine learning (ML) based models for predicting the ultimate strength of rectangular concrete-filled steel tube (CFST) columns under eccentric loading. *Engineering Structures, 276*, 115392. DOI: 10.1016/j.engstruct.2022.115392

Xue, H., Zhao, K., Xia, F., Wang, G., & Shen, A. (2023). Numerical Investigation of the Ultimate Load-Carrying Capacity of Square Concrete-Filled Steel Tube Columns Considering Initial Stresses Generated during Construction. *Buildings (Basel, Switzerland), 13*(11), 2830. DOI: 10.3390/buildings13112830

Yan, J., Ma, C., Kang, B., Wu, X., & Liu, H. (2021). Extreme learning machine and AdaBoost-based localization using CSI and RSSI. *IEEE Communications Letters, 25*(6), 1906–1910. DOI: 10.1109/LCOMM.2021.3058420

Zhang, C., Jin, Z., Feng, G., Tai, Y., Gao, R., & Li, C. (2024). Joint load-bearing behaviors and interface failure mechanism of encased coal samples under uniaxial compression. *Composite Structures, 347*, 118480. DOI: 10.1016/j.compstruct.2024.118480

Zhang, Y., Liu, A., Chen, B., Zhang, J., Pi, Y. L., & Bradford, M. A. (2020). Experimental and numerical study of shear connection in composite beams of steel and steel-fibre reinforced concrete. Engineering Structures, 215, 110707.

Zou, Y., Wang, L., Sun, Z., Pan, J., Chen, M., & Wu, Y. (2023). Experimental and numerical studies of concrete-filled corrugated steel tubular column under axial compression. *Engineering Structures, 276*, 114813. DOI: 10.1016/j.engstruct.2022.114813

Chapter 3
Deployment of IoT Devices, Sensors, and Real-Time Data Systems for Infrastructure Health Monitoring and Proactive Risk Management

Soumen Ghosh
http://orcid.org/0009-0008-3778-6090
Global Institute of Management and Technology, Krishnanagar, India

Parag Chatterjee
http://orcid.org/0000-0003-1243-9582
Global Institute of Management and Technology, Krishnanagar, India

Rupam Hazra
http://orcid.org/0009-0003-0332-2128
Global Institute of Management and Technology, Krishnanagar, India

ABSTRACT

The growing demands of climate resilience, urban expansion, and rapid technological change are transforming civil infrastructure. Traditional models can no longer meet the needs of a dynamic world facing environmental risks and resource constraints. This chapter explores how IoT, sensors, and real-time data enable smart, adaptive systems. Embedded sensors in key assets allow engineers to collect detailed data on structural and environmental conditions, supporting predictive maintenance and

DOI: 10.4018/979-8-2600-3038-7.ch003

ıt © 2027, IGI Global Scientific Publishing. Copying or distributing in print or electronic forms without written permission of IGI Global Scientific Publishing is prohibited. Use of this chapter to enerative artificial intelligence (AI) technologies is expressly prohibited. The publisher reserves all rights to license its use for generative AI training and machine learning model development.

early fault detection. AI and digital twin technologies enhance this data, offering simulations and decision tools to optimize infrastructure performance. Ethical and cybersecurity concerns are also addressed to ensure data integrity and system resilience. Through global case studies and updated educational strategies, the chapter highlights the need to train engineers with digital and interdisciplinary skills. Smart infrastructure presents a forward-looking solution to improving sustainability, safety, and urban resilience in the 21st century.

1. INTRODUCTION

Civil infrastructure forms the backbone of modern society, supporting transportation, water supply, energy distribution, and public safety. However, the increasing complexity of global challenges—ranging from extreme climate events and rapid urbanization to aging infrastructure and limited public funding—has exposed the vulnerabilities of traditional infrastructure systems. Historically designed for static conditions and long service lives, these systems are often ill-equipped to respond to dynamic stressors or provide early warning of failures. In response, the civil engineering field is undergoing a digital transformation, marked by the adoption of the Internet of Things (IoT), real-time data systems, and advanced sensor technologies. IoT enables continuous, remote monitoring of physical infrastructure through interconnected devices that collect and transmit data on a wide range of performance indicators such as stress, vibration, corrosion, and temperature. These data streams, when processed with artificial intelligence and analytics, provide valuable insights for predictive maintenance and risk mitigation. Additionally, digital twins—virtual replicas of physical assets—allow engineers to simulate various operational scenarios and optimize system behavior. These innovations are redefining infrastructure as smart, cyber-physical systems that interact with their environments, users, and managers in real time. This chapter delves into the deployment of IoT and sensor-based monitoring systems in infrastructure, illustrating their benefits, challenges, and transformative impact on how infrastructure is designed, operated, and maintained. It also highlights global case studies and outlines strategies for building capacity in the next generation of engineers to ensure long-term success and sustainability.

2. THE CHANGING LANDSCAPE OF CIVIL INFRASTRUCTURE

The landscape of civil infrastructure is rapidly evolving in response to global challenges such as climate change, rapid urbanization, aging assets, and technological disruption. Traditional infrastructure systems, once built for static conditions and long-term durability, are increasingly proving inadequate in the face of extreme weather events, population pressures, and shifting environmental patterns. This shift has prompted a move toward smart, adaptive, and sustainable infrastructure solutions that leverage technologies like the Internet of Things (IoT), real-time sensors, artificial intelligence, and digital twins. These innovations enable continuous monitoring, predictive maintenance, and data-driven decision-making, transforming civil infrastructure into dynamic, cyber-physical systems. These structures have influenced our urban areas and our relationship with the environment more profoundly than any other advancements of the industrial age(Alehashemi,2017). At the same time, there is growing emphasis on integrating green design, circular economy principles, and nature-based solutions to reduce environmental impact and enhance resilience. The role of the civil engineer is expanding beyond traditional technical domains to encompass systems thinking, sustainability planning, and cross-sector collaboration. Infrastructure is no longer viewed solely as physical assets but as living systems that must interact intelligently with their environment and users. This transformation requires new skill sets, updated regulatory frameworks, and greater interdisciplinary cooperation.

A. The Demand for Resilience and Sustainability

The demand for resilience and sustainability is becoming more pressing as the world faces a growing range of environmental, social, and economic challenges. Resilience focuses on the capacity of systems—whether ecological, economic, or social—to adapt, recover, and thrive after disruptions, such as natural disasters, climate change, or geopolitical crises. Sustainability, meanwhile, emphasizes meeting current needs without compromising the ability of future generations to meet their own, integrating environmental, economic, and social considerations into decision-making processes. Currently, economics and engineering oversee infrastructure processes and products by treating them as supply and demand relationships regarding money and energy(Yanev,2023). As global threats such as climate change, resource depletion, and inequality become more pronounced, organizations and governments are increasingly prioritizing sustainable development practices.

a) Environmental Sustainability

Environmental sustainability Centers on conserving natural resources, reducing pollution, and protecting biodiversity. As ecosystems face unprecedented stress from human activity, promoting practices such as renewable energy adoption, waste reduction, and conservation becomes critical. This ensures that resources are available for future generations while mitigating the impacts of climate change, such as extreme weather events and habitat destruction. By prioritizing environmental sustainability, we can help stabilize ecosystems and maintain the health of the planet, ensuring that natural resources are used responsibly.

b) Social and Economic Resilience

Social and economic resilience involves building systems that support communities in navigating crises while promoting fairness and equal opportunities. This includes strengthening social safety nets, enhancing access to education and healthcare, and promoting economic diversity to mitigate the impact of global shocks. Communities that are economically resilient are better positioned to bounce back from disruptions like recessions or pandemics. Additionally, fostering social equity ensures that vulnerable populations are not disproportionately affected, contributing to a more inclusive and stable society.

B. Climate Change, Urbanization, and Aging Infrastructure

The challenges posed by climate change, urbanization, and aging infrastructure are interlinked, creating complex issues for cities and communities worldwide. Climate change intensifies weather extremes, such as flooding, heatwaves, and storms, which strain existing infrastructure and increase the vulnerability of urban areas. At the same time, rapid urbanization leads to an expansion of cities and the increased concentration of people, placing additional pressure on already overstretched systems like transportation, water, and energy. Numerous cities are currently tackling climate change by cutting down their greenhouse gas emissions(Kessler,2011). Many urban Centers are also grappling with aging infrastructure—roads, bridges, buildings, and utilities that were built decades ago and are no longer suited to modern demands or resilient to environmental stresses.

a) Climate Change and Its Impact on Urban Infrastructure

Climate change significantly impacts urban infrastructure by exposing cities to more frequent and severe extreme weather events. For instance, extreme heat can weaken concrete and steel, reducing the lifespan of structures, while flooding can overwhelm drainage systems, leading to property damage and service outages. Furthermore, climate-responsive urban planning can reduce vulnerability by incorporating nature-based solutions like urban forests and wetlands, which can absorb excess rainwater and cool the environment.

b) Urbanization and the Need for Modernized Infrastructure

As urbanization accelerates, cities are facing a growing demand for more efficient and effective infrastructure systems. The rapid influx of people leads to overcrowded public transport systems, inadequate housing, and strained public services like water supply and waste management. Aging infrastructure in cities, often built to accommodate smaller populations, struggles to cope with the increased demands of modern urban life. Investments in infrastructure modernization, such as the adoption of smart city technologies, renewable energy systems, and integrated transport networks, are essential for improving livability, reducing congestion, and ensuring the sustainability of urban areas.

C. Role of Innovation in Modern Civil Engineering

Innovation plays a pivotal role in modern civil engineering by driving advancements in construction methods, materials, and design principles that improve efficiency, safety, and sustainability. New technologies, such as building information modeling (BIM), robotics, and artificial intelligence (AI), are revolutionizing how infrastructure projects are planned, designed, and executed, allowing for more accurate and cost-effective outcomes. In addition, sustainable design principles are now integrated into civil engineering practices, with innovations in materials like self-healing concrete, recycled steel, and energy-efficient systems helping to reduce the environmental impact of construction projects. In light of the numerous technological developments in recent decades, employers in the labor market are requiring civil engineering graduates to possess greater skills and expertise than in the past(Aparicio & Ruiz-Teran, 2007). The use of drones, sensors, and smart monitoring systems enhances the ability to inspect and maintain infrastructure in real-time, enabling more proactive maintenance and reducing downtime.

a) Technological Innovations in Design and Construction

Technological advancements such as 3D printing, robotics, and automated machinery are reshaping how civil engineering projects are executed. 3D printing allows for the creation of complex structures with reduced material waste, and robotic systems can assist in tasks ranging from construction to demolition with greater precision and efficiency. Additionally, the use of BIM enables the integration of data across the entire project lifecycle, from conceptualization to construction, ensuring better coordination among stakeholders and reducing the likelihood of errors or delays.

b) Sustainability and Green Building Innovations

Sustainability is at the heart of modern civil engineering innovations, as engineers develop solutions to minimize environmental impacts and promote energy efficiency. The use of environmentally friendly materials, such as recycled aggregates, low-carbon cement, and green insulation, reduces the ecological footprint of construction projects. Moreover, energy-efficient design, which includes features like passive heating, cooling, and solar power systems, is becoming standard practice in the construction of both residential and commercial buildings.

3. THE INTERNET OF THINGS (IOT) IN CIVIL ENGINEERING

The Internet of Things (IoT) is revolutionizing civil engineering by enabling real-time connectivity, automation, and intelligent decision-making across infrastructure systems. Through networks of embedded sensors, actuators, and connected devices, IoT allows civil infrastructure—such as bridges, roads, dams, buildings, and water systems—to continuously monitor their structural and environmental conditions. The Internet of Things represents a cohesive sector of the emerging internet, formed by the integration of existing technologies and the swift progression of networking (Chandanshive & Kazi, 2017). This constant data stream provides engineers with valuable insights into stress levels, vibrations, temperature changes, material degradation, and external impacts, enabling early detection of faults and proactive maintenance. IoT technologies also facilitate integration with Geographic Information Systems (GIS), Building Information Modeling (BIM), and cloud platforms, supporting digital twins and remote infrastructure management. By combining physical assets with digital intelligence, IoT transforms static infrastructure into smart, responsive systems capable of adapting to changing conditions in real time. This shift enhances safety, operational efficiency, and sustainability while reducing long-term costs and downtime.

A. Fundamentals of IoT Architecture

The fundamentals of Internet of Things (IoT) architecture lie in the integration of various devices, sensors, networks, and software to enable the collection, transmission, and analysis of data in real-time. IoT architecture typically consists of three main layers: the perception layer, which includes the physical devices and sensors that collect data from the environment; the network layer, which transmits the collected data through wired or wireless communication protocols; and the application layer, where the data is processed, analysed, and used for decision-making.

As a dynamic entity, the Internet continues to evolve, and we have now entered a period of extensive connectivity through a range of smart devices (i.e., objects) that link to the Internet (Firouzi et al., 2020). This architecture enables IoT systems to monitor and control devices remotely, making it essential for applications ranging from smart homes to industrial automation.

a) The Perception Layer and Data Collection

The perception layer, also known as the sensor layer, is responsible for sensing and collecting data from the physical world. It includes devices such as temperature sensors, motion detectors, cameras, and actuators that monitor variables like humidity, pressure, light levels, and other environmental factors. This layer is critical as it acts as the entry point for all data into the IoT system. Sensors can be embedded in a wide range of devices—from household appliances to industrial machinery—allowing for continuous data collection. The quality, accuracy, and frequency of data gathered at this layer directly influence the effectiveness of the entire IoT system, making it important to choose appropriate sensors for each application.

b) The Network Layer and Data Transmission

The network layer plays a crucial role in connecting the perception layer with the application layer by transmitting the data gathered by sensors to processing systems for analysis. This layer uses a variety of communication protocols, such as Wi-Fi, Bluetooth, Zigbee, LoRa, and cellular networks, to ensure seamless data transfer. It is responsible for ensuring that the data is transmitted efficiently, reliably, and securely to minimize latency and maximize the integrity of the data. The network layer also handles issues related to bandwidth and congestion, especially when large amounts of data are being transmitted.

B. IoT Device Ecosystems for Infrastructure Monitoring

IoT device ecosystems for infrastructure monitoring involve the deployment of interconnected sensors and devices to monitor and manage the health, performance, and safety of critical infrastructure such as bridges, roads, pipelines, and buildings. Nonetheless, recent methods in the big data age have infrequently been applied to the monitoring of operational ecosystems, especially concerning delicate ecosystems located in remote regions (Li et al., 2019). These ecosystems consist of a range of sensing technologies, including vibration sensors, temperature and pressure gauges, and cameras, which collect real-time data to detect potential issues like cracks, wear, or structural fatigue. The data collected by these IoT devices is transmitted through a network layer to centralized systems for analysis, allowing engineers and maintenance teams to make informed decisions regarding repairs, upgrades, or replacements. With the ability to provide continuous, real-time monitoring, IoT ecosystems enhance the resilience and safety of infrastructure by providing early warnings of failures and reducing the need for manual inspections.

a) Real-Time Monitoring for Early Detection of Issues

IoT ecosystems in infrastructure monitoring enable real-time detection of anomalies or defects in structures, helping identify issues before they escalate into costly repairs or safety hazards. For instance, sensors placed in bridges or dams can monitor vibrations, temperature changes, or deformations, which could signal impending structural issues. Similarly, in roads and highways, IoT sensors can detect cracks or uneven surfaces, triggering alerts that enable maintenance teams to address problems promptly. Continuous data collection allows infrastructure operators to assess conditions constantly, providing an up-to-date picture of the infrastructure's health.

b) Predictive Maintenance and Resource Optimization

IoT device ecosystems can optimize the maintenance and management of infrastructure through predictive analytics, reducing both operational costs and the likelihood of unexpected failures. By analyzing data trends over time, machine learning algorithms can forecast potential equipment breakdowns or system failures, prompting maintenance only when necessary, rather than adhering to fixed schedules. This not only saves costs associated with unnecessary inspections or replacements but also ensures that resources are directed toward critical areas in need of attention.

C. Communication Protocols and Network Topologies

Communication protocols and network topologies are fundamental components of IoT systems, enabling devices to connect, exchange data, and work together effectively. Communication protocols define the rules and standards for data exchange between devices, ensuring that the data is transmitted reliably, securely, and efficiently. A network topology refers to the layout of nodes, typically switches, routers, or features of software switch/routers, and their connections within a network, which is often depicted as a graph (Mamat et al., 2019). Common protocols used in IoT networks include MQTT (Message Queuing Telemetry Transport), CoAP (Constrained Application Protocol), and HTTP, each serving different needs depending on factors like bandwidth, power consumption, and latency.

a) Communication Protocols in IoT Systems

Communication protocols are crucial for the smooth operation of IoT networks, as they enable devices to communicate with each other and with central systems. For example, MQTT is often used in IoT environments where low bandwidth, low power consumption, and real-time messaging are required. It operates on a publish-subscribe model, which reduces the need for devices to maintain constant connections, thereby conserving energy. On the other hand, CoAP is designed for constrained devices and networks, such as those with limited processing power or storage, and is optimized for use in machine-to-machine (M2M) communication.

b) Network Topologies and Their Impact on IoT Performance

The network topology dictates how devices in an IoT ecosystem are connected and how they communicate with each other, directly affecting the performance, reliability, and scalability of the system. A star topology connects all devices to a central hub, making it easy to manage but potentially a single point of failure if the hub goes down. A mesh topology, where devices are connected to multiple others in a decentralized manner, offers greater fault tolerance and redundancy, ensuring continuous operation even if one or more devices fail. Tree topologies combine elements of both star and mesh, offering scalability while maintaining organized connections.

4. SENSOR TECHNOLOGIES FOR INFRASTRUCTURE HEALTH MONITORING

Sensor technologies play a critical role in infrastructure health monitoring by enabling the continuous assessment of structural integrity, environmental conditions, and operational performance. These sensors—ranging from strain gauges, accelerometers, and Fiber optic sensors to temperature, humidity, corrosion, and displacement sensors—are strategically embedded or mounted on infrastructure assets such as bridges, tunnels, buildings, and pipelines. They collect high-resolution data in real time, detecting early signs of wear, stress, deformation, or damage that could compromise safety or performance. Advances in wireless sensor networks (WSNs), low-power electronics, and energy-harvesting technologies have made these systems more efficient, scalable, and cost-effective, even in remote or hard-to-access locations. Over the past twenty years, a significant number of studies have been carried out to assess the advancements in research and development of optical fiber sensing technology, along with the use of optical fiber sensors for monitoring different types of engineering structures (Ye et al., 2014).

A. Types of Sensors: Structural, Environmental, Geotechnical

Structural sensors assess the physical condition of infrastructure by identifying problems like strain, stress, vibration, or displacement in bridges, buildings, and tunnels. Environmental sensors measure external conditions like temperature, humidity, air quality, and rainfall, which can affect the performance and durability of infrastructure systems. Geotechnical sensors are used to assess subsurface conditions, such as soil pressure, groundwater levels, and ground movement, helping engineers predict and prevent failures related to foundation shifts or landslides. Techniques such as computed tomography scanning, environmental scanning electron microscopy, microfluidics, particle image velocimetry, and transparent soils are now being used to investigate how geomaterials respond to various environmental changes in laboratory conditions (Soga et al., 2019).

a) Structural Sensors and Their Applications

Structural sensors play a vital role in monitoring the health and stability of civil infrastructure by capturing physical changes and stress responses. Common types include strain gauges, accelerometers, and displacement **sensors**, which measure how materials deform or shift under load. For example, strain gauges can detect the elongation of steel beams in a bridge, indicating potential fatigue or overloading. Accelerometers help monitor vibration frequencies in tall buildings or towers, which is crucial for assessing stability during earthquakes or heavy winds.

b) Geotechnical Sensors for Subsurface Monitoring

Geotechnical sensors are essential for monitoring ground behavior and subsurface conditions that can impact the safety of infrastructure such as dams, tunnels, and highways. For instance, piezometers help track fluctuations in water tables that could lead to slope instability or foundation weakening. Inclinometers detect lateral soil movement, which is especially important in landslide-prone or seismically active areas. By providing early warnings of geotechnical risks, these sensors enable engineers to implement preventive measures, ensuring the long-term safety and reliability of critical infrastructure.

B. Wireless Sensor Networks (WSNs) and Smart Nodes

Wireless sensor networks (WSNs) are networks that utilize sensors to track physical or environmental conditions(Dener,2017). Wireless Sensor Networks (WSNs) and smart nodes are integral to modern infrastructure monitoring systems, enabling real-time data collection and communication without the need for complex wiring. A WSN consists of a distributed network of spatially separated sensor nodes that gather environmental or structural data and wirelessly transmit it to a central processing unit. Each smart node in the network typically includes a sensor, microcontroller, communication module, and power source, allowing it to process data locally before transmitting relevant information. These systems are highly scalable, flexible, and cost-effective, making them ideal for monitoring hard-to-reach or expansive infrastructure, such as bridges, tunnels, and pipelines.

a) Components and Functionality of Smart Nodes

Each node integrates a sensor unit (for capturing data), a microcontroller (for local data processing), a wireless transceiver (for communication), and a power unit, often powered by batteries or energy harvesting methods like solar panels. The local processing capability reduces the volume of data sent over the network by filtering out irrelevant or redundant information, thus conserving energy and bandwidth. These nodes can operate collaboratively, relaying data from one to another in mesh or star topologies, depending on the system design. Their modular and autonomous nature allows for easy deployment and maintenance, especially in remote or hazardous locations.

b) Advantages of Wireless Sensor Networks in Infrastructure Monitoring

WSNs offer several advantages over traditional wired monitoring systems, particularly in terms of flexibility, scalability, and cost-efficiency. Because they eliminate the need for extensive cabling, WSNs are easier and cheaper to install, especially in large or geographically dispersed infrastructure projects. They can be deployed incrementally, allowing the network to grow with the monitoring needs. WSNs also enable real-time, 24/7 monitoring, which enhances the early detection of structural issues and improves response times to potential failures.

C. Calibration, Durability, and Deployment Techniques

Calibration, durability, and deployment techniques are critical factors in ensuring the accuracy, reliability, and long-term performance of sensors used in infrastructure monitoring systems. Calibration involves adjusting sensors to produce accurate readings by comparing their outputs to known standards, which is essential for maintaining measurement precision over time. Durability refers to the sensor's ability to withstand harsh environmental conditions such as temperature fluctuations, moisture, corrosion, and mechanical stress—especially important in outdoor and industrial applications. Effective deployment techniques ensure that sensors are strategically placed and securely installed to collect relevant and reliable data without interference or damage.

By utilizing this combined approach, it becomes feasible to assess durability metrics and/or mechanical characteristics, as noted by various researchers (Villain et al., 2018).

a) Importance of Calibration in Sensor Accuracy

Calibration is essential for ensuring that sensors deliver accurate and consistent data throughout their operational life. Without proper calibration, even high-quality sensors can drift over time due to environmental conditions, aging, or mechanical wear, leading to inaccurate or misleading readings. Regular calibration—performed before deployment and periodically during operation—helps maintain the integrity of the monitoring system. For example, in a structural health monitoring system, mis calibrated strain gauges could fail to detect critical stress changes, resulting in safety risks.

b) Deployment Techniques for Long-Term Durability

Effective deployment techniques enhance the durability and functionality of sensors by ensuring optimal placement, protection, and connectivity. This includes selecting appropriate mounting hardware, sealing against moisture or dust, and using materials resistant to corrosion and physical damage. For instance, sensors placed on bridges or roads must be installed in locations where they are exposed to relevant forces but protected from excessive wear or impact. Deployment may also involve considerations for power supply—such as integrating solar panels or long-life batteries—and ensuring strong wireless connectivity for data transmission.

5. REAL-TIME DATA SYSTEMS AND DIGITAL TWINS

Real-time data systems and digital twins are redefining infrastructure management by enabling dynamic, responsive, and intelligent decision-making throughout an asset's lifecycle. Real-time data systems collect continuous inputs from IoT sensors embedded in infrastructure to monitor parameters such as load, vibration, temperature, and environmental stress. This data is transmitted via wireless or edge computing networks to centralized platforms for immediate analysis, supporting timely interventions and condition-based maintenance. When integrated into a digital twin—a virtual replica of a physical infrastructure asset—this real-time data allows engineers to simulate performance, predict failures, and optimize operations under various scenarios. Digital twins serve as interactive models that mirror the live status and behavior of infrastructure, enhancing planning, diagnostics, and stakeholder communication. These systems can also incorporate weather forecasts, traffic data, and historical performance trends to enable predictive analytics and scenario planning. By utilizing this combined approach, it becomes feasible to assess durability metrics and/or mechanical characteristics, as noted by various researchers (Uhlemann et al., 2017).

A. Data Acquisition, Storage, and Transmission

In contemporary wireless communication, several technologies primarily offer solutions for wireless data transmission networks, including GSM, CDMA, 3G, and Wi-Fi (Jun et al., 2009). Data acquisition, storage, and transmission are foundational processes in any IoT-based infrastructure monitoring system, ensuring that sensor data is accurately collected, securely stored, and efficiently transmitted for analysis. Data acquisition involves gathering real-time information from various sensors deployed across infrastructure elements, capturing parameters such as stress, temperature, vibration, or soil movement. Once collected, this data must be stored in a reliable and scalable system—either locally (on edge devices) or remotely (in cloud or server-based databases)—to support future retrieval and analysis.

a) Efficient Data Acquisition Techniques

Efficient data acquisition is essential to capture accurate and relevant information while minimizing energy consumption and processing overhead. Techniques such as event-driven sampling, where sensors only collect data when a specific threshold is crossed, help reduce unnecessary data collection and conserve battery life. Advanced methods like edge processing allow smart sensors to filter and pre-process data before transmission, reducing the load on network bandwidth and central systems. Time-synchronized acquisition ensures that data from different sensors can be accurately correlated, especially in systems monitoring dynamic behavior like vibration or stress responses.

b) Secure and Scalable Data Storage and Transmission

Scalable storage and secure transmission are crucial for managing the vast volumes of data generated by sensor networks over time. Cloud-based platforms offer flexible and expandable storage solutions, supporting long-term data retention and advanced analytics, while edge storage provides faster access to recent data and ensures continuity during network outages. For data transmission, encryption protocols and secure channels such as TLS/SSL are essential to protect sensitive information from cyber threats. Redundancy in transmission paths and backup storage mechanisms also enhance system reliability, especially in mission-critical infrastructure like dams or tunnels.

B. Integration of SCADA, Cloud, and Edge Computing

Integration of SCADA, cloud, and edge computing is transforming infrastructure monitoring by enabling faster data processing, real-time control, and scalable analytics. SCADA (Supervisory Control and Data Acquisition) systems are traditionally used for real-time monitoring and control of industrial processes, offering centralized data visualization and decision-making. However, by integrating cloud computing, vast amounts of historical and real-time data can be stored, processed, and analyze at scale, enabling long-term trend analysis, AI-driven insights, and remote accessibility. In the diverse array of technologies transforming the industry, edge computing and cloud integration have arisen as essential resources with significant transformative potential (Narayana, 2024). Edge computing complements this by processing data closer to the source—directly on or near the sensors—reducing latency and bandwidth usage while ensuring faster local responses to critical events.

a) Role of SCADA in Real-Time Monitoring and Control

SCADA systems are foundational for industrial and infrastructure monitoring, providing real-time data acquisition, visualization, and control capabilities. They collect data from sensors and programmable logic controllers (PLCs), allowing operators to monitor system conditions and respond immediately to alarms or changes. SCADA interfaces offer dashboards, trends, and control options that are essential for day-to-day operations in sectors like water management, power distribution, and transportation. Although traditionally localized, modern SCADA systems can now be linked to cloud and edge platforms for greater flexibility and reach.

b) Benefits of Cloud and Edge Computing Integration

Cloud and edge computing greatly enhance the scalability and intelligence of infrastructure monitoring systems. Cloud computing allows for the centralized collection and storage of vast data streams from multiple sites, supporting complex analytics, machine learning, and global access to system dashboards. In contrast, edge computing enables data processing directly at the source—such as in smart sensors or gateways—allowing critical decisions (like issuing alerts or activating failsafe's) to occur almost instantly. This dual approach minimizes network congestion and ensures rapid local responses, even during connectivity issues.

6. AI AND ANALYTICS FOR INFRASTRUCTURE MONITORING

Artificial Intelligence (AI) and advanced analytics are transforming infrastructure monitoring by enabling automated, accurate, and predictive assessments of structural health and performance. By analyzing vast amounts of real-time and historical data collected from IoT sensors, AI algorithms—such as machine learning, deep learning, and neural networks—can detect patterns, identify anomalies, and forecast potential failures with high precision. These systems can differentiate between normal fluctuations and signs of deterioration, helping engineers prioritize maintenance and allocate resources more efficiently. For example, AI can process vibration data to detect early signs of fatigue in a bridge or analyze temperature and moisture patterns to predict corrosion in pipelines. A noteworthy development is the incorporation of wireless sensor networks (WSNs), which lower installation expenses and simplify processes, allowing for remote monitoring and distributed data analysis (Plevris & Papazafeiropoulos, 2024). In combination with computer vision and image processing, AI can also interpret drone or satellite imagery to assess damage in inaccessible areas. Additionally, analytics dashboards provide actionable insights through visualizations and decision-support tools, enhancing transparency and operational oversight.

A. Machine Learning for Pattern Recognition and Anomaly Detection

Machine learning (ML) for pattern recognition and anomaly detection is a powerful tool in infrastructure monitoring, enabling systems to analyze large volumes of sensor data to identify trends, behaviours, and irregularities. By training algorithms on historical and real-time data, ML models can recognize normal operating patterns and quickly flag deviations that may signal structural issues, equipment malfunctions, or environmental threats. These models continuously learn and improve over time, allowing for more accurate and timely predictions as more data is collected. Utilizing machine learning methods for intrusion detection enables the automatic creation of a model using the training dataset, which consists of data instances defined by a collection of attributes (features) and corresponding labels (Omar et al., 2013).

a) Pattern Recognition for Operational Insights

Pattern recognition using machine learning involves identifying consistent trends or behaviours in data streams, such as vibration patterns in bridges, load distributions in buildings, or temperature fluctuations in pipelines. By understanding what "normal" looks like for a particular asset, ML models can help operators optimize performance and spot inefficiencies. For example, in HVAC systems within a smart building, ML can learn daily usage cycles and suggest adjustments to reduce energy consumption. In structural health monitoring, repeated stress patterns detected by sensors can inform design modifications or load redistribution.

b) Anomaly Detection for Early Warning Systems

Anomaly detection algorithms are trained to flag unusual data points or behaviours that deviate from expected patterns, which could indicate emerging faults or safety risks. This is particularly useful in critical infrastructure such as dams, tunnels, and highways, where early signs of deterioration or failure may not be visible through traditional inspections. For instance, a sudden spike in strain or tilt readings from a sensor embedded in a bridge may indicate a weakening joint or foundation issue. ML models can detect these anomalies in real time and trigger alerts, enabling immediate investigation or intervention.

B. Predictive Maintenance and Lifecycle Forecasting

Predictive maintenance and lifecycle forecasting use data analytics and machine learning to anticipate equipment failures and optimize the long-term management of infrastructure assets. Instead of relying on scheduled maintenance, predictive maintenance analyses real-time sensor data to identify early signs of wear, fatigue, or malfunction, allowing for timely interventions before problems escalate. Lifecycle forecasting extends this approach by projecting how infrastructure components will perform over time, factoring in usage patterns, environmental conditions, and material degradation. Decisions based on data in the context of predictive maintenance management (PdM) are made by observing the variables of industrial equipment and their relationship with the statistical patterns of past data(Vallim Filho 2022). These techniques enhance asset reliability, reduce maintenance costs, and minimize unplanned downtime, all while improving safety and operational efficiency.

a) Benefits of Predictive Maintenance in Infrastructure Systems

Predictive maintenance allows engineers to detect potential issues before they lead to equipment or structural failures, reducing emergency repair costs and service disruptions. For example, vibration sensors on a bridge may detect abnormal patterns indicating joint fatigue, prompting targeted maintenance rather than full-scale inspections. This approach also avoids the inefficiencies of time-based maintenance, which may result in unnecessary servicing or missed signs of failure between inspection intervals. By focusing resources only where needed, predictive maintenance increases efficiency and frees up budgets for critical upgrades. Additionally, it improves safety by preventing accidents caused by undetected infrastructure weaknesses.

b) Lifecycle Forecasting for Long-Term Planning

Lifecycle forecasting provides infrastructure managers with a data-driven understanding of how assets age, degrade, and perform over time, supporting more strategic capital planning and resource allocation. Using historical data, material properties, and real-time monitoring, forecasting models can predict when an asset will require major repair, replacement, or reinforcement. For example, in a water pipeline network, forecasting tools can predict the deterioration of pipe sections based on corrosion rates and flow dynamics. This helps authorities prioritize investments, avoid sudden failures, and ensure continuous service delivery.

7. CYBERSECURITY AND ETHICAL DATA MANAGEMENT

Cybersecurity and ethical data management are essential components of modern infrastructure systems, especially as civil engineering increasingly relies on IoT devices, real-time data streams, and digital platforms. With infrastructure assets connected to the internet and continuously transmitting sensitive data, they become vulnerable to cyber threats such as hacking, data breaches, ransomware, and system sabotage, which can have severe consequences for public safety and operational continuity. Ensuring robust cybersecurity involves implementing encryption, secure communication protocols, intrusion detection systems, and regular vulnerability assessments across both hardware and software layers. The rapid advancement of digital integration has simultaneously revealed fundamental weaknesses in data privacy, cybersecurity, and ethical governance(Bayya,2022). Equally important is the ethical management of data, which includes ensuring data privacy, ownership rights, informed consent (when applicable), and compliance with relevant regulations like GDPR or local data protection laws. Engineers and planners must adopt transparent data governance frameworks that define who collect, accesses, stores, and uses the data—balancing innovation with responsibility.

A. Cyber Threats in Connected Infrastructure

Cyber threats in connected infrastructure are a growing concern as more critical systems—such as power grids, transportation networks, and water supply—become integrated with IoT, cloud, and digital technologies. These infrastructures are now vulnerable to a wide range of cyberattacks, including ransomware, data breaches, denial-of-service (DoS) attacks, and unauthorized remote access. A successful cyberattack can disrupt operations, damage physical assets, compromise public safety, or even lead to national security threats. Many infrastructure systems still rely on outdated or unpatched software, making them soft targets for sophisticated attackers. As connectivity increases, protecting infrastructure from cyber threats has become just as important as ensuring its physical integrity. Nevertheless, critical infrastructures that rely on IoT technology present an attractive target for cybercriminals. These unique systems are becoming increasingly susceptible to cyber vulnerabilities and are facing a growing number of cyberattacks (Djenna et al., 2021).

a) Common Types of Cyber Threats in Infrastructure

Infrastructure systems face several types of cyber threats, such as malware, phishing attacks, insider threats, and man-in-the-middle (MITM) attacks. Malware can infiltrate SCADA systems or IoT devices, disrupting operations or stealing sensitive data. Phishing emails may trick employees into revealing login credentials, granting attackers access to critical systems. MITM attacks can intercept data between devices and control Centers, altering commands or corrupting sensor data. Each of these threats can compromise system reliability, delay response actions, or lead to severe service outages.

b) Consequences of Cyberattacks on Public Infrastructure

The consequences of cyberattacks on infrastructure can be far-reaching, affecting both operational continuity and public safety. For example, an attack on a city's traffic management system could cause gridlock, accidents, or hinder emergency services. In a more severe case, a compromised power grid could lead to blackouts, impacting hospitals, transportation, and communication networks. Financial costs also escalate rapidly, including recovery expenses, reputational damage, and potential legal liabilities. Beyond the immediate damage, these attacks erode public trust in smart infrastructure systems. As such, cybersecurity must be treated as a core element of infrastructure design, operation, and governance.

B. Data Ownership, Integrity, and Privacy

Data ownership, integrity, and privacy are central concerns in the management of smart and connected infrastructure systems, especially as they rely on continuous data collection from a wide range of sources. Data ownership determines who controls and has the right to use or share the collected information, which becomes complex in multi-stakeholder environments involving governments, private contractors, and technology providers. Data integrity ensures that the data remains accurate, complete, and unaltered throughout its lifecycle—crucial for maintaining trust in automated systems and decision-making processes. Meanwhile, data privacy involves protecting personally identifiable information (PII) and sensitive operational data from unauthorized access or misuse, aligning with ethical standards and regulatory frameworks like GDPR or national privacy laws. Balancing these three aspects is essential for fostering public trust, enabling secure data sharing, and maintaining the reliability of digital infrastructure systems.

a) Defining Data Ownership and Access Rights

Clear definitions of data ownership are vital in determining who can access, manage, and benefit from data collected by infrastructure systems. In many cases, multiple parties—such as city governments, sensor manufacturers, and third-party service providers—may be involved, raising questions about who truly "owns" the data. Without well-defined agreements, there is a risk of disputes, data misuse, or restrictions that hinder innovation and collaboration. Establishing transparent data governance policies, ownership contracts, and access controls ensures accountability and proper usage. It also facilitates ethical data sharing across departments or organizations, enabling broader insights without compromising control.

b) Ensuring Data Integrity and Safeguarding Privacy

Maintaining data integrity involves implementing safeguards such as encryption, checksums, and secure logging systems to detect tampering, loss, or corruption. Integrity is crucial for systems where incorrect data could lead to dangerous decisions—such as faulty sensor readings causing a bridge to be deemed safe when it's not. At the same time, protecting privacy means minimizing the collection of unnecessary personal data, anonymizing datasets, and enforcing strict access controls. For example, smart traffic systems should track congestion patterns without storing drivers' identities. Combining technical safeguards with ethical practices ensures that data remains both reliable and respectful of individual rights.

8. SMART INFRASTRUCTURE IN PRACTICE: GLOBAL CASE STUDIES

Smart infrastructure is no longer a theoretical concept but a practical reality, as demonstrated by a growing number of global case studies that showcase the successful deployment of IoT, sensors, and real-time analytics in civil engineering projects. In South Korea, the Seohae Bridge features an extensive network of sensors that track the structural integrity and environmental factors, facilitating proactive maintenance and minimizing the likelihood of failure. Similarly, in the United States, New York City's water distribution system uses smart meters and leak detection sensors to improve efficiency and minimize water loss. In Europe, the Netherlands has implemented smart flood defense systems that integrate real-time weather data, tide levels, and sensor input to autonomously control barriers and pumps. India's Smart Cities Mission has incorporated intelligent traffic systems, air quality monitoring, and infrastructure dashboards to manage urban growth more sustainably. At the same time, Japan employs artificial intelligence and robotics for the inspection of tunnels and the monitoring of infrastructure that can withstand earthquakes. These case studies highlight how smart infrastructure enhances safety, efficiency, and resilience while reducing operational costs and environmental impact. They also demonstrate the importance of cross-disciplinary collaboration, public-private partnerships, and strong governance in scaling and sustaining smart infrastructure solutions worldwide.

A. Monitoring of Bridges, Highways, and Tunnels

Monitoring of bridges, highways, and tunnels is vital for ensuring public safety, optimizing maintenance, and extending the lifespan of critical transportation infrastructure. Using IoT sensors, real-time data can be collected on parameters such as vibration, strain, displacement, temperature, and traffic loads. These insights help engineers detect early signs of structural fatigue, corrosion, or ground movement, allowing for predictive maintenance rather than reactive repairs. Continuous monitoring also enables quick response to natural disasters, heavy usage, or accidents, reducing the risk of catastrophic failures. With smart infrastructure technologies in place, asset managers can make informed decisions to prioritize resources, minimize downtime, and improve transportation reliability.

a) Structural Health Monitoring in Bridges

Bridges are subject to dynamic loads, weather impacts, and aging, making structural health monitoring (SHM) crucial for early detection of damage or wear. Sensors such as strain gauges, accelerometers, and displacement transducers are embedded into critical areas to track real-time performance under traffic or environmental stress. For example, unusual strain patterns may indicate fatigue cracks forming in a steel beam. SHM systems allow engineers to monitor load-bearing components continuously and plan maintenance proactively, avoiding sudden closures or costly emergencies. These systems are especially valuable in aging infrastructure, where deterioration may not be visible until advanced stages.

b) Geotechnical and Environmental Monitoring in Tunnels and Highways

Tunnels and highways are exposed to environmental and geotechnical risks such as ground settlement, water ingress, and temperature fluctuations. Monitoring tools like piezometers, inclinometers, and environmental sensors help track soil pressure, slope stability, humidity, and gas concentrations. For instance, in long tunnels, gas detectors and temperature sensors can prevent fire hazards and ensure proper ventilation. Highways built on unstable terrain can benefit from subsurface movement detection to prevent landslides or pavement failure. Integrating this data with GIS systems allows for spatial analysis of risk zones, enhancing situational awareness and enabling timely interventions.

B. Smart Water and Wastewater Systems

Smart water and wastewater systems utilize IoT sensors, data analytics, and automated controls to monitor, manage, and optimize the flow, quality, and treatment of water resources. These systems provide real-time visibility into parameters like water pressure, flow rate, contamination levels, and energy consumption—enabling efficient operations and rapid response to leaks, blockages, or contamination events. Smart technologies also help utilities reduce water loss, detect aging infrastructure problems early, and ensure compliance with environmental regulations. By integrating sensor data with SCADA, cloud, and GIS platforms, operators can make informed, data-driven decisions to improve service delivery and environmental sustainability. Overall, smart water systems play a key role in conserving resources, reducing operational costs, and ensuring public health and safety.

a) Leak Detection and Pressure Monitoring

IoT-enabled leak detection systems and pressure sensors are critical for identifying and addressing water loss in distribution networks. These tools detect abnormal pressure drops or flow patterns that indicate leaks, pipe bursts, or valve failures—often before they become visible or cause major damage. For example, acoustic sensors can pick up the sound of leaking pipes underground and send real-time alerts to maintenance teams. By addressing leaks early, utilities save millions of Liters of water, reduce energy costs for pumping, and prevent infrastructure damage. Automated pressure monitoring also helps optimize water delivery and prevent pipe stress in high-demand zones.

b) Wastewater Quality Monitoring and Treatment Optimization

Smart sensors in wastewater systems monitor key parameters like pH, chemical oxygen demand (COD), turbidity, and toxic gas levels to ensure proper treatment and discharge. Real-time monitoring allows operators to adjust chemical dosing, aeration, or filtration processes automatically, improving efficiency and compliance with environmental standards. For instance, a sudden drop in oxygen levels in a treatment tank could trigger adjustments in aeration systems to maintain biological activity. These optimizations reduce energy consumption, improve effluent quality, and lower operational costs. In addition, historical and real-time data can be used for predictive maintenance of treatment equipment, minimizing downtime and extending asset lifespan.

9. BUILDING CAPACITY FOR THE FUTURE OF CIVIL ENGINEERING

Building capacity for the future of civil engineering requires a fundamental shift in education, training, and professional development to prepare engineers for the demands of a rapidly evolving, technology-driven landscape. As digital tools such as IoT, AI, digital twins, and data analytics become integral to infrastructure planning and management, civil engineers must acquire interdisciplinary skills that blend traditional engineering principles with competencies in computer science, data management, cybersecurity, and systems thinking. Academic institutions need to revise curricula to include hands-on experience with smart technologies, sustainability practices, and collaborative problem-solving across sectors. Lifelong learning and upskilling programs are essential for existing professionals to stay current with emerging tools and regulatory changes. Moreover, fostering a mindset of innovation, ethical responsibility, and resilience is critical to address the complex challenges of climate change, urbanization, and resource scarcity.

A. Interdisciplinary Education and Workforce Training

Interdisciplinary education and workforce training are essential to equip engineers, technologists, and decision-makers with the diverse skillsets needed for modern, smart infrastructure systems. As civil engineering integrates with digital technologies like IoT, AI, cybersecurity, and data analytics, traditional training must evolve into more cross-disciplinary programs. Professionals now need to understand not only structural mechanics but also programming, network protocols, sensor systems, and ethical data handling. Universities, vocational institutes, and industry programs are increasingly developing hybrid curricula and hands-on training modules that blend engineering fundamentals with digital literacy. This evolution in education ensures that the future workforce can design, deploy, and manage resilient, intelligent infrastructure in an increasingly connected world.

a) Bridging the Gap Between Engineering and Digital Technology

A major goal of interdisciplinary training is to bridge the gap between traditional civil engineering and emerging digital technologies. Engineers must now grasp coding basics, data interpretation, sensor integration, and cyber-physical systems to effectively implement smart infrastructure. For example, a bridge engineer may need to understand how to interface with wireless sensor networks to interpret load data or detect anomalies. Offering coursework in computer science, cybersecurity, and IoT as part of engineering programs empowers graduates to work collaboratively with software developers, data scientists, and IT professionals. This fusion of knowledge results in more agile, innovative infrastructure solutions.

b) Hands-On Workforce Development and Industry Partnerships

Practical, hands-on training is critical for preparing the workforce to operate and maintain smart infrastructure systems. Industry partnerships with universities and technical institutes help align training programs with real-world requirements through internships, apprenticeships, and certification programs. For instance, utility companies or smart city projects may offer field-based learning where trainees install and monitor IoT devices, analyze sensor data, or simulate infrastructure failures. These experiences enhance technical proficiency and foster problem-solving under realistic conditions. Strong collaboration between academia and industry ensures that the workforce remains up-to-date with evolving technologies and regulatory expectations.

B. Upskilling Engineers for Digital Systems and AI Integration

Upskilling engineers for digital systems and AI integration is increasingly vital as infrastructure management becomes more reliant on automation, real-time data, and intelligent analytics. Traditional engineering roles are evolving, requiring knowledge in areas like machine learning, IoT platforms, data visualization, and cloud-based control systems. Engineers must learn to work with smart tools that not only monitor performance but also predict failures, optimize operations, and automate decision-making. Upskilling through professional courses, online certifications, and industry workshops helps bridge the knowledge gap and fosters innovation in infrastructure design and maintenance. This transition enables engineers to take on hybrid roles that combine domain expertise with digital fluency, making infrastructure systems more efficient, adaptive, and resilient.

a) Training in AI and Machine Learning Applications

Engineers must understand how AI and machine learning can enhance infrastructure monitoring, fault detection, and predictive maintenance. This includes learning how to train models using sensor data, interpret AI-generated insights, and apply these tools to improve decision-making. For example, civil engineers may use AI to identify crack patterns in concrete from image data or to forecast bridge deterioration based on load history and environmental conditions. Training can be delivered through modular programs that blend theory with practical projects, giving engineers hands-on experience with real-world datasets. These skills not only improve individual capability but also lead to smarter, data-driven infrastructure systems.

b) Adoption of Digital Twin and IoT Platforms

Familiarity with digital twin technology and IoT platforms is crucial for engineers working on modern infrastructure. Engineers must be able to interact with digital replicas of physical assets, integrate live sensor data, and simulate performance under different scenarios. Upskilling programs can teach how to use tools like BIM (Building Information Modeling) integrated with IoT dashboards for live monitoring and lifecycle analysis. For instance, understanding how a digital twin updates in response to structural stress can help engineers plan maintenance with greater accuracy. Mastering these platforms allows engineers to shift from reactive maintenance to proactive, predictive strategies that extend asset life and reduce costs.

10. CONCLUSION

The deployment of IoT devices, sensors, and real-time data systems represents a pivotal advancement in the evolution of civil infrastructure, enabling a shift from reactive maintenance models to proactive, predictive, and intelligent asset management. Through continuous monitoring and data-driven insights, these technologies empower engineers to detect faults early, optimize maintenance schedules, and enhance the safety and longevity of critical infrastructure. Real-time data integration with AI and digital twins creates a dynamic feedback loop that facilitates informed decision-making and rapid response to environmental or operational stressors. However, this digital transformation is not without challenges. It necessitates robust cybersecurity infrastructure, ethical data governance, and systemic thinking to address risks related to data breaches, technological failures, and unintended consequences of automation. Moreover, it requires a significant investment in human capital—through education, upskilling, and cross-disciplinary collaboration—to ensure that engineers are prepared to navigate and lead within this rapidly evolving landscape. Global case studies demonstrate that smart infrastructure is both feasible and impactful, offering models for scalability and replication. As cities and nations strive to meet climate targets, enhance resilience, and build sustainable futures, the integration of IoT and real-time systems into civil infrastructure stands out as a strategic imperative. By embracing innovation and building capacity, civil engineering can play a central role in shaping a safer, more connected, and sustainable world.

REFERENCES

Alehashemi, A., Mansouri, S. A., & Barati, N. (2017). Urban infrastructures and the necessity of changing their definition and planning Landscape infrastructure; a new concept for urban infrastructures in 21st centur. *The Monthly Scientific Journal of Bagh-e Nazar, 13*(43), 5–18.

Aparicio, A. C., & Ruiz-Teran, A. M. (2007). Tradition and innovation in teaching structural design in civil engineering. *Journal of Professional Issues in Engineering Education and Practice, 133*(4), 340–349. DOI: 10.1061/(ASCE)1052-3928(2007)133:4(340)

Bayya, A. K. (2022). Advocating Ethical Data Management and Security. *International Journal of Scientific Research in Computer Science* [IJSRCSEIT]. *Engineering and Information Technology, 8*, 396–417.

Chandanshive, V. B., & Kazi, A. M. (2017, March). Application of Internet of Things in Civil Engineering construction projects-A State of the Art. In *Proceedings of the 11th INDIACom, 4th International Conference on Computing for sustainable global development* (Vol. 4, pp. 1836-1839).

Dener, M. (2017). WiSeN: A new sensor node for smart applications with wireless sensor networks. *Computers & Electrical Engineering, 64*, 380–394. DOI: 10.1016/j.compeleceng.2017.06.027

Djenna, A., Harous, S., & Saidouni, D. E. (2021). Internet of things meet internet of threats: New concern cyber security issues of critical cyber infrastructure. *Applied Sciences (Basel, Switzerland), 11*(10), 4580. DOI: 10.3390/app11104580

Firouzi, F., Farahani, B., Weinberger, M., DePace, G., & Aliee, F. S. (2020). Iot fundamentals: Definitions, architectures, challenges, and promises. In *Intelligent internet of things: from device to fog and cloud* (pp. 3–50). Springer International Publishing. DOI: 10.1007/978-3-030-30367-9_1

Jun, C., Sun-Zheng, Y. U., & Jing-Li, L. (2009). The design of a wireless data acquisition and transmission system. *Journal of Networks, 4*(10), 1042–1049.

Kessler, R. (2011). Stormwater strategies: cities prepare aging infrastructure for climate change.

Li, X., Zhao, N., Jin, R., Liu, S., Sun, X., Wen, X., Wu, D., Zhou, Y., Guo, J., Chen, S., Xu, Z., Ma, M., Wang, T., Qu, Y., Wang, X., Wu, F., & Zhou, Y. (2019). Internet of Things to network smart devices for ecosystem monitoring. *Science Bulletin, 64*(17), 1234–1245. DOI: 10.1016/j.scib.2019.07.004 PMID: 36659604

Mamat, H., Ibrahim, B. H., & Sulong, M. P. (2019, November). Network topology comparison for internet communication and IoT connectivity. In *2019 IEEE Conference on Open Systems (ICOS)* (pp. 1-5). IEEE. DOI: 10.1109/ICOS47562.2019.8975702

Narayana, S. S. (2024). Synergizing edge computing and cloud integration for enhanced efficiency in the oil and gas industry. *J. Comput. Eng. Technol, 15*, 63–71.

Omar, S., Ngadi, A., & Jebur, H. H. (2013). Machine learning techniques for anomaly detection: An overview. *International Journal of Computer Applications, 79*(2). Advance online publication. DOI: 10.5120/13715-1478

Plevris, V., & Papazafeiropoulos, G. (2024). AI in structural health monitoring for infrastructure maintenance and safety. *Infrastructures, 9*(12), 225. DOI: 10.3390/infrastructures9120225

Soga, K., Ewais, A., Fern, J., & Park, J. (2019). Advances in geotechnical sensors and monitoring. In *Geotechnical fundamentals for addressing new world challenges* (pp. 29–65). Springer International Publishing. DOI: 10.1007/978-3-030-06249-1_2

Uhlemann, T. H. J., Schock, C., Lehmann, C., Freiberger, S., & Steinhilper, R. (2017). The digital twin: Demonstrating the potential of real time data acquisition in production systems. *Procedia Manufacturing, 9*, 113–120. DOI: 10.1016/j.promfg.2017.04.043

Vallim Filho, A. R. D. A., Farina Moraes, D., Bhering de Aguiar Vallim, M. V., Santos da Silva, L., & da Silva, L. A. (2022). A machine learning modeling framework for predictive maintenance based on equipment load cycle: An application in a real-world case. *Energies, 15*(10), 3724. DOI: 10.3390/en15103724

Villain, G., Garnier, V., Sbartaï, Z. M., Derobert, X., & Balayssac, J. P. (2018). Development of a calibration methodology to improve the on-site non-destructive evaluation of concrete durability indicators. *Materials and Structures, 51*(2), 40. DOI: 10.1617/s11527-018-1165-4

Yanev, B. (2023). The supply and demand of infrastructure robustness, resilience and sustainability. *Građevinski materijali i konstrukcije, 66*(3), 157-168.

Ye, X. W., Su, Y. H., & Han, J. P. (2014). Structural health monitoring of civil infrastructure using optical fiber sensing technology: A comprehensive review. *TheScientificWorldJournal, 2014*(1), 652329. DOI: 10.1155/2014/652329 PMID: 25133250

Chapter 4
Connecting the Virtual and Real Worlds:
Digital Twins for Smart Cities and Urban Development

G. L. Swathi Mirthika
http://orcid.org/0000-0002-4591-7811
SRM Institute of Science and Technology, Kattankulathur, India

B. Sivakumar
http://orcid.org/0000-0003-1790-1780
SRM Institute of Science and Technology, Kattankulathur, India

Gowtham Rajendiran
http://orcid.org/0000-0002-7175-0576
SRM Institute of Science and Technology, Kattankulathur, India

Soshya Joshi
http://orcid.org/0000-0001-9791-0534
SRM Institute of Science and Technology, Kattankulathur, India

ABSTRACT

Virtual simulations become smart city and urban planning applications with digital twin technology. The essential stages from virtual to real-world deployment are studied in this urban digital twin development research. Integrating sensor, satellite, and municipal record data creates a digital city model. Virtual models allow simulation and modelling. It helps stakeholders imagine traffic management and sustainability. Through real-time monitoring and input, the digital twin correctly depicts the city. Urban planners and legislators can utilise current facts to decide. Machine learning-based predictive analytics and optimisation identify patterns and

DOI: 10.4018/979-8-2600-3038-7.ch004

nt © 2027, IGI Global Scientific Publishing. Copying or distributing in print or electronic forms without written permission of IGI Global Scientific Publishing is prohibited. Use of this chapter to generative artificial intelligence (AI) technologies is expressly prohibited. The publisher reserves all rights to license its use for generative AI training and machine learning model development.

optimise urban development. Digital twins offer transparent stakeholder decision-making and community participation. Digital twins guide physical implementations, and feedback loops update virtual models. Digital twins help cities solve problems, protect the environment, and service residents. Virtual and real worlds joined by digital twins improve urban designs.

1. INTRODUCTION

In the past few years, pertaining to smart cities, both in their general scope and limited senses the word has been subject to escalating forceful treatment and has into our psyche with growing intensity a variety of different angles. It is connected to the development of technology that smart cities are being built. The progression of technology elevates the standard of living of the people and, in many respects, makes it easier for the occupants to carry out their daily tasks, both for the inhabitants of today and for those who will come after them. Gartner has identified digital twins as one of the top 12 critical technology trends for the year 2022. Additionally, public organizations have acknowledged digital twins as an impactful instrument for urban development and civic administration. The objective of this chapter is to give our perspectives on the value of digital twins for this particular purpose, with a particular emphasis on the opportunities that exist in the cities, a model for the development of a digital twin that is suitable for its intended use, and the common needs for putting the technology into practice.

It is prompted to inquire about the defining characteristics of a "smart" city. A city is considered intelligent when it effectively combines technological advancements and the ecological balance to improve the competence of all its procedures. The ultimate goal is to achieve sustainable development, ensure the safety and health of its residents, and increase the overall standard of living for the population, as well as the immediate surroundings and the environment (Anthopoulos et al., 2023).

DTs were initially developed by the aircraft industry of the United States defense sector. Recently, its scope has been broadened to encompass transportation, industrial manufacturing, intelligent education, and several other industrial areas. They possess the capacity to provide valuable contributions to a wide range of applications, encompassing simulation, monitoring, evaluation, prediction, optimisation, and control. There exists a close correlation between DT technology and a number of other technologies. For the purpose of bringing about the digital transformation of businesses, DT is considered to be an essential technology, and it is also a popular technology that is of interest to both the business world and academic institutions.

It has only been in the past few years that the concept of DT has garnered a significant amount of attention, despite the fact that it has been existing for multiple years. The academic exchange presentations that took place at the "3rd Academic Conference on Digital Twin and Intelligent Manufacturing Services" in 2019 additionally boosted the rational of researchers from a variety of fields on ideas and technology associated to DTs. A year ago, the inaugural International Conference on Digital Twin Technology and the inaugural International Workshop on Model-driven Engineering with Digital Twin took place. Both of these events were held in United States.

2. CONCEPT OF DIGITAL TWIN

2.1 The Conceptual Model

To accurately establish a digital model, it is imperative that there is no automated data interchange among the real model and the computerized version. A digital model is a virtual image or representation of a tangible thing that is either already existing or expected to exist in the near future. Some examples of digital models include, but are not limited to, blueprints for buildings, designs for products, and the development of new products. The absence of any kind of automatic data exchange amid the digital model and the physical system is the maximum crucial characteristic that distinguishes it from other systems (Bauer et al., 2021). This indicates that once the digital model has been established, any modification that is done to the physical object will not have any effect on the digital model in any given circumstance. Illustration of a digital model can be found in Figure 1.

2.2 The Shadow of Digital

A digital tracker is a depiction of an article that exists in the digital realm and has a flow that is only in one direction between the computerized and physical worlds.

There is a direct correlation between a change in the state of the physical object and a change in the digital object, rather than vice versa. Another example of a digital shadow is shown in Figure 1.

2.3 The Digital Twin

The term "Digital Twin" refers to the situation in which data travels between a physical thing that already exists and a digital object, and the two objects are fully integrated in both directions. In the event that a modification is affixed to the tangible entity, it will automatically result in a modification to the digital object, and vice versa.

It is possible to detect the common misconceptions that are found in the literature with the assistance of these three definitions. On the other hand, there are a number of misconceptions that have been observed, and they are not restricted to having only these particular examples. There is a common misunderstanding that digital twins must be a perfect three-dimensional representation of a physical object. This is one of the prevalent fallacies. On the other hand, there are some people who believe that a digital twin is essentially a 3-D blueprint.

Figure 1. Combination stages of digital twin

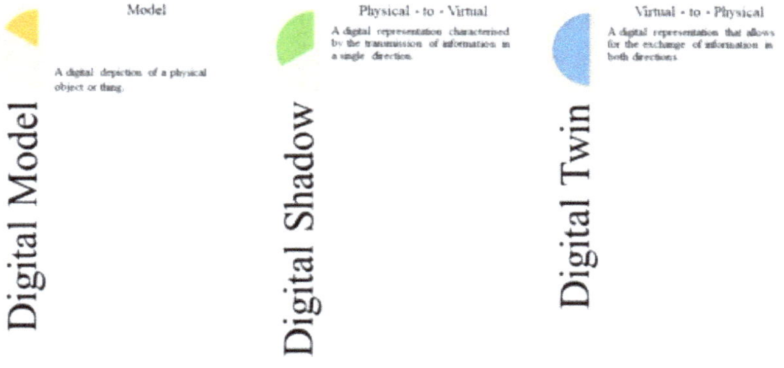

Figure 1. Combination Stages of Digital Twin

An early instance of Digital Twin technology was employed by NASA engineers in 1970 to resolve and rescue Apollo 13. They used a simulant, which replicated the knowledge unit, along with a distinct replica of the module's electric system. NASA engineers successfully executed the procedure in less than two hours, thereby preserving the exists of the three space traveler on board. This was an exceptional initial use of this technology, and the expertise has only advanced further since then. Currently, NASA use Digital Twins to advance the development of cutting-edge vehicles and planes (Boschert & Rosen, 2016).

The notion of Digital Twins is not novel. Nevertheless, the development of Digital Twins takes accelerated significantly in current years, transitioning from a mere idea to a tangible realityIntegrating of Digital Twins with supplementary technologies such voice capabilities, augmented reality, IoT, and artificial intelligence (AI) is expected. Digital Twins were included in Gartner's ten most significant Technology Trends for 2017, indicating their substantial importance. Based on Gartner's projection, over 50% of prominent industrial firms are expected to use Digital Twins in essential business applications by 2021. Industry research conducted by Markets and Markets has predicted a substantial increase in the use of Digital Twin technology in the following years (Charitonidou, 2022). This increase can be attributed to the increasing interest of the manufacturing industry in cost reduction and enhancement of supply chain operations.

3. DT TECHNOLOGY COMPONENTS

There are three primary components of Digital Twins: data collecting, data modelling, and data submission. Digital Twin use four technologies to gather and store live data, acquire information for generating important insights, and construct a digital representation of a tangible object. The mentioned technologies encompass the Internet of Things (IoT), Artificial Intelligence (AI), Extended Reality (XR), and Cloud. Furthermore, Digital Twin employs a specific technology, which varies in its degree of utilization based on the type of application.

3.1 The Internet of Things (IoT)

is an extensive network of linked entities. Connections may occur among items, individuals, or between individuals and objects. The primary technology employed in all applications of Digital Twins is the Internet of Things (IoT). Based on a forecast, more than 90 percent of all Internet of Things (IoT) systems are expected to have the capability of Digital Twinning by 2027.

The Internet of Things (IoT) use sensors to collect data from tangible objects in the actual environment. Data transmitted by the Internet of Things (IoT) is used to create a digital duplicate of a physical object. Subsequent analysis, modification, and optimisation of the digital replica is possible. Digital Twin applications rely on the Internet of Things (IoT) to constantly update data and create a real-time virtual replica of a physical object. Thus, every Digital Twin application utilises the Internet of Things as its primary technology.

3.2. Cloud Computing

Cloud computing refers to the delivery of remotely hosted services via the Internet. The system efficiently saves and provides access to data over the Internet.

Cloud computing facilitates the use of data processing hardware and cloud-based data storage systems for Digital Twins.

Cloud computing enables the storage of enormous amounts of data from Digital Twin in a virtual Cloud, allowing for convenient access to the necessary information from any location. Cloud computing allows for the efficient reduction of computation time for complicated systems and the resolution of challenges related to storing vast quantities of data.

3.3 Artificial intellect (AI)

Computational artificial intelligence (AI) is a branch of computer science that seeks to emulate the essential elements of human intelligence to create a novel kind of intelligent machine capable of engaging with humans in a manner resembling human-to-human intelligence. Artificial intelligence (AI) research includes the domains of Robotics, image recognition, and spoken language recognition.

Neural networks, machine learning, deep learning, and expert systems are integral elements of artificial intelligence (AI) that have the potential to augment digital twin modelling. Through the use of sophisticated analytical tools, artificial intelligence (AI) can autonomously examine data gathered from digital twins and provide significant insights, forecast future results, and issue recommendations for avoiding any problems (Dani et al., 2023).

3.4. Extended Reality (XR)

is a thorough concept that includes immersive technologies, namely Virtual Reality (VR), Augmented Reality (AR), and Mixed Reality (MR). Through the integration of the physical and virtual realms, these technologies have the capacity to broaden our understanding of reality. Extended reality (XR) technology facilitates the generation of digital depictions of objects, therefore enabling the coexistence and immediate interaction between digital and physical entities. In order to facilitate user interaction with digital content, Digital Twins utilise extended reality (XR) capabilities to generate digital representations of physical objects.

3.5. Production Industry,

next recognised application of Digital Twin is in the industrial industry. Manufacturers strive to develop methods to trace and supervise products in order to optimise efficiency and reduce costs, therefore fuelling their motivation. This elucidates the reason why Digital Twins are exerting the more significant impact in this particular setting. Manufacturing is rapidly adopting Digital Twins because of the crucial role they play in facilitating connectivity in the development of smart cities. The present expansion is in line with Industry 4.0, the fourth industrial revolution, which utilises device connection to facilitate the usage of Digital Twins in manufacturing operations. In real-time, the Digital Twin can deliver updates on machine performance and feedback from the production line.

This empowers producers to proactively anticipate difficulties at an earlier stage. Digital Twins optimise device connection and feedback, resulting in significant enhancements in reliability and performance. Integration of AI algorithms with Digital Twins can enhance precision by storing extensive amounts of data for the purpose of performance and forecast analysis (Dani et al., 2023). Establishing a Digital Twin environment for the purpose of product testing and real-time data analysis in production conditions could prove to be quite advantageous. One prominent application of Digital Twins is in the automotive industry, exemplified by Tesla. Developing a Digital Twin of an engine or automobile component can facilitate numerical simulation and data analysis. artificial intelligence improves testing precision by doing.

3.6 Live vehicle data analytics predicts present and future component performance.

Multiple uses of Digital Twins exist within the building industry. Digital Twins have the capability to be utilised at every phase of the construction process of a building or structure. This approach has practical applications in the development of smart city infrastructures and as a tool for real-time prediction and monitoring. Applications of Digital Twins and data analytics can enhance precision in predicting and controlling buildings and structures, by implementing virtual modifications in a physical manner.

The utilisation of the Digital Twin enables construction teams to implement algorithms in real-time prior to the creation of the actual structure, therefore enhancing the precision of simulations. Real-time simulation, as opposed to static blueprint models, is a common objective in the field of Digital Twins study. Despite their usefulness, the predictability and learnability of these models are limited by their absence of real-time parameters. The Digital Twin can employ machine and deep learning methodologies for the purposes of learning and monitoring.

3.7 Medical care

Applications of Digital Twin technology extend to the healthcare sector as well. Advancements in enabling technology have had an unequalled impact on healthcare, since what was previously deemed unattainable is now attainable. The Internet of Things (IoT) has benefitted from the affordability and ease of deployment of its gadgets, resulting in a notable surge in connection. The increasing interconnectedness is consistently improving the viable application of Digital Twin technology in the healthcare sector. A forthcoming application entails the development of a Digital Twin of a human object, enabling a live analysis of the physical structure. One current practical application is the use of a Digital Twin to simulate the effects of particular drugs. Another use entails using a Digital Twin to precisely plan and carry out surgical operations.

Furthermore, similar to other healthcare applications, the utilisation of a Digital Twin allows researchers, doctors, hospitals, and healthcare providers to replicate environments customised to their specific needs, either in real-time or for future progress and applications. Furthermore, the Digital Twin can be employed in tandem with AI algorithms to augment prediction capacities and enable well-informed decision-making. Many healthcare applications do not directly interface with the patient but nevertheless provide valuable contributions to ongoing care and treatment. Hence, these systems make a vital contribution to the provision of patient care. Though still in its nascent phase, the use of digital twins in healthcare has vast potential applications. These applications span from using digital twins for bed management to integrating them into large-scale wards and the general administration of hospitals.

The capacity to mimic and performance in real-time is crucial in healthcare, as it can determine whether a person lives or dies. The Digital Twin can also aid in the anticipation of maintenance needs and continuous overhaul of medical equipment. The utilisation of the Digital Twin in the medical setting has the capacity, in conjunction with Artificial Intelligence (AI), to provide life-saving determinations by analysing both current and past data.

This document identifies many applications of a Digital Twin, highlighting the similarities in their intended purpose. It demonstrates how predictive maintenance may be applied to both industrial plant machines and patient care. Additionally, it demonstrates instances where there is no overlap between applications and the use of Digital Twin is limited to its intended purpose. The progress in artificial intelligence (AI), Internet of Things (IoT), and Industry 4.0 has enabled the development of digital twin applications.

3.8 Some of the other real time use cases are listed below

Real-time monitoring and predictive analysis involve the use of digital twins to continuously monitor a city and forecast future situations, such as potential infrastructure problems or traffic congestion. By adopting a proactive strategy, organisations may do preventative maintenance and allocate resources effectively, so assuring seamless operation.

3.9 Urban Planning and Design

Conduct simulations to assess the effects of new developments or policy changes prior to making investments. Evaluate stream of traffic flow, energy consumption, and environmental influence by virtually testing designs.

3.10 Enhancing design efficiency and reducing expensive

Enhanced Operational Efficiency: Unceasing monitoring and real-time data analysis provide predictive maintenance, allowing for the identification of probable equipment faults before they actually happen. This minimises the amount of time that a system is not operational and decreases the need for expensive maintenance.

Citizen Engagement and Participation: Interactive platforms enabled by digital twins enable citizens to access up-to-date information, report problems, and offer feedback, hence enhancing citizens' involvement and participation. This promotes openness and cooperation between citizens and city authorities.

Resilience and Emergency Response: Utilize simulations to replicate emergency situations such as natural disasters or security threats, thereby strengthening the ability of cities to withstand and recover from such events. Digital twins assist cities in comprehending potential consequences, optimizing reaction strategies, and efficiently allocating resources.

3.11 Data Security and Privacy

Ensuring belief and confidentiality are crucial in smart cities, particularly regarding data. Therefore, maintaining the security of data is of utmost importance for digital twins. State-of-the-art encryption, robust access restrictions, and sophisticated anonymization methods are utilised to safeguard citizen data and thwart unauthorised admittance. By following this approach, citizens can have assurance that their data is managed in a responsible manner and utilized for the betterment of the municipality.

3.12 Integration with Emerging Technologies

Digital twins are not merely static duplicates; they continuously adapt and progress in parallel with state-of-the-art technology. Artificial intelligence (AI) and machine learning (ML) enhance data analysis by revealing concealed designs and tendencies, leading to improved managerial.

Interoperability and Standards: The use of open standards and frameworks is essential for ensuring smooth integration across various digital twins and urban systems. Association between organizations and industrial consortia fosters additional innovation and advancement, guaranteeing compatibility.

3.13 The concept of digital twin and computer simulation

The modelling and replication serve as the foundation for the production of a digital duplicate and the understanding of digital transformation (DT) of the object or system that is being examined. It is a simulation that is based on the real world that constitutes the ultimate digital twin model. Because of this, it can be concluded that DT would be unable to exist in the long run if it were not for the advances in simulation and modelling technologies. DT that is processed on the basis of the model and its simulation makes it possible for it to creation, its continuing development, and its subsequent application for the user's benefit are all included (Gupta et al., 2021).

DT is not merely a digital model, despite the fact that it uses digital models; this is the primary distinction between the two. "The distinction between the two digital twin and a simulation are highly dependent on the magnitude of the simulation: A simulation is normally used to investigate a single process; however, a digital twin can be used to conduct any number of meaningful simulations, which allows it to investigate several processes simultaneously. (How the year 2023). The development of DT results in the creation of a virtual environment that concurrently focusses on the virtual simulation of numerous processes. This results in the development of a more robust environment for the purposes of practice and research. On the other hand, this is hardly the end of the disparities. Simulations offer limited opportunities for evaluating the performance of the system, and they only take into account the system's actual physical components to a limited level. This is not even taking into consideration the fact that the conditions of the processes being researched are continually changing. There is a two-way flow of information between the real system and its virtualization in the form of DT, which is made possible by the design of DT, which is based on the concept of a data link with reality. Not only does DT take into account the current condition of the real item, but it also takes into account the state of the real object in the past. This allows for the formulation of predictions, which may be taken into consideration simultaneously in key decisions.

Data and information regarding the object or system that is being researched are provided by DT through the implementation of a variety of technologies, including recording stimuli, data, and information recording. By utilizing the technology that was discussed, we are able to comprehend a wide variety of sensors that have been installed, as well as drones, diagnostic devices, and mobile applications. Additionally, the system processor receives recorded data from the sensors and other technologies of the object that is being studied. This data is then exchanged with the object that is being investigated.

DT data is provided in real time by recording and continuously updating data, which is accomplished by sensors. When the computational capacity that comes with the DT virtual environment is combined with the ability to analyze several problems simultaneously from a wide variety of domains, it is conceivable to do so.

As a result, the primary distinction between simulation and DT is in the content, breadth, and method of acquiring data, as well as the subsequent processing of those data. In most cases, a simulation is constructed on the basis of the starting assumptions of a pre-specified set of parameters and the different scenarios that are formed. Its launch and research of the future evolution of the real object can be repeated and altered according to the preferences that are wanted, as well as when modifications are made to the parameters. The visualization in DT is continuously modified in accordance with the data that is received in real time, which has the potential to produce a higher potential for improving the processes that are being researched.

4. DIGITAL TWIN IN SMART CITIES (SC)

The concepts of Internet of Everything (IoE) and Automation of Everything (AoE) combine people, organizations, lives, processes, data, and things into a solid and coherent structure known as Cyber-Physical Systems (CPSs). This structure aims to create a smarter and more connected globe through synergy. Smart city (SC) is a city that integrates physical infrastructures, Information Communication Technology (ICT) infrastructures, social infrastructures, and business infrastructures in order to utilize the combined intelligence of the city (Liu et al., 2021). The primary goals of establishing Smart Cities (SCs) can be summarized as follows: i) Facilitating the integration of distributed services and resources in a synergistic manner. ii) Enhancing existing public services and providing new citizen-centric, user-driven, and demand-oriented services. iii) Monitoring cities using user-friendly visualization tools. iv) Enabling near real-time services for end-users and/or smart actions. v) Promoting sustainability through optimized services. vi) Improving the quality of life and livelihoods of citizens. vii) Driving economic development, innovation, and global competitiveness in cities.

For more information about the technological architecture of supply chains (SCs) including communication networks. Additionally, provides further information on real-world use cases of SCs.

The main components that facilitate accurate sensing and effective autonomous action are: i) active involvement with all stakeholders, ii) utilisation of edge IoT devices and citizens to gather data and interact with the environment intelligently by leveraging advanced communication technologies, iii) implementation of edge/fog platforms, iv) utilisation of cloud platforms that include cloudlets, and v) integration of intelligent domains both within its own system and with national and global intelligent domains. The additional intelligent areas in which SC is integrated to facilitate a collaborative implementation, leading to the advancement of a cluster of SCs and subsequently intelligent regions and intelligent nations.

The metaverse is a concept that refers to virtual and supplemented immersive 3D spaces that are democratized, decentralized, and driven by users. It allows for a more tangible connection between the virtual and physical worlds, enabling people who are not physically together to interact through their avatars and have a wide range of experiences. The metaverse, which combines virtual and physical realities, aims to create virtual worlds that are as detailed as the actual world. However, it is still in its early stages and requires significant development.

Open-source metaverse development platforms are currently in high demand and are anticipated to accelerate the creation of several unexplored metaverse experiences. In order to achieve truly persistent and immersive computing that can be accessed by billions of people in real time, a 1000-fold increase in computational efficiency is necessary compared to the current state of the art. This surpasses the growth predicted by Moore's law, which is expected to only result in a 10-fold increase over the next five years through hardware improvements. However, this high computational requirement can be addressed through energy-efficient computing, better algorithms, and improved architectures, as suggested by Intel. For instance, Ethereum plans to reduce their energy consumption by around 99% by transitioning to a mechanism called proof of stake, which relies on energy-efficient computing and processing. The metaverse is currently limited by constraints like as resources, processing capacity, edge intelligence, and sensory devices. As a result, it has not yet achieved its goal of complete immersion, materialisation, and interoperability.

Blockchain is a form of Distributed Ledger Technology (DLT) that operates as a decentralised Peer-to-Peer network. It securely keeps a digital ledger in a distributed manner. Smart contracts enhance the functionalities of blockchain technology. They are programmable codes that can digitise and automate the terms and conditions of an agreement among different entities. These contracts are deployed on the blockchain. Smart contracts offer several benefits including automation, access control, establishment of trust, and removal of the need for third-party involvement in execution. Blockchain can be employed in urban metaverse settings as a means of providing digital evidence of asset ownership.

5. URBAN APPLICATIONS OF THE METAVERSE AND UPCOMING TRENDS

We are utilizing technology to address our urban challenges and improve the quality of life for our citizens. What is the process for implementing the metaverse to fulfil this objective? This section investigates the existing and future contributions of the metaverse to the urban ecosystem and its potential relevance in future urban metaverse planning. The implementation of intelligent urban living and working through virtualization has already commenced in many cities in diverse manifestations. Seoul, as the initial metaverse Smart City, has spearheaded several urban metaverse initiatives. Santa Monica is the inaugural American city to participate in the metaverse (Lv et al., 2022).

The integration of the metaverse with smart city applications is experiencing rapid growth as a component of intelligent urbanization.

The use cases of the real-world urban metaverse that are currently in the early stages of development. These use cases demonstrate how cities are actively participating in the metaverse. In addition, the specific goals, patterns, and requirements that the public and commercial sectors are aiming for in the future urban metaverse. This table provides insights into how cities can interact with the metaverse in the coming years and decades. Several municipal administrations, particularly in Chinese towns, have recently unveiled their intentions to convert their smart city programs into metaverse settings. There is currently no initiative being undertaken by a representative.

The objective of this project is to build and propose a comprehensive urban metaverse space development called MetaOmniCity. To assist city managers in visualizing and subsequently creating this environment using clearly defined and mutually accepted rules.

Because it is so widely used in a variety of contexts, DT also has a significant number of applications in smart cities. It is possible to use digital technology to make use of smart maps and geospatial virtual infrastructure in order to contribute to the general prosperity of cities and the people who live in them.

According to Pramotedham, the Chief Executive Officer of Esri Singapore, "only with a digital twin can a government agency effectively analyse what can be done with data to improve the lives of citizens, economic opportunities, and revitalise the wider community." Through the utilisation of cyber-physical systems, the digital twin makes it possible to create a comprehensive representation of a virtual city environment.

Because of this, the digital twin is able to respond to new issues and enable the solution of problems with infrastructure and transport networks. The primary capability of the digital twin is therefore to mirror various functions of the city through data that is available in real time. The level of modernisation, as well as the level of building and implementation of smart city technologies, are all factors that determine the availability of imported data in DT.

Laser scanners, cameras, sensors that capture changes in weather conditions or flood activity, global positioning system technologies, and several mobile application platforms in smartphones are some examples of the technologies that have been launched for the purpose of data collecting. These technologies are generated in a variety of forms and formats.Each and every piece of information that is gathered is first saved in the cloud, and then it is transmitted to the digital version of the Smart City Digital Twin.

Utilising the level of smart technologies that have been introduced, DTs make it possible to collect information about the site and its operation characteristics in a single information database. This makes it easier to quickly solve any technical problems or incidents that may occur while the site is being operated (these may include leaks, power interruptions, failure of various systems, including ventilation, fire, and so on). Due to the fact that, in many instances, this sector is not developed enough to fulfil the requirements of delivering automated information provision, the digital twin was launched for industrial companies as well as transportation infrastructure.

Utilising DT in Smart Cities offers the benefit of virtually controlling the status of the city's physical system by forecasting and optimising its behaviour in real time. This is accomplished through the use of DT. Through the utilisation of virtual models or machine learning, it is feasible to enhance productivity and guarantee continuous monitoring by means of the capture of data in real time. It is possible to utilise the DT test platform to verify many scenarios in order to select the variation that is the most appropriate for improving business decisions, enhancing system performance, and controlling the physical system. The detection of incorrect behaviour in the analysed instance of emergencies is one of the ways that DT contributes to the improvement of safety and the acceleration of resilience.

Digital transformation makes it possible to do a study of the detection, mitigation, and resilience of technology in a virtual environment prior to the actual deployment of the technology. A number of things are made possible by the information included in DT, including the detection of attacks and the restoration of the system state by utilizing data from the DT virtual environment. DT can also be utilized to create a "what if" scenario, which is another example. The study of the system's response and a more accurate risk assessment are the outcomes that come about as a consequence of testing the development of the situation in unexpected circumstances.

The digital twin theory is being implemented in various sectors of Smart Cities, as shown in Figure 2. These fields include:

- Agriculture
    - o Natural resource distribution
    - o Greenhouse production involves the cultivation of plants in a controlled environment, typically a structure made of glass or plastic.
    - o On the other hand, designing a livestock farm entail creating a layout and infrastructure that is suitable for raising and managing animals.
- City transportation
    - o Ensuring safety when driving and understanding the functioning of railway turnouts.
- Urban health management
    - o Aquatic purification
    - o Medical infrastructure
- Security
    - o Urban disaster management
    - o Urban infrastructure defect analysis

The Figure.2 which outlines the key sub-regions on which the world focusses most while generating DT applications in Smart City, provides a closer look at the applications that are being developed in specific areas. In Smart City, DT is increasingly permeating the aforementioned regions and provides a multitude of benefits because of this. By participating in the planning and decision-making processes associated with the Smart City through the use of DT, the benefit consists of linking an active approach to improving the Smart City.

Urbanisation and building planning are the two most important areas that make use of DT in Smart City programs. One of the advantages of subterranean utility networks is that they provide information that may be utilised for the purpose of optimising land usage and developing spatial arrangements. In its most basic form, DT has the potential to lessen the amount of uncertainty and risk that is involved in the planning and design of the development of infrastructure in urban areas. DT is utilised for a wide variety of activities, including the elimination of disturbances to the environment and subsurface services, as well as the enhancement of safety in the processes of building and engineering works.

Figure 2. Digital twin applications in smart cities

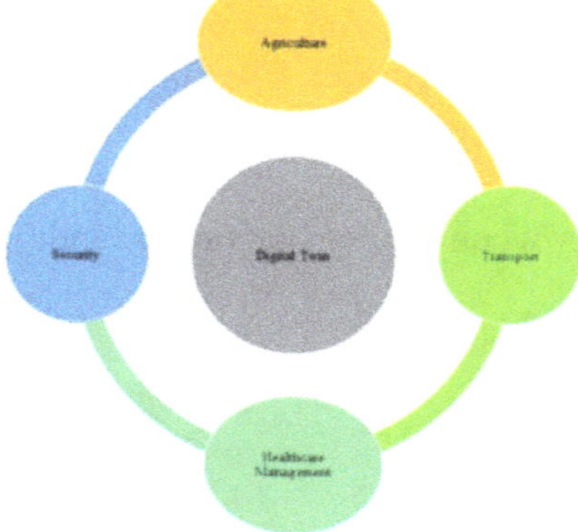

Figure 2. Digital Twin Applications in smart Cities

The proper management of traffic flows is vital in terms of safety, particularly because of the increase in the number of vehicles that are involved in road traffic and the need to preserve its fluidity. There is also the possibility that the potential of DT might be utilised to improve the safety of transportation and infrastructure. The interaction between communication systems and automation systems is made possible by DT. In the event of an emergency and the subsequent management of traffic by intelligent traffic systems, it would be feasible to plan the redirection of traffic away from the area that is experiencing the threat with the assistance of DT (Mandal, 2024).

DTs are able to detect the occurrence of a possible extraordinary event by revealing anomalies, which in some cases could strengthen the resistance of threatened elements of critical infrastructure and improve the response rate to the crisis situation. This is made possible and made possible by the availability of a large amount of historical data.

In the event that a hazardous material is released into the environment, it is important to take into consideration the possibility of incorporating new technologies into the risk assessment. The information obtained from these technologies might be utilised in disaster planning in order to assess the potential areas that could be affected by the weather.

As a result of the fact that the development of DT is still in the process, which is what the forecasts suggest, it is assumed that its application inside the Smart City will continue to grow. For the purpose of advancing the implementation of DT in smart cities, it is essential to raise awareness, engage in open dialogue, and exchange thoughts, perspectives, and possible proposals with one another.

5.1 Collecting big data: Examining the structure and organisation of urban data from a topological standpoint in the context of critical data studies:

The authors Dietmar Offenhuber and Carlo Ratti state in the preface to their book, Decoding the City: Urbanism in the Age of Big Data, that the phrase "big data" refers to the availability of a huge amount of information that can be interpreted by machines. They are referring to a quantitative approach in the field of social science that is known as computational social science. This approach is centred on the analysis of large amounts of data rather than traditional methods. In addition, they concentrate a particular emphasis on the development of the subject of network science, which is primarily concerned with the analysis of intricate networks. The move from spatial perspectives on urban data to topological viewpoints is the subject of the problems that have been raised within the field of network science, and in particular within the work of Manuel Castells. In the discussions that have been taking place about the topological perspectives on urban data, Michael Batty and Manuel Castells have been playing a leading role since the beginning. In the topic of critical data studies, one of the most important questions that needs to be answered is about the methods that are used to acquire large amounts of data, as well as the influence that the local conditions that were present during their formation have on our research. The following are some more concerns that are at the core of the critical data studies: first, the investigation of the degree to which sense data are an operational component of an economic system; second, the inquiry into which social groups benefit from the production of big data; and third, the assessment of the extent to which sense data are an operational component of an economic system (Attaran & Celik, 2023).

The concept of "local reading" is one that Yanni Alexander Loukissas has investigated in his work, which includes his article titled "Taking Big Data apart: Local readings of composite media collections," which was published in 2017 in the journal Information, Communication & Society, as well as his recently published book titled "All Data Are Local: Thinking Critically in a Data-Driven Society, " which is of great significance for this endeavour to take into account the principles that have been mentioned above. When it comes to gaining a deeper understanding of the shortcomings of "digital universalism," the following concepts, from which Loukissas departs in the aforementioned book, are helpful: Data are complicated attachments to place; data are obtained from a variety of sources; data and algorithms are inextricably entwined; interfaces recontextualise data; and data are indexes to local knowledge. Here are some of the characteristics of the data.

5.2 The challenges of digital generalization

There has been a growing number of complaints levelled against the concept of "digital universalism" during the past several years. Particular focus is placed on the perils and difficulties that its ambition to obtain a comprehensive representation and knowledge of things involves. The primary argument that is made in the critiques is that "digital universalism" is a "myth." The shortcomings of the concept known as "digital universalism" are connected to the fact that, on the one hand, it is founded on theoretical assumptions, and, on the other hand, it makes use of a restricted number of variables and processes. In the same vein, another problem of the concept of "digital universalism" is the fact that the data and information that it is founded on are necessarily selected.

The concept of "digital universalism" is founded on the erroneous belief that "once online, all users could be granted the same agencies on a single network, all differences could dissolve, and everyone could be treated alike." This is the foundation of the myth. Digital universalism could become a new kind of colonialism in which practitioners at the "periphery" are made to conform to the expectations of a dominant technological culture." Certain researchers, such advocate for a "de-Westernization of critical data studies" as a means of challenging the concept of "digital universalism (Ramu et al., 2022).

In the definition of the Internet of Things, "the networked interconnection of everyday objects, which are often equipped with ubiquitous intelligence". In the section of his book titled "The Stack: On Software and Sovereignty" that is devoted to the the observation that for Stack urbanism, mobilization precedes and supersedes settlement. This observation by Bratton, which states that well before smart cities evangelism, the modernist call for a more intense technologization of design's disciplinary doxa, blending urban and cybernetic programs, was a predominant discourse," is helpful in gaining an understanding of the fact that the evangelism that is associated with the discourse surrounding smart cities is not something new but rather has a long history.

Big data ideology refers to the pursuit of gathering comprehensive information on a certain topic, while minimising the significance of the data's sources, and presuming that data alone can fully replace other forms of knowledge.

5.3 Understanding urban data: Transitioning from technical methodologies to socio-technical viewpoints

Within the domain of smart cities, a crucial transition that warrants attention is the transition from the technological viewpoint to the socio-technical viewpoint.. For instance, various researchers, such as Simon Joss, Frans Sengers, Daan Schraven, Federico Caprotti and Youri Dayot have thrown light on the idea that the concept of smart city should be tied to an effort to expose 'multiple aspects beyond an infrastructure-technology focus'."14" The concept of "data universalism" is one that is prevalent in the ongoing discussions about big data and is pertinent to the questions that are discussed in this article. In their article titled "Big Data from the South(s): Beyond Data Universalism," Stefania Milan and Emiliano Trer´e argue that the term "data universalism" is a fallacy that relates to the propensity to adapt the cultural variety of technology innovations in the Global South to the ideals set forth by Silicon Valley.a 15 "The main problem with data universalism is that it is asocial and ahistorical, presenting technology as something operating outside of history and of specific sociopolitical, cultural, and economic contexts," Milan and Trer´e argue in their critique of the "hyperbolic narratives of the "big data revolution."16 They argue that this is the primary issue with data universalism.a 17 The topic "how does datafication unfold in countries with fragile democracies, flimsy economies, and impending poverty?" was one of the most important questions that they aimed to answer.16 In order to move beyond the concept of "data universalism," it is imperative that we take into consideration the unequal access to the technologies and data that make it feasible to create smartcities and digital twins on an urban scale. The authors Simon Joss, Frans Sengers, Daan Schraven, Federico Caprotti, and Youri Dayot have shed light on the "competitive dynamics created between world cities posited as "model" smart cities and various second- and third-tier "follower" cities." This is in reference to the issue of this uneven connection to the technologies and data.

5.4 Urban scale digital twins and the construction of tangible links between the virtual and physical realms

"Digital twin" is a word that is used to refer to the digital representation that enables full data sharing. It can contain models, simulations, and algorithms that describe its counterpart as well as their characteristics and behaviour in the actual world. A digital depiction of a physical process, person, location, system, or object is referred to as a "digital twin." When it was first introduced, the phrase "digital twin" was used in the context of the manufacturing industry. In the early 2000s, to use this term, which describes digital simulation models that are designed to operate in tandem with real-time processes. The term "digital twins" refers to the concept of digital duplicates constructed from physical elements. The utilization of technical advancements such as sensing, processing, and data transfer makes it possible to create digital twins. In addition to its application in the subject of urban analytics, digital twins are also utilized in the field of computational social sciences within the discipline of social science.

It is anticipated that the number of urban digital twin deployments would surpass 500 by the year 2025. It has been said by Michael Batty that " The notion of the digital twin has originated from the depiction of the urban environment in relation to its tangible resources.Furthermore, it is important to note that the digital twins can be updated in accordance with changes in the environment. The tangible manifestations of the abstract notions. The new feasibility of this is attributed to the integration of the virtual world with the physical world. It is important to keep in mind that "ideal digital twin would be identical to its physical counter-part and have a complete, real-time dataset of all information on the object/system".

This is the fundamental concept that underpins the construction of digital twins, and it is that concept that we need pay attention to in order to comprehend it. Recently, the perception of an urban scale digital twin has taken centre stage in the realm of urban planning, and more specifically, in the arena of smart city design. This is a relatively recent development. In their essay titled "Smart city based on digital twins". And also discuss the implications of this relationship. In the introduction to their study, they describe the term "digital twin" as a "simulation process that makes full use of physical models, sensors, historical data of operation, etc. to integrate information of multiple disciplines, multiple physical quantities, multiple scales, and multiple probabilities". In addition to this, they bring to light the fact is that the contemporary discussions around the concept of digital twins are considered by a variety of different ways in which this idea is defined. They make the observation that "a consensus definition has not yet been formed," which is a characteristic statement.

The bi-directional mapping link between physical space and virtual space is a shared interest that serves as the common denominator connecting the different meanings of the term under consideration. This essay emphasises the critical importance of the premise that the development of digital twins is based on the aim to create "real-time connections between the virtual and the real."22. Within the context of digital twins, the digital models serve the dual function of watching, identifying, and comprehending the physical environment, while simultaneously aiming to manipulate and alter it. The primary utility of urban scale digital twins is in their capacity to observe and analyse urban activities, enabling the development of design solutions that are both more efficient and environmentally sustainable. The data that can be monitored using digital twins at the urban scale may include a diverse range of features related to the urban environment dynamics. The parameters encompass metrics related to traffic and transportation, electricity production, utility provisioning, water supply, and waste management, among other factors. “Smart city based on digital twins” refers to the software application called “Smart City TrafficBrain.” Based on the use of digital twins, this program collects vast quantities of data related to journey paths and “real-time dynamic traffic information.” Scholars such as Martin Mayfield have underscored the significance of urban-scale digital twins in providing a comprehensive approach to urban and infrastructure design.25The research on smart cities and the theories related to big data are intricately connected to the discussions around urban scale digital twins because of their strong correlation. The essay titled “Critical Questions for Big Data: Provocations for a Cultural, Technological, and Scholarly Phenomenon” was authored by Anah Boyd and Kate Crawford and was published in the scholarly journal Information, Communication & Society. The paper titled “Smart city based on digital twins” highlights that urban scale digital twins aim to create an intricate integrated system connecting the physical world and the virtual domain, enabling mutual mapping and interaction in both directions.

Generation of extensive urban big data is an ongoing activity essential for the development of metropolitan-scale digital twins. This capability is facilitated by the deployment of “many sensors and cameras strategically placed throughout the city, together with the digital subsystems progressively developed by the municipal management departments.” Advanced technologies like as the Internet of Things (IoT), cloud computing, big data, and artificial intelligence (AI) are employed in the development of urban-scale digital twins. An academically significant research initiative is known as DIGITbrain. This project, financed by the European Commission under the Horizon 2020 Research and Innovation Programme (grant agreement number 952,071), involves a consortium of 36 partners. Its objective is to support small and medium-sized manufacturing firms in Europe in creating artificial intelligence-based digital twins. Prior to the emergence of the employment of artificial intelligence and vast volumes of real-time data, apps that were supposed to assess characteristics crucial to urban planning decisions performed significantly worse than they do now. In a highly resource-efficient manner, a flexible local adaptation to short-term expected vehicle arrivals demonstrates exceptional performance. Furthermore, artificial intelligence has not yet made significant advancements in the domain of traffic signal control. The Future Cities Laboratory is now doing a project named “Engaging Mobility,” supervised by Dr. Pieter J. Fourie. This project aims to analyse present and future urban mobility issues by conducting travel behaviour research, applying big data-driven modelling, and proposing solutions. This effort presents a compelling case that is relevant to the topics addressed in this paper (Sharma et al., 2022).

6. DISCUSSION

There are a big number of firms, including Meta, Microsoft, NVIDIA, Intel, Apple, and Samsung, amongst a great number of others, that are making substantial investments in the research and development of developing technologies for the metaverse. These corporations are among the most prominent and enormous corporations. The commercial activities of other prominent firms, including as Nike and Coca-Cola, are also being relocated into the metaverse environment in order to generate immersive experiences for their customers. These experiences will enable customers to interact with the products of these corporations in a manner that will offer them with a high quality of experience. There will be a greater number of individuals who are able to immerse themselves in virtual worlds that are additionally enriched with reality as a result of this tremendous participation with the metaverse, which will elevate this three-dimensional elevation of the linear Internet (also known as the second generation of the Internet) to the next level. This merger would result in the formation of new business models, and a large number of established businesses would gain significant benefits from the enhanced immersive involvement that would emerge from this merger. The union of the corporate world and the metaverse would result in the creation of new business models. There is a strong indication that the economic value of the digital virtual worlds that are employing blockchain technology will significantly increase in the years to come, which offers a significant amount of further economic growth.

A notable and accelerating trend is the exponential growth in digital ownership of high-fidelity virtual assets on Web 3.0 (Tao et al., 2019). An increase in the number of metaverse applications driven by progressively advanced metaverse technology is expected in the near future. These developments will result in a transformation of the way company operations are conducted inside the urban environment. Decentraland (2020) and HorizonWorlds (2021) are contemporary three-dimensional virtual world platforms built on Ethereum, which are user-controlled and serve as exemplary case studies of global communities that facilitate global connectivity. Furthermore, these platforms provide users with the chance to experience a sense of joint participation in a digital environment by enabling them to utilise their virtual currency for the trading of virtual assets, such as lands. To access a diverse selection of decentralised virtual landscapes that offer a multitude of immersive experiences, readers are advised to visit the OpenSea website5. These worlds are intricately crafted with a fluid futuristic architectural design and are categorised in a diverse range of options. These worlds are developed all around the world for the purpose of exploration. These worlds are not clones of real worlds because they do not represent any physical reality. Furthermore, there is no data transmission between the ecology of the physical city and these worlds. Another way of putting it is that these virtual worlds do not have any physical enhancements (Therias & Rafiee, 2023).

There is no text provided. In addition to their physical likeness to the real world, the potential advantages of virtual environments, which enhance the ease and productivity of our busy urban lives, will appeal to a growing number of people. Consensus-based optimal futuristic smart contracts can be created using smart contract development tools and metaverse software and resources.To accomplish this objective, the MetaOmniCity framework is specifically developed to establish a system that streamlines the deployment of Urban Mobility as a Service (UMaaS) in Smart City (SC) settings. The integration of residents into this ecosystem results in an augmentation of the degrees of freedom, so fostering a multitude of prospects and a more auspicious future for cities and citizens within a more democratic society. This is achieved through the globalisation of the metaverse, which eliminates borders and promotes collaborative synergy. Urban metaverse worlds are expected to integrate with our everyday surroundings through several interoperable metaverse windows.

In the future, there will exist a plethora of alternative approaches to engage with virtual surroundings. From a neuroscientific perspective, motor neurone prosthetics, Deep Brain Stimulation (DBS), Brain Implant Technologies (BIT), and Brain Computer Interface (BCI) with bidirectional communication are expected to function as assistive devices for manipulating digital smart environments by utilising the electrical activity of the brain.

These technologies are expected to enable individuals, particularly those with disabilities such as Parkinson's, to immerse themselves in metaverse worlds and turn science fiction into reality through the decoding of brain signals. These technologies have the potential to restore independence and promote equality among individuals, notwithstanding the presence of ethical concerns and cynicism (Waqar et al., 2023).

To create a virtual social setting for two individuals, several things are necessary. Firstly, there must be highly realistic avatars that accurately depict their appearance, including clothing, hair, and skin tones. These avatars should be generated in real time using sensor data that captures the physical characteristics of real-world objects, gestures, and audio. Additionally, there needs to be a fast and efficient data transfer system that can handle large amounts of information with minimal delay. Lastly, a persistent model of the environment is required, which can include both real and simulated elements. Consider the challenge of tackling this problem on a large scale, simultaneously serving hundreds of millions of customers. It becomes evident that our current computing, storage, and networking infrastructure is unable to support this ambition. The demanding need for resources can be reduced through the implementation of energy-efficient computation, improved algorithms, and enhanced architectures, as reiterated.

In order to address this issue, MetaOmniCity has been developed with several UMaaS models that are specifically designed to provide benefits. These models include users/avatars and objects that are directly tied to the models. This approach helps to reduce the demand for resources (Wang, 2020).

Research on Digital Twin solutions in manufacturing focusses on production planning and control, which is crucial for integrating all data within a production system. The literature examines the application cases of Digital Twins in the field of supply chain management. The utilisation of use cases in the construction and healthcare sectors is also experiencing an increase. In the construction business, the notion of Digital Twin, along with the use of mobile devices and wearables on a construction site, can provide a more accurate representation of the actual construction progress compared to the original plan, at any given moment (White et al., 2021; Yao et al., 2023).

Furthermore, it aids in reducing the occurrence of errors and the need for reworks by facilitating the timely transmission of current information to the field. Digital Twin technologies facilitate the identification of latent disorders, enable experimentation with treatment options, and enhance surgical planning in the healthcare industry. Acquiring a precise and comprehensive human body model will assist medical professionals in identifying latent diseases, conducting therapy trials, and enhancing surgical planning. Current scientific efforts are focused on the creation of Digital Twins to analyse the human body, and significant progress has been made. The Living Heart initiative is a widely employed technology for the purposes of clinical diagnosis, testing, development of medical devices, and education and training. Digital Twin technologies have created the first lifelike virtual model of a human organ, including elements such as blood circulation, mechanics, and electrical properties. The swift development of Digital Twin solutions in industrial sectors including manufacturing, supply chain, life sciences, agriculture, and energy may be mostly ascribed to the progress made in artificial intelligence (AI), Internet of Things (IoT), and cloud computing technologies.

AI-powered Digital Twins are utilised to replicate intricate real-world systems (Yang et al., 2022). It utilises data collected by IoT devices to acquire knowledge and operate in conjunction with actual industrial systems, consistently suggesting areas for improvement and assisting in tactical decision-making. It also aids in optimising systems design to enhance efficiency and save expensive redesigns during implementation.

Predictably, the growing requirement for automation in several sectors will fuel the robust demand for the Digital Twin platform in the next years. In the aftermath of the epidemic, Digital Twin solutions are poised to exert an increasingly substantial influence across diverse sectors. The benefits of implementing a Digital Twin system are vast and have not yet been comprehensively explored. Notwithstanding the challenges presented by data quality and security, increasing power and storage requirements, and integration with current infrastructures, Digital Twin solutions are effectively propelling a much-advanced digital transformation to enhance the planet for humanity. In the future, Digital Twins will progressively expand their application to a broader spectrum of use cases and industries. The proposed solutions would incorporate supplementary technologies, such as augmented reality (AR), in order to offer a fully immersive experience. Furthermore, they will leverage artificial intelligence skills to augment connections, offer valuable insights, and facilitate sophisticated analytics. Moreover, technological progress has enabled us to use Digital Twin solutions, therefore obviating the need for physical examination of the physical object concerned.

CONCLUSION

The concept of digital twins has gained broadening and diversification to include urban digital twins as a conceptual foundation for developing intelligent cities. Universal Design Technologies (UDTws) may not always be developed in conjunction with their physical counterparts; they may be developed independently or incorporated at a later stage. Incorporating intelligent elements into existing urban infrastructure, coupled with certain pre-existing provisions, for the purpose of strategic planning. The NGSI-LDin formation model is applicable for representing mathematical data related to digital information. The subject matter concerns Italian twins and the implementation of the NGSI-LD API, which serves the objectives of information discovery and retrieval. Utilising open-source components for the management of digital twin data. The digital twin framework can be developed by drawing upon the FIWARE ecosystem, which is based on NGSI-LD. This framework incorporates the reactive, speculative, and predictive dual functionality concepts. An inherent challenge in developing UDTws as a component of previous. The primary objective of present smart city initiatives is to extract and integrate. Quantitative data acquired from established sources of comprehensive information. For example, the UDTw can incorporate sensors. Artificial intelligence techniques like as Weakly-supervised machine learning holds great potential. The foundation for this task and will remain the central point of our attention. Engaging in research endeavours. We are confident that UDTws will create new opportunities for improvement. Enhancements and cooperation with the control community. In this context, UDTs and autonomous systems engage in interaction mutually identify and anticipate each other for the purpose of recognition and future forecasting of the entity study the environment and the conditions of dynamic systems utilising technology. Techniques of artificial intelligence and simulations. Autonomous systems within a given context include autonomous vehicles, unmanned aerial vehicles, and robotic systems in intelligent urban areas. Additionally, computing and communication systems have the capability to adjust in real-time according to the User-Defined Task workflow idea.

REFERENCES

Anthopoulos, L., Janssen, M., & Weerakkody, V. (2023). Introduction to the Special Issue on Citizen Centricity in Smart Cities. *Digital Government: Research and Practice*, *4*(3), 1–4. DOI: 10.1145/3624556

Attaran, M., & Celik, B. G. (2023). Digital Twin: Benefits, use cases, challenges, and opportunities. *Decision Analytics Journal*, *6*, 100165. DOI: 10.1016/j.dajour.2023.100165

Bauer, M., Cirillo, F., Fürst, J., Solmaz, G., & Kovacs, E. (2021). Urban digital twins–a fiware-based model. *Automatisierungstechnik*, *69*(12), 1106–1115. DOI: 10.1515/auto-2021-0083

Boschert, S., & Rosen, R. (2016). Digital twin—the simulation aspect. *Mechatronic futures: Challenges and solutions for mechatronic systems and their designers*, 59-74.

Charitonidou, M. (2022). Urban scale digital twins in data-driven society: Challenging digital universalism in urban planning decision-making. *International Journal of Architectural Computing*, *20*(2), 238–253. DOI: 10.1177/14780771211070005

Dani, A. A. H., Supangkat, S. H., Lubis, F. F., Nugraha, I. G. B. B., Kinanda, R., & Rizkia, I. (2023). Development of a smart city platform based on digital twin technology for monitoring and supporting decision-making. *Sustainability (Basel)*, *15*(18), 14002. DOI: 10.3390/su151814002

Gupta, M., Kumari, M., & Ved, C. (2021). Digital Twin in Agriculture Sector: Detection of Disease Using Deep Learning. In *Digital Twin Technology* (pp. 47-60). CRC Press.

Liu, M., Fang, S., Dong, H., & Xu, C. (2021). Review of digital twin about concepts, technologies, and industrial applications. *Journal of Manufacturing Systems*, *58*, 346–361. DOI: 10.1016/j.jmsy.2020.06.017

Lv, Z., Chen, D., Feng, H., Singh, A. K., Wei, W., & Lv, H. (2022). Computational intelligence in security of digital twins big graphic data in cyber-physical systems of smart cities. [TMIS]. *ACM Transactions on Management Information Systems*, *13*(4), 1–17. DOI: 10.1145/3522760

Mandal, S. (2024, March). A Privacy Preserving Federated Learning (PPFL) Based Cognitive Digital Twin (CDT) Framework for Smart Cities. *Proceedings of the AAAI Conference on Artificial Intelligence*, *38*(21), 23399–23400. DOI: 10.1609/aaai.v38i21.30400

Ramu, S. P., Boopalan, P., Pham, Q. V., Maddikunta, P. K. R., Huynh-The, T., Alazab, M., Nguyen, T. T., & Gadekallu, T. R. (2022). Federated learning enabled digital twins for smart cities: Concepts, recent advances, and future directions. *Sustainable Cities and Society*, *79*, 103663. DOI: 10.1016/j.scs.2021.103663

Sharma, A., Kosasih, E., Zhang, J., Brintrup, A., & Calinescu, A. (2022). Digital twins: State of the art theory and practice, challenges, and open research questions. *Journal of Industrial Information Integration*, *30*, 100383. DOI: 10.1016/j.jii.2022.100383

Tao, F., Sui, F., Liu, A., Qi, Q., Zhang, M., Song, B., Guo, Z., Lu, S. C.-Y., & Nee, A. Y. (2019). Digital twin-driven product design framework. *International Journal of Production Research*, *57*(12), 3935–3953. DOI: 10.1080/00207543.2018.1443229

Therias, A., & Rafiee, A. (2023). City digital twins for urban resilience. *International Journal of Digital Earth*, *16*(2), 4164–4190. DOI: 10.1080/17538947.2023.2264827

Wang, Z. (2020). Digital twin technology. In *Industry 4.0-impact on intelligent logistics and manufacturing*. IntechOpen. DOI: 10.5772/intechopen.80974

Waqar, A., Othman, I., Almujibah, H., Khan, M. B., Alotaibi, S., & Elhassan, A. A. (2023). Factors influencing adoption of digital twin advanced technologies for smart city development: Evidence from Malaysia. *Buildings (Basel, Switzerland)*, *13*(3), 775. DOI: 10.3390/buildings13030775

White, G., Zink, A., Codecá, L., & Clarke, S. (2021). A digital twin smart city for citizen feedback. *Cities (London, England)*, *110*, 103064. DOI: 10.1016/j.cities.2020.103064

Yang, B., Lv, Z., & Wang, F. (2022). Digital twins for intelligent green buildings. *Buildings (Basel, Switzerland)*, *12*(6), 856. DOI: 10.3390/buildings12060856

Yao, J. F., Yang, Y., Wang, X. C., & Zhang, X. P. (2023). Systematic review of digital twin technology and applications. *Visual Computing for Industry, Biomedicine, and Art*, *6*(1), 10. DOI: 10.1186/s42492-023-00137-4 PMID: 37249731

Chapter 5

Integrating IoT and AI for Smart Urban Sustainability and Green Infrastructure

Rituraj Jain
http://orcid.org/0000-0002-5532-1245
Chandigarh University, India

Ashish Sharma
http://orcid.org/0009-0007-6644-6362
JIET Universe, India

Nausheen Khilji
JIET Universe, India

Rakesh Saxena
Jai Narain Vyas University, India

Sitesh Kumar Singh
http://orcid.org/0000-0002-7108-0808
National University of Science and Technology, Oman

ABSTRACT

This chapter discusses the importance of IoT and AI in the realization of sustainable urban planning through fast urbanization, climate change, and resource shortage. It looks at smart systems in urban environments that can use data in real time using IoT sensors and AI-partnered analytics to streamline resource consumption, better monitor the environment, and advance quality of life. Such uses can be intelligent

DOI: 10.4018/979-8-2600-3038-7.ch005

nt © 2027, IGI Global Scientific Publishing. Copying or distributing in print or electronic forms without written permission of IGI Global Scientific Publishing is prohibited. Use of this chapter to generative artificial intelligence (AI) technologies is expressly prohibited. The publisher reserves all rights to license its use for generative AI training and machine learning model development.

traffic systems, smart grids and green waste management. Climate resilience is enabled through the integration of deep tech with green infrastructure (such as urban forests and green roofs). Privacy of data, cybersecurity, and ethics are other issues covered in the chapter, which call for strong policies and multidisciplinary solutions in smart city development that is sustainable and fair.

1. INTRODUCTION TO SMART URBAN SUSTAINABILITY AND GREEN INFRASTRUCTURE

Currently, smart urban sustainability and green infrastructure concepts are taking place to build environmentally sound, efficient and livable cities. These approaches allow using innovative technologies, sustainable design principles and ecological solutions to resolve urban issues and enhance the quality of living of residents.

Urban sustainability, being term of smart urban sustainability, stresses the use of technology and data driven way to optimally utilize resources, minimize environmental influence, and strengthening urban services. This idea includes energy efficiency, waste, transportation, and taking care of public services of the city.

Natural and semi-natural features have been planned and integrated into a wide range of ecosystem services on a city scale. Therefore, this infrastructure includes parks, green samizdat, urban forests, green roofs, rain gardens, and permeable pavements. They include as well and are termed as green infrastructure. Mitigation of climate change and environmental impacts on urban environment, improvement of air and water quality, handling of stormwater, and increase of biodiversity through green infrastructure are the key.

Information and communication technologies (ICTs) and green infrastructure are one with each other, to produce efficient, livable and environmentally responsible cities as smart urban sustainability. Part of compact sustainable cities, green infrastructure consists of integrated networks of natural and semi natural areas providing ecosystem services (Sugandha et al., 2022) . The goals of this approach are to respond to the challenges imposed by urbanization at a rate of rapidity, climate change, and resource scarcity.

Urbanization it is rapidly increasing, a pressure intensified for urban resource and infrastructure, exacerbating climate change and the sustainability challenges (Nguyen et al., 2024) . The combination of cities as critical points for contributing to global warming and cities as very vulnerable to the effects of global warming (Gómez-Villarino et al., 2020) is completely novel and new to global science. Although the predicted rise in 2.5 billion urban dwellers by 2050 further compounds these environmental issues (Van Oorschot et al., 2024), those living in cities are increasingly becoming the focus of proslumg cleaners. Furthermore, as Nguyen et al.

(2024) pointed out, over fifty percent of the global population lives in regions prone to natural disasters which indicates that there is need for formulating techniques to manage disaster risk alongside the sustainable development goals (SDGs).

ICTs and green infrastructure are used in the smart cities to manage resources, promote sustainability, and improve quality of life of the residents to overcome these challenges. Technologies including IoT, AI, and the Metaverse integration brings in transformative potential to develop sustainable urban environment (Adaramola et al., 2024) . Although addressing such issues such as data privacy, costs, regulatory frameworks, and the need for inclusive, socially sustainable development is the prerequisite for successful implementation of smart urban sustainability (Shah, 2023).

IoT and AI is emerging as key enablers of developing the smart and sustainable cities. Being able to monitor urban systems, optimize operations, and improve the citizen experience for any city, from transportation to energy to environment, to public safety, is made possible by combining features provided by IoT with those offered by AI (Hoang, 2024).

Interestingly, these technologies are also associated with challenges. For a wider implementation of the smart city solution based on IoT, there are several issues that need to be addressed including data privacy, security, standardization, interoperability, scalability, sustainability (Abdel Aal, 2023). In addition, these technologies themselves are not environmentally bulletproof, as they require energy use and electronic waste (Bibri et al., 2023).

Thus, the integration of IoT and AI in the strategic way will make urban management and urban service delivery more effective and greatly promote the goal of sustainable development. The use of these technologies could allow cities to use data to make smarter decisions, streamline administrative processes, optimize resources, etc. thus combating urbanization challenges to provide a better living condition to the citizens of a city. But their implementation needs to be considered from the point of view of ethical, regulatory and environmental issues, to avoid irresponsible and unfair utilization.

As shown in Figure 1, the integration of IoT and AI technologies for smart urban sustainability is described. Real time data collected from urban systems by such IoT devices as sensors or smart meters is entering. The data is forwarded to central systems where AI analytics are applied and used to predict insights and make decisions. An environment leading to more efficient and sustainable urban environments is applied to various applications in smart city, such as energy management, traffic control, waste management and environmental monitoring, based on the optimized decisions.

Figure 1. Integration of IoT and AI for smart urban sustainability

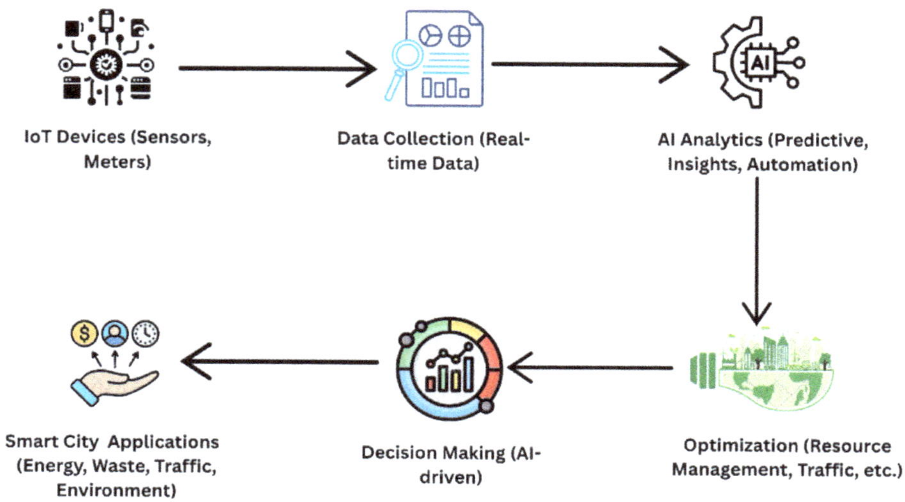

2. THE ROLE OF IOT IN URBAN SUSTAINABILITY

Smart urban systems, integrated in a city, that optimize the usages of the resources, make a life better for its inhabitants through IoT. The network consists of preexisting analog physical devices such as vehicles and home appliances with unprecedented levels of data to and from the devices embedded with electronics, software and sensors.

Bearing in mind the urban environments, IoT devices are a collection points of very numerous data, covering information in real time about diverse parts of the life and infrastructure of the city. These are the devices that monitor and measure such critical parameters as energy consumed in buildings, activities in waste process management, levels of air quality in the urban landscape, and patterns of traffic. The information creates a continuous stream of data that city planners, policymakers, and administrators have to have a complete and up to date picture of what is a city doing.

2.1 Types of IoT devices used in urban sustainability

As IoT devices, environmental sensors capture the needs of urban sustainability and smart grid/meter sensors offer an intelligent means to its deployment. For example, there are sensor used to monitor not only air but also water quality in smart cities. These sensors are capable of detecting different pollutants and the real time data collected from these sensors is useful for analysis and decision making (Ullo & Sinha, 2020). For example, sensors are used by air quality monitoring systems to measure CO2, alcohol, smoke, NH3, Benzene and NO2 (Kumar et al., 2024). Likewise, IoT based sensors are used in monitoring of water quality to assess surface and groundwater quality at various locations (Ighalo et al., 2020).

Another important type of IoT devices employed in the urban sustainability is the smart grid and smart meter. The efficiency on energy management that these devices allow is especially favorable for the development of Nearly Zero Energy Buildings (NZEB) embedded with them (Martínez et al., 2021). In particular, smart meters provide remote monitoring and control of load for promoting adoption of new concepts like smart grids and automation of power systems (Marahatta et al., 2021). Further integration of IoT and Green Fintech in smart cities leads to better resource management, that is, of energy, water, waste through interconnected sensors and real time data processing, (Adaramola et al., 2024).

Automated traffic management IoT systems monitor and resolve pending delays to trains (Razmjoo et al., 2022). These systems can employ sensors and other devices to track traffic patterns, optimize signal timings, and provide updated information to travelers. Such intelligent systems that manage traffic have emerged as an effective way of combating urban traffic congestion which in return helps reduce time and fuel consumption as well as emissions (Kushwaha, 2024).

IoT devices are widely deployed for monitoring and improving the collection processes in waste management. Real time monitoring of waste level and bin location are done by the use of smart bins equipped with proximity sensors, ultrasonic sensors and GPS modules (Sallang et al., 2021) . In this way, these devices can inform waste management centers when bins need emptying, to facilitate better routes. A few of the advanced systems involved cameras and deep learning models for the waste segregation and classification (Sallang et al., 2021). For example, an integrated system with Raspberry Pi 4 powered by SSD MobileNetV2 Quantized model was used for (Sallang et al., 2021) to detect and categorize waste into various compartments.

The integration of IoT with Green Fintech also provides a number of additional benefits in urban sustainability. This configuration supports the running of dynamic pricing models and the provision of financial incentives for sustainable practices in smart cities (Adaramola et al., 2024). Nevertheless, there are challenges, including cost and privacy, and possible regulatory frameworks that must be overcome for the widespread adoption of this. An overview of various IoT devices being used in the Urban Sustainability Initiatives is presented in Table 1. They provide the monitoring and optimization of such resources as air quality, water quality, energy use, waste management, and traffic flow.

Table 1. Overview of different IoT devices employed in urban sustainability initiatives

IoT Device Type	Function
Air Quality Sensors	Detect pollutants such as CO2, NH3, and benzene in urban environments.
Water Quality Sensors	Monitor real-time water contamination levels for quality assessment.
Smart Meters	Track and optimize energy usage in residential and commercial buildings.
Waste Management Sensors	Optimize waste collection routes using real-time bin status.
Traffic Sensors	Improve traffic flow by adjusting signal timings dynamically.

2.2 Applications of IoT in sustainability

The information pertaining to urban sustainability reflects improved resource management, minimal environmental impact and overall livability of a city through the Internet of Thing (IoT) systems. The context of one such is the IoT based intelligent systems which include optimization of traffic flow and reduction of congestion and emissions making use of real-time information derived from sensors, GPS and connected vehicles. Smart public transport systems thus benefit IoT by using this technology to provide commuters with real time information and updates regarding public transport services so as to reduce delays and improve traffic conditions on smart traffic lights that are adjusted dynamically based on fixed events.

Efficient energy distribution and conservation constitutes another crucial area for which IoT based smart grids come to play through real time monitoring of energy consumption so as to optimize distribution and waste elimination. Predictive maintenance in power infrastructure allows power infrastructure to fail predictively, preventing power failures, with the consumers able to monitor and adjust these changes with smart meters in order to track and respond to them.

Real time environmental monitoring is also done by IoT as the sensors are continuously reporting the air and water quality and again detecting the pollutants and providing data for corrective measures. Greenhouse gas emission monitoring can be performed in a smart way to lend support to city-wide sustainability initiatives and to their policy makers to implement suitable climate policies.

IoT enables systems for smart water and water waste management to monitor for leaks, optimize distribution of water, or treat wastewater more efficiently in order to optimize usage of water resources. Similar is smart waste bin equipped with sensors that relay when they are full so collection services know when to come and recycle with fewer errors. Integration of IoT in these domains allows cities to build more sustainable, efficient and environmentally friendly urban ecosystem. In Table 2 This table summarizes key IoT applications in different sectors of urban sustainability. Traffic management, energy efficiency, environmental monitoring, water management, waste management are some areas, in which IoT is utilised for making urban ecosystems sustainable and efficient.

Table 2. Key IoT applications across various sectors of urban sustainability

Application	IoT Implementation
Traffic Management	Smart traffic lights, real-time congestion monitoring, adaptive signals.
Energy Efficiency	Smart grids, predictive maintenance, real-time energy consumption tracking.
Environmental Monitoring	IoT sensors for air and water pollution control, climate monitoring.
Smart Water Management	Leak detection, distribution optimization, wastewater treatment.
Waste Management	Smart bins, route optimization, AI-driven waste classification.

3. THE ROLE OF AI IN SUSTAINABLE URBAN DEVELOPMENT

AI is essential to smart cities in that it serves as the glue to build upon the development and management of the city with data in mind. With AI, great amounts of data collected in urban environments can be analyzed, using means of descriptive and predictive models to serve as base for Smart City applications (Cesario, 2023). By integrating Internet of Things (IoT) with AI technologies, ubiquitous data collection and automatic analysis comes to the urban management and service delivery (Hoang, 2024).

However, for an effective urban planning like in smart cities, it is necessary for AI driven data analytics and predictive modeling. With these technologies we can do proactive planning, planning is able to respond to the challenge as it evolves (Gkontzis et al., 2024). To demonstrate this, Ayasrah et al. (2024) show that the Multi Modal Stacked VAR LSTM model, that combines Vector Autoregression and Long Short Term Memory neural network, is capable of predicting accurate Carbon Footprint for urban environment. For example, devices and systems that generate real time data from embedded sensors in the urban infrastructure can also be analyzed by the AI based predictive maintenance models to identify and anticipate the failure patterns, and to reduce maintenance costs by as much as 30% and to avoid almost 92% of unplanned failures (Alqasi et al., 2024).

Office of Urban Sustainability in artificial intelligence for traffic performance, traffic management, energy forecasting and energy demand management, is a revolution. The smart traffic management using AI POWERED systems have proven to be good in making the traffic traffic path in the urban cities, an example from one study showed a 32.94% improvement on the heavy traffic and the result of such study shows a positive productivity. Real time traffic data, network characteristics and individual travel pattern are used in these systems in formulating intelligent routing strategies that minimize travel times and negate environmental costs (Dikshit et al., 2023).

The AI driven solutions for traffic optimization includes the utilization of drone and machine learning algorithm for the improvement of traffic flow, alleviation of traffic congestion and guaranteeing traffic safety (Boopathi, 2024). Predictive insight into traffic patterns are offered by these technologies so that the data active decision can be made regarding the allocation and sustainability of resources. AI based demand forecasting has also been proved to offer a promising energy management in terms of reducing energy consumption, decreasing waste and increasing sustainability in manufacturing processes (Rojek et al., 2024).

Traffic optimization as well as energy management prove to be very useful AI applications for urban sustainability. AI technologies can help the cities to approach more efficient transportation systems, lower energy consumption, and develop more sustainable urban space. Although, there are challenges to be addressed in data privacy, algorithm bias and ethical considerations to ensure the use of AI for the purpose of shaping sustainable cities (Yigitcanlar et al., 2020).

Using AI models for pollution detection and control has been proved to be very effective. In this work, environmental hazards are being monitored and are early detected using machine learning algorithms, sensor networks and satellite imagery (Adefemi et al., 2023). AI technologies enable reading out of their tests, providing forecast accuracy, pollution source detections, as well as air and water quality monitoring. These improvements serve to boost environmental monitoring by providing a greater understanding, forecasting, as well as mitigation of environmental risks. For example, spatially varying concentrations of air pollutant can be predicted using AI/ML and environmental and health risks assessed by adding air pollution data to predict such different exposure levels within communities (Krupnova et al., 2022).

Smart governance and policy making using AI is transforming urban management and decision making process. By capitalizing on AI driven analytics, cities can gather insights, gain ground and react to the urban behavior dynamic, and therefore the policy can be predictive maintenance of infrastructure, energy pattern optimization and traffic management. By joining internet of things (IoT) devices with AI, ML, and DL models, the data can now be used for data driven city govern to make smart grids, self sustaining urban transportation networks and effective public service mechanism. Furthermore, it empowers developing countries in addressing their traditional development barriers with the use of innovative, cost effective and scalable solutions (Farahani & Ghasemi, 2024). Table 3 includes key techniques of AI used in urban sustainability. Its uses include optimisation of urban systems, resource management optimisation, optimisation of decision making processes, among others, using AI technologies such as predictive analytics, machine learning, deep learning and natural language processing.

Figure 2 shows the PESTEL-based framework showing how AI/IoT enabled system can help foster urban sustainability by managing governance, monitoring the environment, resource productivity, predictive analytics, and ethical implications. The figure highlights the intersection of six dimensions that are important in shaping sustainable and resilient urban ecosystems, including Smart Governance, Resource Optimization, Dynamic Insights, Predictive Analytics, Pollution Control and Ethical Considerations. Smart governance outlines the notion of transformation in its policy-making, which is based on the principles of data-driven decision-making, whereas smart governance optimizes its usage of resources, as it claims cost efficiency in resources allocation in cities. Dynamic insights record the behavioral reactions in real time and predictive analytics identify how AI-driven systems can predict urban demands. Pollution control plays its role in emphasizing the necessity of continuous control of the environment, and ethical concerns play an important role in the challenges of fairness and privacy of data. All these constitute a systematized approach to the capacity of newer technologies to achieve a balance between innovation and social responsibility in the creation of greener, smarter cities.

Figure 2. PESTEL framework for smart urban sustainability

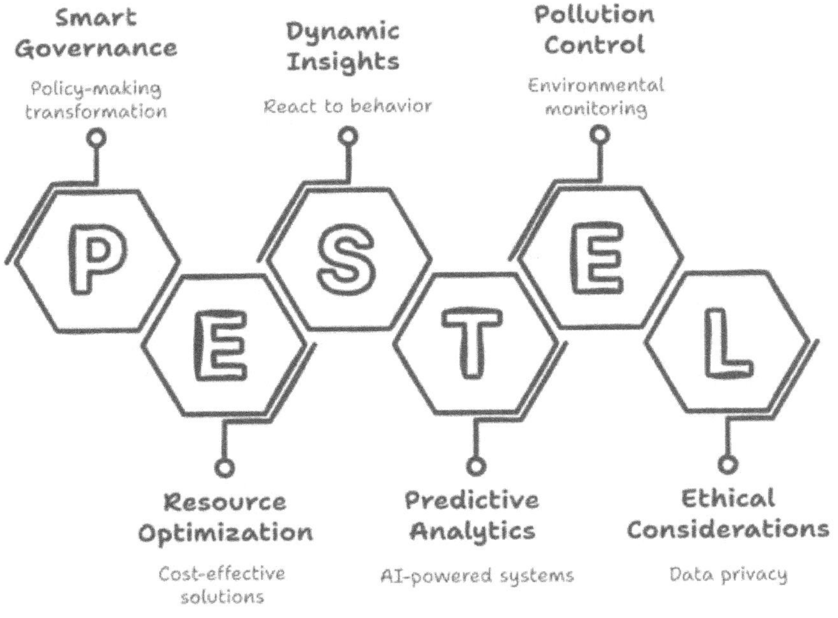

Table 3. Key AI techniques applied in urban sustainability

AI Technique	Application Area
Predictive Analytics	Forecasting energy demand, waste generation, traffic congestion.
Machine Learning Models	Enhancing waste sorting, air pollution detection, traffic control.
Deep Learning	Pattern recognition in environmental data, smart surveillance.
Natural Language Processing	AI-driven urban planning tools, automated policy drafting.

4. SYNERGY BETWEEN IOT AND AI FOR SMART CITIES

IOT coupled with AI for the real time data processing and applying the decision making has revolutionized many industries allowing for better and more efficient system of doing things. AI algorithms are used by IoT devices to analyze huge volumes of data to extract valuable insights and make real time (Rajan, 2024) This is the synergy, Artificial Intelligence on Things (AIoT), that devices can access in order to process and analyze assets further, and disrupt industries such as healthcare, industry and agriculture.

Here, AI is very important in data analysis done with the data collected from IoT devices with predictive analytics, image classification, object detection, and we can even have some natural language process. For example in agriculture, AI algorithms process the information that comes flowing in through IoT sensors about disease detection and optimized feeding strategies that leads to a better management of water quality (Huang & Khabusi, 2025). Just like in healthcare, AI and IoT technologies make automation of real time no patient monitoring, medical image analysis and patient based personalised treatment plan possible (Mehta et al., 2023).

There are examples of AI-IoT synergy in almost all sectors. Integrating AI and IoT helps smart tourism destination provide personalized recommendations and real time data collection for better visitor experiences. The AIoT systems do the real time environmental monitoring and then make decisions based on the data to keep the operations efficient and sustainable in aquaculture. SEMAR IoT application server platform shows how AI can be used within IoT systems for predictive analytics, image classification and natural language processing (Panduman et al., 2024). The examples go on to highlight, how this synergy of AI – IoT can address diverse complex problems and fine tune the operations across various industries.

Waste and power management are examples of smart city infrastructure which are undergoing disruption because of integration of IoT and AI technology. Real urban resource monitoring is enabled by the IoT sensors and then used by AI to run predictive analytics and optimization. IoT enabled smart bin that is equipped with ultrasonic sensor to measure the fill level and AI algorithms to reduce the route of collection and automate sorting processes (Nwokediegwu et al., 2024) have been developed for the use of waste management. The synergy of these two systems offers efficiency, savings and sustainability for the urban waste management systems.

According to researchers, the use of these technologies is interestingly different in developed and developing regions. AI in Africa is at a place where it focuses on community engagement (Nwokediegwu et al., 2024), it is concentrating on scalable and adaptable solutions while in the USA AI is giving attention to improving what is already existing in the infrastructure (Nwokediegwu et al., 2024). Furthermore, this incorporation facilitates the security and trust in smart city systems by combining blockchain with IoT and AI, and this proves to be necessary in urban environments that are more and more interconnected, or at least when a perception of being more interconnected exists (Dritsas & Trigka, 2024).

Finally, IoT and AI are combined to transform smart cities in regard to waste management and energy efficiency in closing. These technologies include waste classification models powered by AI (Cao et al., 2020), as well as optimized vehicle routes for waste collection but are contributing to more sustainable and efficient urban environments. Despite these, challenges like data privacy, costs and skills are present to be faced to take the full advantage of these innovations on smart city development.

The figure 3 demonstrates the potential of the combination of IoT and AI in smart cities, making it possible to process data in real-time, perform predictions, and make automatic decisions applicable to smart cities and healthcare, agricultural, tourist, waste-management and energy-system infrastructures. The figure indicates the way that the capabilities of data gathering of IoT integrate with advanced analytical capabilities of AI to create AIoT ecosystems capable of supporting the efficient and sustainable urban management. Every IoT device is gathering real-time data about the environmental conditions, the functioning of the infrastructures, and the activities of citizens, with this data being used to process and perform predictability analytics, image and language processing, and optimization as AI algorithms do this work. The most promising applications are in precision agriculture (disease detection, water management) and healthcare (real-time monitoring, medical image analysis, personalized treatment), but also tourism, aquaculture, or urban services like smart waste management and energy optimization. The figure also indicates globalities: developing areas tend to be concerned with scalability and community involvement whereas developed countries pay much attention to improving the existing infrastructure. AIoT is an efficient solution based on IoT and AI or even AI enhanced with blockchain technologies, which lead to trust and sustainability impacts, but at the same time costs, data privacy, and skills gaps prove to be consequential factors.

Figure 3. Synergy between IoT and AI for smart cities

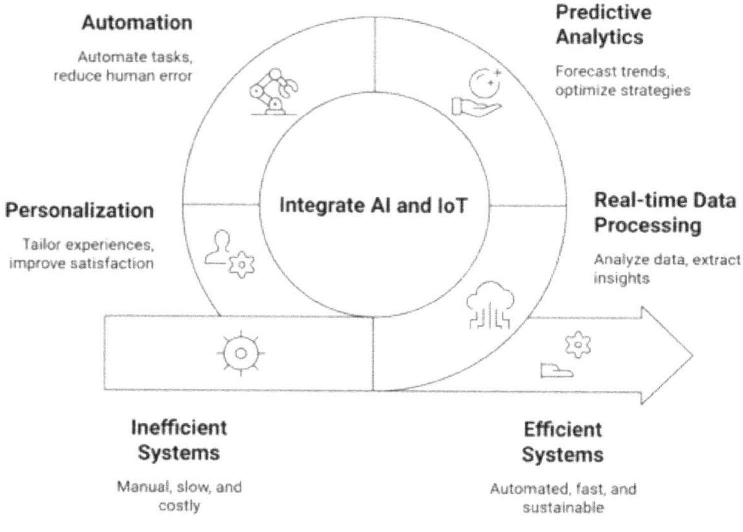

5. SMART ENERGY MANAGEMENT AND RENEWABLE INTEGRATION

5.1 Importance of energy sustainability in urban development

The current use of new technologies coupled with energy smart operations has made renewable energy an integral part of energy sustainability. Nowadays, monitoring, data collection, and IoT-enabled automation can all be done in real-time, thus allowing for more effective load management, demand response, and energy management (Jia et al., 2024).

Smart grids, which are IoT systems, try to reduce the integration and control of renewables, like solar and wind, into the electric grids. A smart grid is one of the vital elements of a smart city. (Mishra & Singh, 2023). In fact, these systems enhance security and reliability of the grid, increase renewables and decrease greenhouse gas emissions and fossil fuel dependency (Kiasari et al., 2024). In incorporating IoT devices into the smart grid framework, energy optimization decisions can be made that lead to decreased energy consumption and increased stability for the grid (S, 2024).

Thus, the concept of energy aware distributed cyber physical system (DCPSs) that merges common nonfunctional requirements with other aspects of sustainability including energy utilization came into being (Cicceri et al., 2023). This assists in the creation of smart grids by facilitating energy self-control and distribution in a manner that is sustainable and efficient.

Table 4 presents various case studies where AI and IoT technologies have been successfully integrated into smart city initiatives, leading to improvements in resource management, waste reduction, and sustainability efforts.

Table 4. Smart energy management systems in urban development

Technology	Application	Impact
Smart Grids	Optimizing energy distribution	Reduces energy waste and enhances grid reliability
AI-Powered Demand Forecasting	Real-time energy consumption prediction	Minimizes peak loads and optimizes power generation
IoT in Renewable Integration	Integration of solar and wind energy	Improves efficiency of renewable sources
Microgrid Operations	Localized energy resource management	Enhances energy stability at a local level
EV Grid Integration	Smart scheduling for electric vehicle charging	Balances grid load and lowers energy costs

5.2 Case studies of smart cities implementing AI-IoT in energy management

The use of AI and IoT technologies is transforming energy management in smart cities in a way that makes them operate more efficiently and, at the same time, with much lesser resource intake. These technologies are successfully implemented several case studies for the urban energy systems.

The introduction of IoT and Green Fintech in Amsterdam have managed to robustly facilitate the resources through the implementation of dynamic pricing models and financial incentives targeted towards sustainable practices (Adaramola et al., 2024). The results of this approach have increased transparency and economy of energy consumption. Similar to this, Singapore has made use of AI and IoT solutions for the efficiency of energy usage and the embracement of sustainable practices (Adaramola et al., 2024).

In the present day, the combination of AI, IoT, and renewable energy technologies have proven to radically change the power systems of the current era. The energy management system through IoT and the smart grid made possible real time monitoring, data analysis, automation, and more advanced load managements, demand response, and energy storage solutions (Jia et al., 2024). With this integration, grid stability has been greatly improved and there has been a promotion of renewable energy sources.

For example, AI algorithms have been applied to energy consumption patterns optimization based on IoT sensors' real time data for the purpose of minimizing energy usage and enhancing grid stability (S, 2024). For instance, an AI equipped smart grids can predict energy demand fluctuations for more intelligent to distribute energy.

The combination of AI and IoT in microgrid operations has improved management and optimization of the local energy resources within the microgrid. This has been extremely helpful for the deployment of renewable sources of energy like solar and wind into the existing grid structure (Camacho et al., 2024). They assist in forecasting renewable energy production, as well as controlling the energy storage system which ensures that power is produced and supplied stably.

Advancements of AI-IoT applications have been on electric vehicle (EV) grid integration. AIs are used to determine charging schedules that take grid load and energy prices into consideration while the IoT devices allow for feedback communication between EVs and the grid (Szpilko et al., 2023). Interestingly, although these case studies offer encouraging results, it is still difficult to take care of areas like data privacy, interoperability and regulatory adaptation (S, 2024). Furthermore, the various uses of AI algorithms in different renewable energy scenarios identify ingenious for more commonly used approaches (Necula, 2023).

Table 5 presents various case studies where AI and IoT technologies have been successfully integrated into smart city initiatives, leading to improvements in resource management, waste reduction, and sustainability efforts.

Location is a vital factor in the comprehension of the work of AI-IoT systems in various urban areas. The developed countries like Singapore and the Netherlands focus on efficiency and optimization of the available infrastructure whereas developing cities like Lahore and Nairobi focus on growth, cost affordability, and community planning. The results obtained after the incorporation of region-specific case studies not only counterchecks the proposed framework but also outlines situational barriers and levers. Such comparison enhances this chapter because it allows showing how the geography directly affects adoption and outcomes.

Although AI-IoT applications in energy management reduce the load on the grid, enhance renewable integration and provide predictive models of demand, they also come with a certain package of concerns. It is Energetically demanding to incorporate variable sources of renewables into the system as it involves expensive grid interconnection and forecasting models. There are also data privacy and cybersecurity concerns about relying on IoT sensors and AI-driven analysis when there is sharing of sensitive consumption patterns across the networks. In addition, the prohibitive expenses of commercialization, especially at scale, may not make sustainable energy-based theories, technologies and innovations accessible in developing settings, which as well, causes inequalities in access to sustainable energy-based theories, technology and innovations. By acknowledging the potential and the constraints it will give a sounder picture of the reality of how energy systems can be transformed without over-promising and creating impossible deadlines.

Table 5. Case studies of AI-IoT integration in smart cities

City	AI-IoT Application	Key Outcomes
Amsterdam	Green Fintech for dynamic pricing and resource management	Increased transparency and economic sustainability
Singapore	AI-IoT solutions for energy efficiency and sustainability	Enhanced energy optimization and reduced emissions
Beijing	AI-powered waste management optimization	25% reduction in waste collection trips, 30% fewer waste overflow cases
Lahore	IoT-enabled waste collection with dynamic route optimization	32% improvement in route efficiency, 29% lower emissions
United States	AI-driven waste management for underserved communities	Predictive analytics for waste collection and recycling optimization

6. AI AND IOT FOR GREEN INFRASTRUCTURE AND ENVIRONMENTAL MONITORING

6.1 IoT sensors for real-time environmental monitoring

More effective environmental monitoring and management is taking place due to the combination of AI and IoT technologies improving green infrastructure, which covers natural and engineered systems with environmental benefits.

Environmental monitoring of green infrastructure is enabled by IoT sensors, where they can be deployed as real time sensors. They can continuously monitor vital markers including temperature, pH, dissolved oxygen and salinity (Huang & Khabusi, 2025). The IoT connected sensors allow remote monitoring of weather, soil conditions, as well as crop health in an agricultural setting. To facilitate collection of sensor data, its display, and analysis on the same platform (Panduman et al., 2024), the SEMAR (Smart Environmental Monitoring and Analytics in Real-Time) IoT application server platform has been developed. By integrating the IoT sensors with the AI algorithms, predictive insights for water quality management, diseases detection and optimization of resource utilization are provided (Huang & Khabusi, 2025).

However, recently, looking at the sensing technologies using organic materials showed one of the promising ways of smart environmental systems integrated with IoT network platforms (Song et al., 2022). Furthermore, Unmanned Aerial Vehicles (UAVs) are investigated for the sake of improving the efficiency of the environment monitoring solved with the IoT technology by flying closer to IoT devices and transmitting data (Alsamhi et al., 2021).

Table 6 highlights different environmental concerns and how AI and IoT technologies contribute to their monitoring and management, ranging from air and water quality to flood prediction and biodiversity conservation.

Table 6. AI and IoT applications in environmental monitoring

Environmental Concern	IoT Implementation	AI Role
Air Quality Monitoring	Sensors detect pollutants like CO2, NH3, NO2	Predictive analysis for air pollution trends
Water Quality Management	Real-time water quality assessment using IoT	AI algorithms detect contamination and optimize water distribution
Flood Prediction	AI-driven flood prediction models using remote sensing	Predicts flood risks and enhances early warning systems
Biodiversity Conservation	AI-based habitat assessment, species recognition	Machine learning models analyze species distribution

6.2 AI-IoT applications

Smart water management, flood control and biodiversity monitoring with AI and IoT applications have brought about the invention of new and unique solutions to the environmental problems.

AI-IoT systems help improve operational efficiency as well as resource conservation in intelligent water management and conservation systems. With these technologies you are able to monitor the water quality in real time, for the health of infrastructure, and optimize water distribution. IoT sensors and AI algorithms used to process data from different sources, including ground gauges and remote sensing tools, to approximate a complicated relationship among hydrological processes and enhance water assets the board the board administration (Chang et al, 2023). Water applications on microcontrollers, sensors, and communication modules become the driver for IoT implementations for water conservation.

In this way, AI based systems bring in early warnings of flood and also help to improve the preparedness for this type of natural disaster. AI algorithms leverage data from several sources to determine when these floods may occur in an attempt to help in the planning of evacuation and mitigation strategies. Integration with environmental science would yield some innovative approaches to the issues of climate change such as predictive modeling for climate impacts and sustainable resource management (Shuford, 2024).

AI and IoT are very important in wildlife tracking, habitat assessment, and species identification for biodiversity monitoring and ecosystem protection. AI algorithms analyze camera trap videos, drone photos, and sound recordings to determine the number of people living within a certain area, especially where deforestation and new or endangered species are found. With these technologies, real time environmental monitoring has been made possible and is attributed for conservation and biodiversity analyses (Alotaibi & Nassif, 2024).

The fusion of AI, IoT, and environmental science has become the solution to tackle the global environmental issues. Yet, such technologies have a number of challenges like data quality, algorithmic bias and infrastructure limitations, which need to be overcome to fully extract the potential of these technologies (Chisom et al., 2024) . Future research then should concern about improving the quality of data, broadening of interdisciplinarity collision, and of ethical consideration in artificial intelligence directed environmental management.

Safety, health and environment (SHE) factors are also to be mentioned along with sustainability and governance among the dimensions to be considered. AI-IoT systems in smart cities are not only about efficiency in the use of resources but also about a safer living environment, healthier communities and reduced environmental hazards. Examples would be AI-supported real-time air quality forecasting to prevent respiratory diseases, IoT-based flood early-warning systems that will make people safe, and AI-supported waste segregation that reduces the risk of contamination. Embedding SHE Integrates an appreciation of smart urban ecosystems in a more comprehensive way.

7. SMART WASTE MANAGEMENT SYSTEMS

7.1 Role of AI-IoT in optimizing waste collection routes

AI and IoT technologies help in reducing collection routes for smart waste management system. Real time fill level and waste production trend data is given by IoT enabled smart bins with ultrasonic and weight sensors (Hussain et al., 2024). It is this information that gets transmitted to central systems through various such communication technologies as Wi-Fi, cellular networks and long range communication networks (Vishnu et al., 2021) .

This data is processed by AI algorithms to predict which waste will be produced where and how efficient collection routes could be. According to Yao et al. (2024), in Beijing, they integrated air driven route optimization and real time monitoring powered by IoT in which waste collection trips were reduced by 25%, and dropped waste overflow cases were by 30%. The same is manifested by a study of Lahore, Pakistan that positioned route efficiency improvement by 32%, fuel consumption and emissions by 29% using dynamic route optimization algorithms (Addas et al., 2024).

By combining AI and IoT technologies, a multi level decision making process incorporating bin status and coordinated with routing is made possible to optimize the routing problem (Ghahramani et al., 2022). The approach solves the problem such that bins will be processed based on priority to eliminate errors and ensure that accuracy is high while collecting waste (Alqahtani et al., 2020). Integrating these technologies will not only improve the overall effectiveness of the operations, but also contribute to environmental conservation and economic sustainability measured by a net present value of $3.5 million in the coming decade in Beijing (Yao et al. 2024).

7.2 Real-world implementations of AI-IoT waste management solutions

AI-IoT waste management solution has been empowered with real world implementations and have shown improvements in terms of efficiency, sustainability, and cost effectiveness into various cities around the world.

The integration of AI and IoT into waste management system of Beijing showed significant improvement during the years. The use of AI meshed routes and IoT supported real-time monitoring increased efficiency, allowing reductions of 25% in waste collection trips and 30% in waste overflow incidents (Yao et al, 2024). Along with increasing the operational efficiency of these systems, advancements in technology will also contribute toward environmental sustainability by decreasing fuel use and emissions. The economic analysis indicated a Net Present Value of USD 3.5 million for a 10 year period which suggests that such implement should have a financial viability.

A series of pilot studies was performed in Lahore, Pakistan in 10 different locations using a comprehensive IoT enabled waste management system alone yielding more than 200 million data points. In addition, these benchmarks facilitated an increase of up to 32% In route efficiency, 29% off fuel consumption and emissions, 33% hike up in waste processing throughput, as well as 78% drop in waste vehicle maintenance cost (Addas et al., 2024). Everything is for the purspose of capturing real, operational, economic and sustainable value.

In fact, the second was implemented on an IoT Based Urban Waste management System. A supervisory system which takes care of human activities and helps in the waste management processes was developed as an IoT based system. A Long Short Term Recurrent Neural Network was used to perform the aforementioned analysis with a further installment to a cuckoo search algorithm to optimally find the parameters to analyze waste type, truck size and waste source. In the system, the standard approach of bin processing with regularized error of 0.16 and a maximum accuracy of 98.4 percent in the processing time of 15 minutes was emphasized (Alqahtani et al 2020).

The author in the case of the United States has explored the potential of AI driven waste management systems for underserved communities. However, this concept paper without alteration is a framework of applying such AI functionalities as predictive analytics, optimization algorithms and IoT sensors to increase waste collection, recycling efficiency and help bring about environmental sustainability in these places (Nwokediegwu & Ugwuanyi (2024)).

AI-IoT solution's realizations in these applications indicate that it has the power to transform waste management. These are the significant improvement of operational efficiency, environmental sustainability and cost effectiveness. Despite which, there is still need for cautious implementation in regard to particular factors, infrastructure shortcomings, and community participation. While smart implementation of these technologies is already catching on in cities, the development of these technologies as well as their implementation can only get better as cities adopt them.

Regardless of the positive results in terms of AI-IoT-powered waste handling systems, there are a series of practical obstacles, which hamper its implementation efforts. The predicative routes and sensor networks usually require a significant initial expenditure, and the necessity of a well-developed network connection may eliminate rural or under-serviced territory of the city. Problems of data quality and algorithm bias on waste classification can hamper efficiency when not taken into consideration carefully. Also, the attachment of AI algorithms to municipal systems presupposes the involvement of qualificated workers who can be absent in most local administrations. Such difficulties point to the necessity to combine technological advancement with institutional capacity-building, regulatory assistance, and bottom-top involvement to achieve a leveled and sustainable end.

8. CHALLENGES AND ETHICAL CONSIDERATIONS IN AI-IOT IMPLEMENTATION

If smart cities integrate Artificial Intelligence (AI) and Internet of Things (IoT), there will be many benefits of itself but many ethical and challenge raised. The problem is the underlying problem of data privacy and security. The volumes of real time data generated by sensors, cameras, smart devices, is depended on these data to make smart city projects successful. This data is susceptible to be cyber attacked, unlawfully accessed and misused if not secured properly, it represents a concern to personal privacy and surveillance. To mitigate such risks, encryption must be as robust as possible, data must be well stored and security must be as stringent as possible.

The other big challenge is for AI in decision making to be ethical. The processed data is vast and reflects enormous amounts of various types of activities throughout our cities, towns, and, therefore, necessary to be processed by AI algorithms in order to enable decisions, for example, in traffic control, law enforcement, or resource allocation. Yet even though human oversight isn't in operation during the AI decisions, it can lead to unintended consequences during these sensitive areas like predictive policing or mental health care. These concerns can be mitigated in the case of establishing clear ethical guidelines and human intervention in key decisions.

The high costs and infrastructure challenges of deployment of AI-IoT also constitute a great barrier. Solutions for the smart city are actually quite costly in having to invest so much in sensors, AI powered analytics, additional data management systems. Affordability and resource allocation will become an issue for developing nations in particular. These financial restraints can be overcome by public private partnerships and scalable solution.

Additionally, AI bias and lack of transparency in decision-making present risks. Even if the data that trains AI models is biased, these models can reinforce the existing inequalities, and in the event of incorrect quantification, and subsequently unfair outcomes, in urban management. It is possible to maintain transparency in use of AI algorithms and run regular auditions to minimize bias and build public trust.

To accomplish this while keeping the cities sustainable, cities would have to adopt strategies like regulatory framework, AI ethics committees and open data policies. But with the right approach to AI and IoT integration, it can avoid the extremes that will increase the risk. Table 7 highlights the major challenges faced in AI-IoT adoption for smart urban development, along with proposed solutions. Addressing issues such as data privacy, high costs, AI bias, and cybersecurity risks is essential for the successful implementation of AI-IoT technologies in smart cities.

A further critical look at the socio-ethical and governance behaviour of the integration of AI-IoT is needed to develop long-term sustainability. In addition to environmental and economic efficiency, sustainability requires an impeccable ratio of harmony, responsibility, and inclusion as the social pillar of this concept. When I.A. is trained on biased data, there is a risk that instead of promoting equality it will in fact perpetuate the inequalities that existed in the first place, and a lack of transparency in decision-making can contribute to the loss of trust by those individuals the I.A. will affect. These risks can be addressed by bias audit and explainable AI models as a way of strengthening algorithmic justice. The other key considerations are strong data governance models including open data policies, federated learning models that respect privacy, and blockchain-based data governance systems that permit greater transparency. Through incorporating such ethical and governance values, cities will be able to ensure that smart technologies not only optimise, but also protect the rights of citizens, deliver equity, and improve trust in urban innovation.

Table 7. The major challenges faced in AI-IoT adoption for smart urban development

Challenge	Description	Proposed Solution
Data Privacy	Risk of unauthorized data access and surveillance.	Strong encryption, regulatory compliance, transparent data policies.
High Costs	High capital required for IoT infrastructure and AI models.	Public-private partnerships, scalable low-cost solutions.
AI Bias & Transparency	Algorithmic bias in AI decision-making.	Bias audits, diverse data sets, explainable AI models.
Cybersecurity Risks	Vulnerability to cyberattacks on smart city networks.	AI-based anomaly detection, blockchain for data security.

9. FUTURE TRENDS AND POLICY RECOMMENDATIONS

Emerging technologies can play a big role in future AI-IoT integration in smart cities for the sake of efficiency, security, and scalability. Edge computing and decentralized AI are one of the key progress — they allow data processing near to the source, lowering latency, as well as a reliance on a centralized cloud system. This improved is for real time decision making in the applications like traffic management, energy distribution and environmental monitoring. The other disruptive technology is Blockchain which ensures trust and tamper proof distribution of the data among IoTs in a smart city ecosystem. Another advantage of federated learning is that they allow the training of their AI models even on multiple decentralized devices without having to share sensitive data, leading to better privacy as well as using the distributed intelligence.

However, strong policy frameworks at government levels would help to maximize the benefits of AI-IoT in urban sustainability. Data privacy, cybersecurity, ethical deployment of AI are to be taken up with appropriate regulations to ensure protection of citizens without hampering innovation. They also should establish interoperability standards for various AI-IoT solutions in cities to facilitate seamless integration.

An effective way in which governments and urban planners can aid in the adoption of AI-IoT is by playing this crucial role. Such investments in smart infrastructure, funding for AI research, connecting with private sectors and academia can bring this process to a smooth pace. And governments must also assist ensuring unequal access to smart city solutions, to avoid divides between the different socio-economic groups.

Smart urban sustainability is seen the way ahead towards AI-IoT enabled cities by being ‘energy efficient’, ‘environmentally conscious’, and ‘citizen centric’. Renewable energy grids optimized via AI, AI to aid in waste management and planning of the city via AI will ensure that there are self-sustainable ecosystems. Future cities are ones which practise responsible governance and innovate for sustainability improving the quality of life of the people that live in them. One of the new things in these cities of tomorrow is the combination of AI and IoT, the cities will be smarter, greener and more resilient.

The capacity building and skill development are crucial in the effective implementation of AI- IoT systems in smart cities. Cities and industries need educated individuals who can work with sensor networks, extract insights out of big data, and embrace ethical ways to use AI. Some examples of implementation consist of special training programs, industry-academia links, and urban upskilling. As an example, digital literacy training on municipal employees and collaborative AI -IoT labs in higher learning institutions could help to fill the skill gap. By emphasizing such measures, one can make sure that technological progress will be reasonable and sustainable.

CONCLUSION

This development in the area of integrating IoT and AI technologies has resulted in the development of smart and a green infrastructure that is in line with urban sustainability. This chapter showed what high potency stems from the alliance between IoT sensors and devices, and AI algorithms to better experience more comfortable, livable and environmentally responsible cities. IoT also enables the ability of real-time data collection across urban systems by networks of connected sensors and devices that deliver high visibility into how the city operates and how its resources are being used as key takeaways. The sheer volume of generated IoT data is too much for the human resources. It becomes too difficult for them to interpret further. This explains how AI systems can analyze massive data to gain useful insights and optimize processes, as well as achieve predictive capabilities. Important integrated urban areas like waste management, energy management, transportation, and environmental monitoring are synergized by IoT and AI. The use of IoT and AI enables smart energy grids to integrate alternative renewable energy sources, balance supply with demand, and consumption. Sustainability is facilitated by the integration of IoT sensing and AI analytic technologies which allow for the real time monitoring of air, water, and ecosystem based on environmental parameters. By utilizing IoT bin sensors, the efficiency of collecting waste in the cities significantly increases through AI-optimized waste collection routes. Nevertheless, there are data privacy, cybersecurity, algorithmic bias challenges along with high costs of implementation barriers. Proper governance and ethical да are needed. The future trends in such AI-IoT include edge computing, blockchain integration, and federated learning to improve the performance of AI-IoT capabilities by satisfying security and privacy. For realizing the full potential of these technologies, the government needs to support with supportive government policies, standards for interoperability and public private partnerships. Therefore, the integration of IoT with AI concepts brings about exciting opportunities of making cities smarter and sustainable. Yet, to realize this potential technological, ethical and governance challenges will need to be confronted collaboratively with policymakers, urban planners, technology providers and citizens. If properly exploited, the AI-IoT synergy can be a strong asset to collectively realize urban sustainability and upscale urban quality of life.

REFERENCES

Abdel Aal, S. I. (2023). Neutrosophic Framework for Assessment Challenges in Smart Sustainable Cities based on IoT to Better Manage Energy Resources and Decrease the Urban Environment's Ecological Impact. *Neutrosophic Systems with Applications*, *6*, 9–16. DOI: 10.61356/j.nswa.2023.33

Adaramola, T., Nwariaku, H., Adigun, O., Arowolo, M., Omole, O., & Wada, I. (2024). Internet of thing integration in green fintech for enhanced resource management in smart cities. *World Journal of Advanced Research and Reviews*, *23*(2), 1317–1327. DOI: 10.30574/wjarr.2024.23.2.2448

Addas, A., Khan, M. N., & Naseer, F. (2024). Waste management 2.0 leveraging internet of things for an efficient and eco-friendly smart city solution. *PLoS One*, *19*(7), e0307608. DOI: 10.1371/journal.pone.0307608 PMID: 39083562

Adefemi, A., Adekoya, O., Adegbite, A., Abatan, A., & Ukpoju, E. (2023). Artificial intelligence in environmental health and public safety: A comprehensive review of USA strategies. *World Journal of Advanced Research and Reviews*, *20*(3), 1420–1434. DOI: 10.30574/wjarr.2023.20.3.2591

Alotaibi, E., & Nassif, N. (2024). Artificial intelligence in environmental monitoring: In-depth analysis. *Discover Artificial Intelligence*, *4*(1), 84. Advance online publication. DOI: 10.1007/s44163-024-00198-1

Alqahtani, F., Al-Makhadmeh, Z., Tolba, A., & Said, W. (2020). Internet of things-based urban waste management system for smart cities using a Cuckoo Search Algorithm. *Cluster Computing*, *23*(3), 1769–1780. DOI: 10.1007/s10586-020-03126-x

Alqasi, M. A. Y., Alkelanie, Y. A. M., & Alnagrat, A. J. A. (2024). Intelligent Infrastructure for Urban Transportation: The Role of Artificial Intelligence in Predictive Maintenance. *Brilliance: Research of Artificial Intelligence*, *4*(2), 625–637. DOI: 10.47709/brilliance.v4i2.4889

Alsamhi, S. H., Afghah, F., Sahal, R., Hawbani, A., Al-Qaness, M. A. A., Lee, B., & Guizani, M. (2021). Green internet of things using UAVs in B5G networks: A review of applications and strategies. *Ad Hoc Networks*, *117*, 102505. DOI: 10.1016/j.adhoc.2021.102505

Ayasrah, F. T., N, K., Alsharafa, N. S., S, S., Sengan, S., & B, S. (2024). Strategizing Low Carbon Urban Planning Through Environmental Impact Assessment by Artificial Intelligence Driven Carbon Foot-print Forecasting. *Journal of Machine and Computing*, 1140–1151. DOI: 10.53759/7669/jmc202404105

Bibri, S. E., Sharifi, A., Alexandre, A., & Krogstie, J. (2023). Environmentally sustainable smart cities and their converging AI, IoT, and big data technologies and solutions: An integrated approach to an extensive literature review. *Energy Informatics*, *6*(1), 9. Advance online publication. DOI: 10.1186/s42162-023-00259-2 PMID: 37032812

Boopathi, S. (2024). *Advancements in Machine Learning and AI for Intelligent Systems in Drone Applications for Smart City Developments* (pp. 15–45). igi global. DOI: 10.4018/978-1-6684-9596-4.ch002

Camacho, J. D. J., Anthony, B., Aguirre, B., Ponce, P., & Molina, A. (2024). Leveraging Artificial Intelligence to Bolster the Energy Sector in Smart Cities: A Literature Review. *Energies*, *17*(2), 353. DOI: 10.3390/en17020353

Cao, B., Lv, Z., Fan, S., Chen, X., & Li, R. (2020). Optimization of Classified Municipal Waste Collection Based on the Internet of Connected Vehicles. *IEEE Transactions on Intelligent Transportation Systems*, *22*(8), 5364–5373. DOI: 10.1109/TITS.2020.2981564

Cesario, E. (2023). Big data analytics and smart cities: Applications, challenges, and opportunities. *Frontiers in Big Data*, *6*, 1149402. Advance online publication. DOI: 10.3389/fdata.2023.1149402 PMID: 37252127

Chang, F.-J., Chen, J.-F., & Chang, L.-C. (2023). Artificial Intelligence Techniques in Hydrology and Water Resources Management. *Water (Basel)*, *15*(10), 1846. DOI: 10.3390/w15101846

Chisom, O., Obaedo, B., Biu, P., Umoh, A., Abatan, A., & Adegbite, A. (2024). Reviewing the role of AI in environmental monitoring and conservation: A data-driven revolution for our planet. *World Journal of Advanced Research and Reviews*, *21*(1), 161–171. DOI: 10.30574/wjarr.2024.21.1.2720

Cicceri, G., Tricomi, G., D'Agati, L., Longo, F., Merlino, G., & Puliafito, A. (2023). A Deep Learning-Driven Self-Conscious Distributed Cyber-Physical System for Renewable Energy Communities. *Sensors (Basel)*, *23*(9), 4549. DOI: 10.3390/s23094549 PMID: 37177753

Dikshit, S., Shahid, M., Thusu, A., Dwivedi, V., & Atiq, A. (2023). The Use of Artificial Intelligence to Optimize the Routing of Vehicles and Reduce Traffic Congestion in Urban Areas. *EAI Endorsed Transactions on Energy Web*, *10*. Advance online publication. DOI: 10.4108/ew.4613

Dritsas, E., & Trigka, M. (2024). Machine Learning for Blockchain and IoT Systems in Smart Cities: A Survey. *Future Internet*, *16*(9), 324. DOI: 10.3390/fi16090324

Farahani, M. S., & Ghasemi, G. (2024). How artificial intelligence plays a role in achieving sustainable development goals? *Sustainable Economies*, *2*(3), 66. DOI: 10.62617/se.v2i3.66

Ghahramani, M., Zhou, M., Molter, A., & Pilla, F. (2022). IoT-Based Route Recommendation for an Intelligent Waste Management System. *IEEE Internet of Things Journal*, *9*(14), 11883–11892. DOI: 10.1109/JIOT.2021.3132126

Gkontzis, A. F., Verykios, V. S., Feretzakis, G., & Kotsiantis, S. (2024). Enhancing Urban Resilience: Smart City Data Analyses, Forecasts, and Digital Twin Techniques at the Neighborhood Level. *Future Internet*, *16*(2), 47. DOI: 10.3390/fi16020047

Gómez-Villarino, M. T., Ruiz-Garcia, L., & Gómez Villarino, M. (2020). Implementation of Urban Green Infrastructures in Peri-Urban Areas: A Case Study of Climate Change Mitigation in Madrid. *Agronomy (Basel)*, *11*(1), 31. DOI: 10.3390/agronomy11010031

Hoang, T. V. (2024). Impact of Integrated Artificial Intelligence and Internet of Things Technologies on Smart City Transformation. *Journal of Technical Education Science*, *19*(1), 64–73. DOI: 10.54644/jte.2024.1532

Huang, Y.-P., & Khabusi, S. P. (2025). Artificial Intelligence of Things (AIoT) Advances in Aquaculture: A Review. *Processes (Basel, Switzerland)*, *13*(1), 73. DOI: 10.3390/pr13010073

Hussain, D. I., Elomri, D. A., Kerbache, D. L., & Omri, D. A. E. (2024). Smart city solutions: Comparative analysis of waste management models in IoT-enabled environments using multiagent simulation. *Sustainable Cities and Society*, *103*, 105247. DOI: 10.1016/j.scs.2024.105247

Ighalo, J. O., Marques, G., & Adeniyi, A. G. (2020). *Internet of Things for Water Quality Monitoring and Assessment: A Comprehensive Review* (pp. 245–259). springer. DOI: 10.1007/978-3-030-51920-9_13

Jia, L., Hu, Z., & Li, Z. (2024). Applications of the Internet of Things in Renewable Power Systems: A Survey. *Energies*, *17*(16), 4160. DOI: 10.3390/en17164160

Kiasari, M., Ghaffari, M., & Aly, H. H. (2024). A Comprehensive Review of the Current Status of Smart Grid Technologies for Renewable Energies Integration and Future Trends: The Role of Machine Learning and Energy Storage Systems. *Energies*, *17*(16), 4128. DOI: 10.3390/en17164128

Krupnova, T. G., Bondarenko, K. A., Tretyakova, V. D., & Rakova, O. V. (2022). Environmental Justice and the Use of Artificial Intelligence in Urban Air Pollution Monitoring. *Big Data and Cognitive Computing*, *6*(3), 75. DOI: 10.3390/bdcc6030075

Kumar, L. R., Kondaiah, T., Goel, L., Sahil Singh, R. K., Dixit, S., Keerthana, S., Sai Akanksha, S., Satyanarayana, K., Bobba, P. B., Perveen, A., & Debnath, S. (2024). Smart Air Quality Monitoring Using IoT. *MATEC Web of Conferences, 392*, 01084. DOI: 10.1051/matecconf/202439201084

Kushwaha, A. (2024). Application of AI and IoT in Traffic Management of Large Metropolitan Cities. *International Journal of Innovative Science and Research Technology (IJISRT)*, 3426–3431. DOI: 10.38124/ijisrt/IJISRT24APR2069

Marahatta, A., Singh, A., Korba, P., Shrestha, A., Gonzalez-Longatt, F., Thapa, A., Shin, S., & Rajbhandari, Y. (2021). Evaluation of a LoRa Mesh Network for Smart Metering in Rural Locations. *Electronics (Basel), 10*(6), 751. DOI: 10.3390/electronics10060751

Martínez, I., Cambra, D., Blanco, T., Casas, R., Trillo-Lado, R., & Zalba, B. (2021). Internet of Things (IoT) as Sustainable Development Goals (SDG) Enabling Technology towards Smart Readiness Indicators (SRI) for University Buildings. *Sustainability (Basel), 13*(14), 7647. DOI: 10.3390/su13147647

Mehta, R., Tanwar, S., Patel, Y., Mishra, S., & Prasad, V. K. (2023). *Evolving technologies: IoT and artificial intelligence for healthcare informatics* (pp. 231–254). institution of engineering technology. DOI: 10.1049/PBHE041E_ch8

Mishra, P., & Singh, G. (2023). Energy Management Systems in Sustainable Smart Cities Based on the Internet of Energy: A Technical Review. *Energies, 16*(19), 6903. DOI: 10.3390/en16196903

Necula, S.-C. (2023). Assessing the Potential of Artificial Intelligence in Advancing Clean Energy Technologies in Europe: A Systematic Review. *Energies, 16*(22), 7633. DOI: 10.3390/en16227633

Nguyen, D. N., Imamura, F., & Usuda, Y. (2024). Gaps in and Opportunities for Disaster Risk Reduction in Urban Areas Through International Standardization of Smart Community Infrastructure. *Sustainability (Basel), 16*(21), 9586. DOI: 10.3390/su16219586

Nwokediegwu, Z., Dada, M., Obaigbena, A., Majemite, M., & Ugwuanyi, E. (2024). AI-DRIVEN WASTE MANAGEMENT SYSTEMS: A COMPARATIVE REVIEW OF INNOVATIONS IN THE USA AND AFRICA. *Engineering Science & Technology Journal, 5*(2), 507–516. DOI: 10.51594/estj.v5i2.828

Nwokediegwu, Z., & Ugwuanyi, E. (2024). IMPLEMENTING AI-DRIVEN WASTE MANAGEMENT SYSTEMS IN UNDERSERVED COMMUNITIES IN THE USA. *Engineering Science & Technology Journal*, *5*(3), 794–802. DOI: 10.51594/estj.v5i3.903

Panduman, Y. Y. F., Sukaridhoto, S., Funabiki, N., Fajrianti, E. D., & Fang, S. (2024). A Survey of AI Techniques in IoT Applications with Use Case Investigations in the Smart Environmental Monitoring and Analytics in Real-Time IoT Platform. *Information (Basel)*, *15*(3), 153. DOI: 10.3390/info15030153

Rajan, P. (2024). Integrating IoT Analytics into Marketing Decision Making: A Smart Data-Driven Approach. *International Journal of Data Informatics and Intelligent Computing*, *3*(1), 12–22. DOI: 10.59461/ijdiic.v3i1.92

Razmjoo, A., Mirjalili, S., Ahmadzadeh, S., Gandomi, A., Memon, S., Astiaso Garcia, D., Rezvani, A., & Mahlooji, M. (2022). An Investigation of the Policies and Crucial Sectors of Smart Cities Based on IoT Application. *Applied Sciences (Basel, Switzerland)*, *12*(5), 2672. DOI: 10.3390/app12052672

Rojek, I., Mroziński, A., Mikołajewski, D., & Macko, M. (2024). Green Energy Management in Manufacturing Based on Demand Prediction by Artificial Intelligence—A Review. *Electronics (Basel)*, *13*(16), 3338. DOI: 10.3390/electronics13163338

S, B. (2024). *Advancements in Optimizing Smart Energy Systems Through Smart Grid Integration, Machine Learning, and IoT* (pp. 33–61). igi global. DOI: 10.4018/979-8-3693-0492-1.ch002

Sallang, N. C. A., Islam, M. T., Islam, M. S., & Arshad, H. (2021). A CNN-Based Smart Waste Management System Using TensorFlow Lite and LoRa-GPS Shield in Internet of Things Environment. *IEEE Access : Practical Innovations, Open Solutions*, *9*, 153560–153574. DOI: 10.1109/ACCESS.2021.3128314

Shah, H. (2023). Beyond Smart: How ICT Is Enabling Sustainable Cities of the Future. *Sustainability (Basel)*, *15*(16), 12381. DOI: 10.3390/su151612381

Shuford, J. (2024). Interdisciplinary Perspectives: Fusing Artificial Intelligence with Environmental Science for Sustainable Solutions. *Journal of Artificial Intelligence General Science (JAIGS) ISSN:3006-4023*, *1*(1), 106–123. DOI: 10.60087/jaigs.v1i1.87

Song, H. W., Jeon, T., Choi, W., & Oh, J. H. (2022). Recent Advances in Smart Organic Sensors for Environmental Monitoring Systems. *ACS Applied Electronic Materials*, *5*(1), 77–99. DOI: 10.1021/acsaelm.2c01315

Sugandha, S., Favaro, P., & Freestone, R. (2022). The social sustainability of smart cities: A conceptual framework. *City, Culture and Society*, *29*, 100460. DOI: 10.1016/j.ccs.2022.100460

Szpilko, D., Naharro, F. J., Gallegos, A. D. L. T., Lăzăroiu, G., & Nica, E. (2023). Artificial Intelligence in the Smart City — A Literature Review. *Engineering Management in Production and Services*, *15*(4), 53–75. DOI: 10.2478/emj-2023-0028

Ullo, S. L., & Sinha, G. R. (2020). Advances in Smart Environment Monitoring Systems Using IoT and Sensors. *Sensors (Basel)*, *20*(11), 3113. DOI: 10.3390/s20113113 PMID: 32486411

Van Oorschot, J., Slootweg, M., Remme, R. P., Sprecher, B., & Van Der Voet, E. (2024). Optimizing green and gray infrastructure planning for sustainable urban development. *NPJ Urban Sustainability*, *4*(1), 41. Advance online publication. DOI: 10.1038/s42949-024-00178-5

Vishnu, S., Srinivasan, S., Anagnostopoulos, T., Senith, S., Ramson, S. R. J., Fan, X., Abu-Mahfouz, A. M., & Kirubaraj, A. A. (2021). IoT-Enabled Solid Waste Management in Smart Cities. *Smart Cities*, *4*(3), 1004–1017. DOI: 10.3390/smartcities4030053

Yao, Y., He, C., Xiao, P., Gong, C., & Weng, J. (2024). AI-powered Strategies for Optimizing Waste Management in Smart Cities in Beijing. *World Journal of Innovation and Modern Technology*, *7*(5), 22–29. DOI: 10.53469/wjimt.2024.07(05).02

Yigitcanlar, T., Desouza, K., Butler, L., & Roozkhosh, F. (2020). Contributions and Risks of Artificial Intelligence (AI) in Building Smarter Cities: Insights from a Systematic Review of the Literature. *Energies*, *13*(6), 1473. DOI: 10.3390/en13061473

Chapter 6
Cognitive AI Framework for Smart Water Management in Buildings:
Integrating Leak Detection, Water Usage Prediction, and Quality Monitoring

T. Grace Shalini
http://orcid.org/0000-0002-0016-3702
SRM Institute of Science and Technology, India

Nayantra Ramakrishnan
http://orcid.org/0009-0000-5682-3717
SRM Institute of Science and Technology, India

P. Bhavesh
http://orcid.org/0009-0003-1327-9020
SRM Institute of Science and Technology, India

S. S. Krishikaa Mathi Bharathi
http://orcid.org/0009-0002-4079-4864
SRM Institute of Science and Technology, India

ABSTRACT

Modern cities face a critical need to handle water efficiently because increasing water demand meets aging building infrastructure. This chapter brings forth an Adaptive Cognitive AI Framework that operates for intelligent water management solutions throughout residential and commercial buildings as well as institutions.

DOI: 10.4018/979-8-2600-3038-7.ch006

ht © 2027, IGI Global Scientific Publishing. Copying or distributing in print or electronic forms without written permission of IGI Global Scientific Publishing is prohibited. Use of this chapter to generative artificial intelligence (AI) technologies is expressly prohibited. The publisher reserves all rights to license its use for generative AI training and machine learning model development.

The system combines immediate sensor information with automated machine learning systems and smart automation elements. This system unites several AI models so LSTM autoencoders analyze unusual water consumption while Prophet models forecast consumption patterns and XGBoost merging with Random Forest algorithms detect water quality fluctuations. With a reinforcement learning-based control layer the system generates decisions through the combination of present-time data and previously acquired wisdom from system of events. The system detects both contamination indicators along with aberrant consumption patterns which leads to automatic preventive measures. The proposed system presents an intelligent solution to address contemporary water issues in the buildings.

1. INTRODUCTION

Contemporary urban communities with their functional elements rely on water as a basic requirement in order to sustain human health. Increasing global freshwater demand along with growing cities has made constructing water management an imperative challenge for residential and commercial and institutional buildings. World Health Organization along with UN-Habitat reported that leakages in water supply networks provide losses of 30–40% of whole quantities of water distributed to consumers resulting in economic loss and resource waste. The phenomenon of water contamination due to pipeline corrosion in addition to environmental contaminants and backflow incidents goes undetected until extensive damage has ensued. The needs of many urban building residents in densely populated urban areas present serious challenges to fundamental water distribution networks to serve optimally.

Although intelligent sensors and automation hardware are more accessible, buildings continue to run with reactive instead of preventive water management systems at the facility level. Typical reactive management systems implement fixed rules and manual evaluation procedures as well as individual detection systems that do not provide an overall scenario perspective. These systems are able to recognize water flow anomalies and produce chemical level alarms but have no capability to interpret these signals along with other parameters or produce suitable instantaneous system responses. The traditional systems do not consider temporal aspects like weekends and holidays along with ignoring behavioral patterns which vary based on building features and location along with seasonal changes. Incorrect functioning of these systems results in both incorrect notifications and system inefficiency in addition to too much human intervention in their processes.

The advent of Cognitive Artificial Intelligence (Cognitive AI) comes across as a sophisticated method of transforming the way buildings deal with water management and control. Cognitive AI offers greater value in comparison to conventional rule-based systems due to the fact that it combines perception capabilities with learning capabilities and reasoning systems and adaptive decision processes. The system has capabilities to recognize unique water patterns in buildings and improves its decision-making via experience-based learning during uncertain conditions. Cognitive AI structures attain building water system performance via integration of real-time sensing technology with machine learning algorithms along with autonomous control logic that facilitate higher accuracy and sustainability.

A newly introduced Cognitive AI Framework in this chapter attempts to transform fundamental smart water management techniques. The platform consolidates different machine learning tools that work with real-time sensor data to perform leak detection and predict water consumption and analyze water quality status. A decision engine in reinforcement learning relies on context reasoning integrated with past outcomes to control system responses such as valve shutdowns and pressure adjustments along with triggering alerts. The feedback mechanism of the system allows users as well as conditions it to make its performance better over time while reducing errors and enhancing system longevity. The solution becomes more practical usage and scalability across various building types and geographic locations since it offers real-time user control via visual dashboards and analytics functionalities.

This article addresses modern-day infrastructure system vulnerabilities by proposing one smart solution that is user needs adaptable hence catering to future infrastructure development. This new system doubles as technological advancement and provides the core shift in smart buildings towards proactive water management systems with a human-centered approach. This article reviews applicable background literature prior to demonstrating current research limitations and proposing the system design with its anticipated impacts on water sustainability in built environments.

2. LITERATURE SURVEY

Various investigations in smart water management have produced improvements through their specific research on leak detection as well as water quality characterization and control system performance enhancement. Boudhaouia and Wira (2018) established fundamental ideas which showed that smart meter analysis combined with night-time usage patterns could detect leaks as they occur in tertiary building infrastructure. Despite their methodologies the system operated through rigid rules that prevented it from efficiently operating in evolving environments. Pesari and Porika (2024) developed a dual-task deep learning method based on LSTM models that conducted leakage recognition together with usage prediction tasks at the same time. Their model was the first system to combine important operational functions yet it did not embrace real-time decision capability or system context understanding.

Acoustic leak detection using TinyML and CNNs on microcontroller-based systems gained refinement when Atanane et al. (2023) applied their implementation. The innovation added intelligence functions at the device level yet its limitations included accepting only sound-based input and insufficient connection of sensor types. Cheng et al. (2025) examined the transfer learning capability between different infrastructure networks by applying pressure and flow datasets which demonstrated the potential to utilize models in new operational settings. Forecasts of future usage and automatic system controls were not integrated into their system design. Mohd Yussof and Ho (2022) conduct a comprehensive evaluation which emphasizes the developing requirement for system coordination and self-operational capabilities in building system leak detection methods. Most existing solutions maintained dependence on non-predictive fragmented analytics as they lacked integrated system design approaches according to their study.

Li et al. (2024) explained how digital twins work in water conservancy operations through simulation-based performance tracking in their optimization models paper. The strategic capabilities for planning and network performance adjustment provide valuable insights but have shown limited success when applied for building-level AI applications or multidimensional applications. The researchers at Balta et al. (2023) succeeded in improving anomaly detection efficiency by using supervised learning with water quality data which produced promising results. The model presented by the authors worked reactively as a prediction system yet failed to include leakage characteristics along with dynamic consumption patterns in its calculation. Through commercial development Phyn (n.d.) utilizes proprietary AI which learns about how different households use water resources at their homes. This system works well yet proves inappropriate for big or institutional facilities due to its restricted scalability and its closed-source design.

Liu et al. (2024) made a substantial progress in data enhancement through their work on LSTM-GAN models that produced synthetic leak information to enhance model effectiveness. The implementation of this method required substantial training expensives but consisted of zero integration with ongoing quality surveillance systems. The authors of Yan et al. (2024) developed a new multi-sensor fusion method by turning acoustic signals into spectrograms that processed through CNNs resulting in better detection accuracy. The effectiveness of this model came at a cost since it lacked predictive features and real-time monitoring feedback systems. The researchers at Patil et al. (2023) developed an IoT pH monitoring system for drinking water safety but its threshold alert system proved insufficient regarding recurrent anomaly detection.

The application of Q-learning by Sharma et al. (2024) for smart building systems provided energy and HVAC integration management yet it did not include specific water control measures. A comparison of XGBoost and Random Forest algorithms as water quality predictors appeared in Haghiabi et al. (2018) where researchers studied dataset-specific performance results. The predictive models from their research delivered satisfactory results when tested in structured conditions yet they did not integrate an automation feedback system for self-learning capabilities or real-time operational control. Jha et al. (2023) developed an IoT platform featuring solenoid valve systems for real-time water flow management that exemplified first-time real-time water actuation methods. Their system did not possess cognitive functions which included the ability to reason or learn from user input or adjust limits dynamically.

Shukla and Verma (2021) used statistical models to investigate seasonal patterns and monthly consumption trends of urban water use systems. Their system lacked ability to adjust to behavior modifications that happened during real time operations. Rao et al. (2023) analysed sustainable design methods that use AI for energy and water efficiency however their research emphasized minimal aspects of water leakage and contamination detection. The authors Thakur et al. (2026) used neural networks to enhance quality control by optimizing chemical dosing in water treatment processes. Their solution implementation excluded both leak detection functions along with predictive warning systems that could signal anomaly escalation.

Prasad and Kulkarni (2022) conducted a thematic survey that investigated AI utilization throughout utilities for water quality assessment and leak identification. Their analysis revealed that current implementations function poorly in separate units so they urged for entire system integration. Zameer et al. (2023) created regression models that processed IoT data to track chlorine levels in real time yet the system lacked complete decision-making capabilities particularity valve control. A comparison between SVM and k-NN and ANN models was conducted by Kapoor et al. (2024) to determine their effectiveness when used for pipeline water contamination detection. The benchmarking results presented by their study proved useful but their work only monitored detection systems without exploring feedback-driven adaptation along with cognitive learning mechanisms.

The existing research shows how different groups made significant strides in leak identification and quality assessments and predictive modeling yet demonstrates the essential requirement for integrated platform-based water infrastructure management at a building level in real-time. Most available solutions fit into three categories: specialized designs, reactive mechanisms and decision-isolated approaches. The existing gaps highlight why an integrated framework as presented in this chapter needs to emerge because it contextualizes anomalies and forecasts risks and autonomously responds while continuously learning from the environment.

Smart water management advanced significantly over the previous decade because of technological advancements between digital sensing and real-time monitoring systems with machine learning functionalities. Multiple research teams along with industrial enterprises have focused their work on creating better system awareness through anomaly detection and early fault diagnostics and water quality surveillance methods. The rise in IoT-enabled infrastructure deployment made possible the development of continuous and granular monitoring systems for water flow pressure and chemical properties. The majority of research along with commercial solutions maintain their single-issue approach by investigating leak detection or quality monitoring rather than developing unified intelligent control systems.

Research focused heavily on leak detection because this subject directly links to water losses combined with infrastructure degradation and monetary expenses. Research has examined three main methods for leak detection by studying pressure-flow analysis together with acoustic signal interpretation along with data-driven classification techniques. The identification of system abnormalities through sensor data relies on the execution of support vector machines along with k-nearest neighbours and decision trees within machine learning systems. The emergence of deep learning technology brought LSTM sequence learners together with CNNs which serve to model the temporal patterns found in water leakages. The advanced models detect minor behavioural shifts within flow operations that thresholding systems would usually overlook. LSTM autoencoders perform reconstruction of normal flow sequences to detect anomalies through errors in the final output. This development brings edge computing benefits through TinyML applications which allow small microcontroller-based artificial intelligence models to run leak detection analyses (Atanane, et al., 2023). This technology delivers much value for areas with minimal resources and distant locations. Most current implementations work as if environments stay static without real-time adaptability features and they perform inadequately across buildings that have different profiles. When applied elsewhere models which receive training through specific building types combined with flow patterns as well as sensor configurations demonstrate poor performance levels because they fail to achieve effective scalability together with transferability and robustness features (Cheng et al., 2025).

Most research about leak detection focuses only on corrective measures. Such detection systems demonstrate excellent performance at locating leaks after their occurrence yet lack predictive capabilities which estimate potential failures from stress buildup and behavioral changes or deterioration indications. Building managers only learn about significant leaks through the system yet this prevents them from conducting proactive maintenance or optimizing system dynamics. This inability to predict time-based patterns specifically creates challenges for understanding anomalous behavioral patterns because it makes it difficult to determine whether the 3 a.m. flow counts as an anomaly or a part of regular cleaning activities. Irrational alarm triggers become more frequent without proper context since it weakens automated system reliability while requiring human manual checks to regain trust.

Research has simultaneously pursued the monitoring along with the analysis of water quality. The dissemination of safe water remains vital in all structures particularly healthcare facilities alongside hospitality along with educational buildings because individuals can immediately experience hazards from contaminants. Multiple water quality indicators including pH values together with residual chlorine testing and turbidity measurement as well as temperature recording form the basic parameters under observation. The classification of water samples between normal and contaminated state depends on supervised learning methods Random Forest and XGBoost and logistic regression that examine sensor data (Balta et al., 2023). By implementing these models, it becomes possible to achieve better detection capabilities while also cutting down the amount of time required for alert notification following contamination events. The identification of natural risk areas based on chemical readings is achieved through clustering techniques that utilize the k-means or DBSCAN methodologies. Most quality-monitoring systems operate with rule-based and static threshold logic for their operations. Only a limited number of systems apply historical patterns and flow dynamics or pipe age data even though these factors powerfully affect chemical stability conditions and microbial reproduction.

Water quality solutions that currently exist process measurements as stand-alone data points while they should analyze these readings as sequential time-based events that are altered by upstream activities and temperature changes and infrastructure state. The system faces restricted forecasting capabilities because of the lack of connection between measured data and anticipated future events and early alert systems. The implementation of predictive modeling to forecast future chlorine level drops or turbidity increases remains underserved at the building level (Zameer et al., 2023). Most installations maintain their boundaries within water distribution networks and treatment facilities while showing minimal progress toward true-time monitoring of in-building smaller environments. The absence of predictive modeling represents an essential deficiency because quality degradation usually starts as pipes reach terminal points where bacterial growth and stagnation occur in unused pipelines.

Time-series forecasting brings widespread use to urban water management and agricultural operations yet presents insufficient application within the individual building domain. Previous implementations of predictive models in water usage forecasting contain simplistic methodology because they use either rolling averages or seasonal trends (Shukla & Verma, 2021) while failing to incorporate occupancy patterns together with weather data and specific event-based variables including holiday and maintenance factors. The simplification of forecasting models results in decreased operational value which hinders their usefulness for present-day decisions and unforeseen situations analysis. Smart building water systems remain sparse when it comes to integrating Prophet and deep LSTM networks despite their potential for solving the forecasting limitations.

Existing literature severely lacks closed-loop intelligence as a system capability which requires active learning and adaptation to future behavior after taking actions. All AI models used in smart water management conduct offline training before installing static configurations for deployment. Such systems fail to improve through continuous learning and their performance remains sub-optimal since they do not receive input from users or the environment. Building performance suffers because of ongoing deterioration which happens mostly when facilities undergo changes in utility requirements or plumbing systems. The current solutions divide presentational, predictive and decision-making functionalities into separate distinct layers. The leak detection device alerts staff yet the decision to shut off valves or contact maintenance remains externally controlled outside of systematic reasoning abilities. Poor system usability and delayed responses as well as unnecessary escalations occur due to fragmented service design.

Most available solutions show limited success in understanding context-related information. The actual usage patterns of water depend on multiple active circumstances which include variations between weekend and weekday usage and seasonality as well as regular cleaning protocols and scheduled events. The majority of AI models operate without incorporating both contextual tags or time-based reasoning elements. The detection systems generate erroneous measurements regarding typical fluctuations or miss piercing through planned operational patterns. The failure to identify scheduled laundry operations after observing high flow during midnight hotel hours creates both mistrust among users and persistent fatigue of managing alerts. It is common for unnecessary contamination alarms to occur when AI systems do not consider elevated chlorine levels during pipeline flushing operations. AI models need contextual understanding the way cognitive architectures can provide because current AI models repeatedly demonstrate failures.

The residential domain has seen commercial manufacturers attempt to solve some unsolved problems in the market. The Phyn home water monitor (Phyn, n.d.) stands out as a notable product because it employs AI technology to develop a water usage profile for homes based on which it adjusts its alert parameters. Such technology shows promise for personalized control through learning processes but the current systems provide restricted observation capabilities. These proprietary systems maintain closed platforms that do not welcome customization nor integration nor academic research on their operations. Commercial and institutional buildings limit the applications of these systems because they handle different amounts of scale combined with complex flow dynamics and different sensor needs. These systems do not support autonomous actions from AI decisions because they need either user verification or cloud-based notification systems to trigger responses.

Years of academic research in federated learning and transfer learning and edge intelligence demonstrate promising directions for the development of smarter water systems. The federated learning method allows buildings to train models together securely while preserving privacy thus improving operational consistency between various situations. The integration of multiple sensors lets systems confirm anomalies through the combination of matched data points—such as acoustic data matched with hydraulic pressure reduction and environmental humidity characteristics that indicate water leaks (Yan et al., 2024). The combination of memory with context and reasoning and adaptation behavior from cognitive computing principles provides the most thorough method forward. Experimental and simulated developments account for most developments but end-to-end cognitive implementations in actual operations are absent.

The collected literature finds that knowledge remains scattered and underbuilt throughout this field. Each research subdomain presents promising progress yet lacks an integrated framework that unifies real-time sensing structures with contextual awareness platforms and adaptive learning mechanisms and predictive forecasting models together with autonomous decision-making capabilities. Most AI technology implementations today exist as either entirely reactive solutions or static computational models which lack complete utilization of present-day AI functionalities. The need exists for advanced systems to go beyond monitoring because they must exclusively provide intelligent interpretation and adaptive responses and autonomous execution — systems which comprehend anomalies and this comprehension enables them to respond appropriately and develop improved capabilities.

The Cognitive AI Framework presented in this chapter incorporates solutions to overcome the current framework shortcomings. The proposed research combines the positive elements of past work by implementing an automated control system that merges IoT sensor inputs with artificial intelligence algorithms and reinforcement learning models in real-time operations. The system functions through self-modular units which offer scalability and suitability for different building infrastructure types. The framework improves its accuracy levels by infusing user-related feedback and contextual learning elements into its operational framework which results in fewer false alarm alerts and provides useful notification messages. This system evolves water management into an intelligent system based on predictive abilities to create sustainable proactive structures that need low maintenance. By accomplishing this objective, the system creates conditions for upcoming building automation systems which will utilize AI-powered management of water systems to both learn and evolve while ensuring extended durability of infrastructure assets.

3. RESEARCH GAPS

The review of existing literature and commercial solutions highlights several critical gaps that persist in the field of smart water management in buildings. Despite the rapid development of IoT technologies and machine learning techniques, current solutions tend to operate in silos—addressing either leak detection, water quality monitoring, or usage analytics independently. A truly cognitive and comprehensive solution that integrates these components remains lacking.

Absence of Unified, Multi-Objective Systems: The majority of current research investigates one application at a time between leak detection alongside quality monitoring. Solutions built this way ignore the interconnected relation between water stream control and consumer patterns alongside safety operations. There exists no integrated framework which manages leak detection and usage prediction alongside real-time quality monitoring systems with intelligence capabilities at the same time.

Lack of Cognitive, Self-Adaptive Learning Models: At present most models use static rule-based or threshold-based techniques that create both false alarm and detection miss issues. Very few studies exist regarding cognitive AI systems which learn from specific building water usage data alongside historical information and environmental changes.

Limited Integration Between AI Detection and Autonomous Response: The installed detection systems lack automated control functions such as valve shutdown and pressure regulation. A cognitive decision engine remains underdevelopment since it needs to identify risks while making time-dependent resilient actions based on historical events.

Inadequate Prediction of Future Anomalies or Failures: The current detection methods already exist but organizations fail to use predictive analytics models to forecast usage patterns and contamination events. Tools that predict water pipe failures as well as overuse occurrences and chemical inequalities would enhance both maintenance scheduling capabilities and water resource preservation measures.

Absence of Contextual Awareness in Decision-Making: Systems that manage buildings or estates today typically overlook vital elements such as daily time schedules and appliance status and weather variations which impact their operational capabilities. The systems fail to separate actual anomalies from normal routines because of this limitation which leads to unwanted false alarm alarms and unnecessary interventions.

No Closed Feedback Loop for System Improvement: Feedback from users or system outcomes is not commonly used to re-train or fine-tune the models. The absence of a continuous learning mechanism prevents systems from becoming smarter and more reliable over time.

Poor Interoperability and Scalability Across Building Types: Many existing systems are either custom-built for residential settings or large-scale utilities, with few scalable solutions for mid-sized commercial, institutional, or hybrid infrastructures. Cross-building learning and generalization remain a gap in most proposed systems.

Addressing these research gaps is critical to the realization of intelligent, resilient, and sustainable water management systems. The proposed framework in this chapter is designed to overcome these limitations through a modular, AI-driven, and feedback-enabled architecture that can adapt and scale across building environments.

4. PROPOSED METHODOLOGY

Figure 1. Data flow of how our proposed idea will work

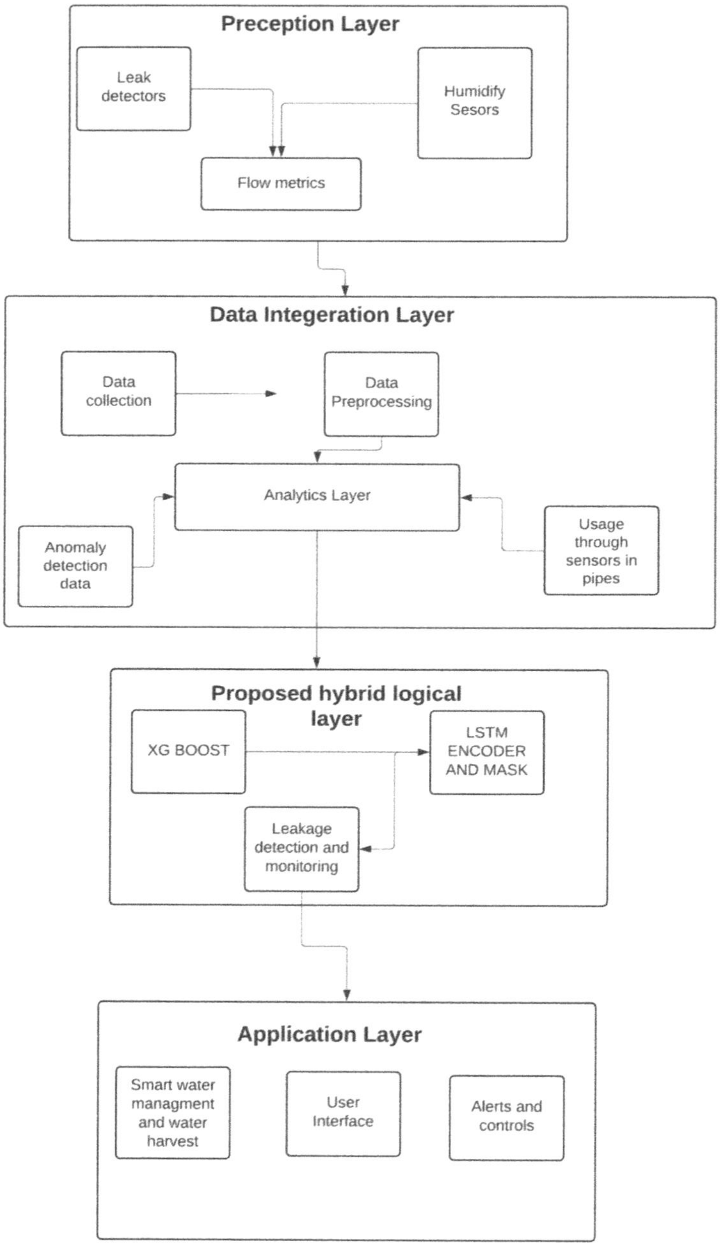

The proposed Cognitive AI Framework for Smart Water Management in Buildings is designed as a multilayered system that mimics intelligent human decision-making processes while managing the complexities of dynamic building environments. It incorporates sensing, preprocessing, intelligent analytics, autonomous decision-making, and user interface capabilities, all connected via a cognitive feedback loop that enhances the accuracy and responsiveness of the system over time. The core strength of this methodology lies in its ability to evolve, contextualize anomalies, and execute real-time actions with minimal human intervention. It is structured into six key architectural layers, each serving a distinct but interconnected function: 4.1 IoT-Based Sensing and Data Acquisition Layer, 4.2 Data Preprocessing and Feature Engineering Layer, 4.3 Cognitive AI Engine, 4.4 Cognitive Decision and Control Layer, 4.5 Dashboard and Alerting Interface Layer, and 4.6 Closed Feedback Learning Loop.

4.1 IoT-Based Sensing and Data Acquisition Layer

The foundational layer functions as the sensory bridge connecting physical water infrastructure with the AI framework. The core component relies on IoT sensors positioned at different points in building plumbing and water storage facilities and pipeline structures and major water consumption areas. The selected sensors cover a full range of necessary data which enables real-time tracking of water system behavior and health. The primary categories of sensors used include flow sensors, which measure the rate of water consumption and help identify sudden or unexpected usage patterns indicative of leaks; pressure sensors, particularly piezoresistive types, which detect fluctuations or drops in pressure that may signal pipe bursts or clogging; acoustic and vibration sensors, which capture ultrasonic or mechanical resonance patterns associated with micro-leaks or flow turbulence; and water quality sensors, including those that detect chlorine residual, pH balance, turbidity, and temperature. Guaranteeing water safety and detecting contamination as early as possible depends on these chemical and physical indicators. Environmental sensors track both temperature and humidity measures which confirm leak situations while distinguishing between genuine leaks and environmental conditions such as water vapor. Multiple sensors use lightweight transmission protocols LoRaWAN and MQTT for wireless data transfer and minimize power usage through optimized communication methods. The sensors either link up with edge-computing gateways that direct them to cloud-hosted servers or they connect to servers for deployment purposes. The multiple sensory inputs generate a detailed water environment picture that allows precise time-based tracking and comprehensive data collection for processing purposes.

4.2 Data Preprocessing and Feature Engineering Layer

The preprocessing pipeline follows raw sensor data collection to optimize machine learning applications through improved data quality and structure design. This pre-processing layer exists to minimize the noise and inconsistency together with missing data which commonly appear within IoT sensor information. During the first stage of preprocessing the application of Kalman filters allows dynamic state estimation while removing temporary noise while Savitzky-Golay filters maintain signal shape during smoothing operations. Then sensors send their measurements to a single timestamp reference point to achieve temporal coordination between different modalities including flow measurements and pressure readings and chemical sensor output. Time-series analysis depends on this step to establish proper event correlation. Normalization involves value scaling which produces uniform ranges to prevent model symptoms caused by dominant features. The process of feature engineering produces high-level metrics from original base data records. The calculation of hourly flow gradients together with minimum night flow thresholds and sudden pressure decay slopes and rolling average turbidity deviations enables the detection of concealed patterns and signals in the data. The feature construction process includes extracting consumption patterns from time-of-day data, weekly patterns and seasonal patterns of consumption. The anomaly pre-screening function includes outlier detection algorithms such as Isolation Forest that uses tree structures for anomaly isolation and DBSCAN which separates events from noise based on density clustering. The cognitive analytics modules receive a structured and clean feature matrix as input from this layer which has been developed through this process.

4.3 Cognitive AI Engine

The system analytic core relies on three parallel operating machine learning modules including Leak Detection Module and Water Usage Prediction Module together with Water Quality Prediction Module. The specific optimized modules within the system form the brain of the cognitive framework by handling distinct water intelligence elements. The Leak Detection Module teaches normal water consumption patterns through time using LSTM (Long Short-Term Memory) Autoencoders. The models receive daily and weekly usage patterns from individual buildings specifically and calculate anomalies through reconstruction error detection. The system marks any significant input variations versus reconstruction results as potential leaks. Users can confirm false positives or missed detections of the model through semi-supervised tuning which results in improved sensitivity following historical feedback. Small leak detection and complex event identification including intermittent and pressure backflow issues can be achieved using this method.

The Water Usage Prediction Module predicts both hourly and daily water demand through time-series forecasting with methods including Prophet and LSTM. The component uses historical flow data together with occupancy schedules along with temperature trends and calendar events to produce context-specific forecasts. The system serves as the leak detection baseline by showing differences between the forecasted usage pattern and actual consumption. This approach establishes that every alert will be assessed in reference to both the time-based patterns and established behavioral patterns of building water usage. Supervised ensemble learning algorithms XGBoost and Random Forest Regression power the Water Quality Prediction Module to generate predictions of vital water quality indicators. The prediction models incorporate chemical reading data together with temperature and pipeline specifications and material type alongside pipeline age and pressure disturbances to forecast turbidity as well as chlorine dissipation and pH imbalance trajectories. Early warning from this module enables proactive action to prevent contamination, pipe corrosion, or health risks. All three modules operate asynchronously and feed their inferences to the decision-making layer, enabling multi-dimensional insight and cross-verification of anomalies.

4.4 Cognitive Decision and Control Layer

Within this framework the independent action and policy development area operates as the most crucial component. Through integration of detection and prediction outputs the system develops suitable control strategies by evaluating current state along with contextual awareness and previous intervention performance. The control logic operates through two functional models that combine rule-based determinism for instant critical responses with an agent using reinforcement learning methods to handle complex scenarios requiring adaptive behavior. The RL agent trains within OpenAI Gym-like environments which provide rewards and penalties to actions including valve closures and pressure reductions and delay of alerts based on their outcome effects. Study of night-time leaks that remain stable helps the agent in learning how to reduce unnecessary shutdowns in analogous future situations. The decision matrix uses risk severity alongside system confidence level together with time of occurrence and sensor convergence and user behavior patterns. The system performs four action types including full main valve closure in severe leak conditions and short-term pressure decrease for managing smaller problems while warnings can proceed with delay followed by automatic diagnostic systems activation. Through dynamic policy evolution the system maintains the ability to evolve control strategies by processing real-time learning data. The system demonstrates robustness to false alarms while managing operational states at different levels through automated control procedures which do not need permanent human oversight.

4.5 Dashboard and Alerting Interface Layer

In order to close the loop between machine intelligence and human stewardship, this layer gives a real-time, interactive dashboard for monitoring, control, and reporting. The dashboard collects and displays data from all modules—presenting flow trends, pressure profiles, quality statistics, alert histories, and forecasted anomalies in a readily understandable manner. It enables facility managers not only to observe the present status of the system but also comprehend why AI-based decisions were made through visualizing model confidence, action histories, and feedback reports. The user interface is multi-user role-based, enabling both administrators and maintenance staff to access the system based on access permissions. Alerts are produced according to severity and confidence levels and are sent via several mediums like SMS, email, and application push notifications. The users can accept, dismiss, escalate, or suppress alerts, with useful feedback logged into the system. Some of the technologies utilized are light-weight React.js frontends for responsiveness, Node-RED for orchestrating logic flow, Grafana for visualizing, and Firebase or MQTT brokers for real-time data streaming. The dashboard is even optimized for mobile use, making remote monitoring accessible even for offsite or field situations.

4.6 Closed Feedback Learning Loop

The closed feedback learning loop constitutes the sixth essential piece which embodies the cognitive aspects of the system architecture. This layer brings adaptiveness to traditional automation systems by enabling continuous feedback processes to refine and recalibrate operational models. User actions like confirmation of alerts and manual control triggers together with incident reports serve as indications which the system uses for learning. Analysis of these signals enables detection model parameter adjustments and demands new algorithm training operations along with RL policy engine improvement. False alarms that occur multiple times during established maintenance sessions will teach the system to disregard alerts that match these situations in future occurrences. When users report missed leaks which the system was unable to detect the initial time the system incorporated those observations into its training data set. The feedback loop tracks environmental changes throughout time including occupancy patterns and seasonal effects and hardware breakdowns which prevents the system from losing relevance or becoming breakable. The loop generates virtual memory for the system by storing operational data which improves the system's ability to maintain reliability and contextual intelligence.

Through the proposed methodology buildings gain access to a complete and highly technical solution for intelligent water management operations. A system that incorporates AI models, real-time sensing, edge/cloud hybrid infrastructure and continuous learning operates as a responsive and scalable intelligent solution. The layered structure of this tool enables modular design features together with cognitive functions that offer adaptability and predictive abilities which make it an assets for building infrastructure development.

Figure 2. Cognitive water leakage detection Ai layers

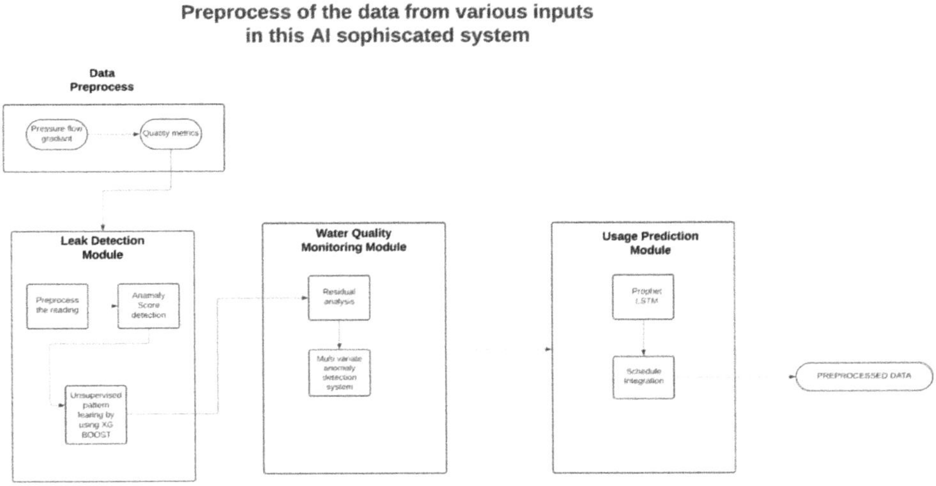

4.7 Technical Summary and System Signature

1. The system receives real-time streams from flow meters along with pressure sensors and acoustic/vibration detectors and water quality sensors and environmental sensors that measure pH, turbidity, chlorine, temperature, humidity and ambient temperature.
2. The system implements Kalman Filter and Savitzky-Golay Filter along with Isolation Forest and DBSCAN algorithms and Normalization methods as well as Feature Engineering techniques that use flow gradients and pressure slopes and night flow plateaus.
3. The application employs an LSTM Autoencoder for leak detection whereas Prophet and LSTM methods predict usage and XGBoost and Random Forest models analyze water quality and a RL Agent from OpenAI Gym optimizes control measures.
4. The system performs actions which include valve shutdown and pressure reduction and anomaly revalidation and maintenance alert triggering which are delayed based on the confidence thresholds.

5. The system produces five main output groups which include leak detection alerts together with usage prediction visualizations and quality risk score assessments and dashboard displays and automated valve actuation logs and user feedback systems.
6. The system uses TensorFlow together with Scikit-learn and Prophet and OpenAI Gym and Node-RED and Grafana and Firebase and React.js and MQTT and LoRaWAN for its framework implementation.
7. The system operates through modular real-time feedback services that utilize a hybrid cloud-edge deployment framework.

Figure 3. Proposed architecture of the calculating an accurate reading with the help of LSTM

5. COMPARATIVE STUDY & TOOLS JUSTIFICATION

The section demonstrates the differences between current methods and the scheduled Cognitive AI Framework through a functional and attribute-based comparison. This section provides rationalization for using particular tools and algorithms within the system because they effectively tackle previously identified research gaps. The decision-making process for tool and framework selection considered multiple factors.

1. The time-series forecasting features of Prophet optimize sufficient performance when applied to periodic consumption data. XGBoost along with Random Forest delivers excellent results for tabular data frameworks which include quality readings while maintaining both interpretability and accuracy rates. The reinforcement learning environment simulation and creation capabilities exist within OpenAI **Gym.**
2. The distributed sensor nodes can exchange information through the communication protocols MQTT and LoRaWAN using an efficient energy-saving method while maintaining low system latency performance.

6. ADVANTAGES OVER EXISTING SYSTEMS

The suggested Cognitive AI Framework for Smart Water Management brings a revolutionary change in the way buildings monitor, forecast, and react to water consumption, leakage, and quality variations. In contrast to conventional systems that tend to work in isolation and are based on fixed thresholds or human intervention, this framework is intelligent, dynamic, and autonomous. Its major advantages lie in the combination of sophisticated machine learning, multi-sensor fusion, predictive analytics, and real-time cognitive reasoning. This segment critically contrasts the suggested method with traditional water management configurations, outlining a conceptual discussion of the differentiating characteristics that make it an improved, forward-looking solution.

Standard leak detection methods base their operation through applying static thresholds to sensor data measurements which include flow rate limits and pressure range settings. The basic threshold methods generate many false alarms since they cannot distinguish between normal high-usage scenarios and actual leakage incidents. Using LSTM autoencoders the proposed framework trains specific time-series patterns for every building. The system learns regular patterns from environmental usage behaviour so it recognises anomalies which hint at leaks while ignoring typical variations in operations. The model improves itself automatically through time by learning adjusted detection thresholds from feedback and training sessions that make it far more reliable in its operations.

The current system lacks expected usage prediction capabilities because it does not have built-in intelligence for forecasting. The methods used mainly depend on basic prediction rules and average estimations that produce limited value for proactive decision processes. A new solution solves this gap through integration of time-series forecasting mechanisms which use Prophet and LSTM models. The predictive tools forecast upcoming water use at different time scales from daily to hourly periods through analysis of external conditions with temperature data along with occupancy schedules and historical patterns and holiday factors. The implemented system detects both high-demand periods and unusual usage trends because of its forecasting capabilities from Prophet and LSTM models which lead to enhanced facility resource management and anomaly detection abilities.

Water quality monitoring systems in traditional setups use manual sampling combined with sensor-generated fixed-alert systems. The detection methods react slowly while remaining limited in number and producing delayed assessments of health risks. The proposed framework advances this process through AI models that evaluate multivariate inputs between pH, chlorine concentration and turbidity alongside temperature values using XGBoost and Random Forests algorithms. The training process involves historical data. Through this model the prediction of water quality changes reaches high levels of accuracy. A predicted decrease in residual chlorine will trigger system recommendations for maintaining pipelines through flushing or chemical adjustments which protect users from bacterial growth. The unexplored field of using predictive water quality analytics for building systems obtains its initial breakthrough through the framework presented.

The framework stands out with its ability to make cognitive choices that form the most significant benefit. Three faults exist in traditional alert systems which prompt the need for manual operator intervention that includes valve closure or maintenance service calls. A manual process enters human mistakes and extended durations when applied to functions that demand automation. The proposed system utilizes OpenAI Gym environments to learn through reinforcement learning which responses generate the best results after detecting operational problems. The system enhances its performance through experience because it identifies the correct actions which stop damage from occurring while avoiding service interruptions. The automated decision process cuts down operator dependencies and delivers time-efficient solutions during crisis situations.

The proposed system delivers its strength through integration of numerous sensor capabilities. Typical existing solutions examine separate parameters like pressure and flow rates independently before drawing conclusions without uniting the measurements. The fragmented data interpretation results in wrong meanings and inaccurate conclusions. The system connects flow meters and pressure sensors to acoustic/vibration monitors together with chemical analysers and environmental data features like temperature and humidity into one data pipeline. When different dimensions are analysed together as a system current and future state becomes clear while noise is eliminated along with anomaly validation for comprehensive water system health presentation. For example, a high flow event accompanied by acoustic anomalies and pressure drops could confidently indicate a leak, while one without pressure or sound changes might be a legitimate usage spike.

Contextual awareness leads to what makes the framework distinct. Water systems operate without considering factors related to specific time periods or situations and grant equivalent treatment to all usage readings. This model operates using time-of-day information together with building occupancy and weekend or holiday status and repair schedule data to create its framework. Through this functionality the system adapts expectations and develops response strategies on the fly. Normal water consumption happens at 7 a.m. weekdays in school buildings but unusual usage becomes noteworthy when it occurs during a Sunday evening. Situational intelligence delivers to model pipelines activates more accurate alerts and makes systems reflect actual operational behavior.

The proposed system introduces new approaches to how users interact with the system. Raw data presentation coupled with basic alert production is the common design of traditional dashboards although these interfaces lack interactive components or explanatory elements. The proposed framework implements a user-friendly visual interface which combines Grafana and React.js frameworks for development. Systems gather real-time water usage patterns combined with future anomaly predictions and expectations about water quality and show previous operational choices to users. Users receive detailed explanations with override choices when alert pop-ups appear in this system. The user can obtain alert reasons and previous resolution actions when triggers occur while viewing historical treatment of similar incidents. The system achieves better trust throughout its workforce through interactive features that lead to active management practices.

The proposed framework offers two primary benefits such as scalability and deployment flexibility. Traditional systems depend on established hardware arrangements and proprietary software applications which become burdens in environmental expansion initiatives and integration processes. By using containerized microservices (Docker cloud-native platforms (AWS IoT Core), and standard REST APIs The system requires low structural modifications to become operational in residential buildings as well as commercial establishments and healthcare facilities and school campuses. The edge processing technology makes core features available to properties that have limited internet access including leak detection and real-time alerts.

The proposed system achieves excellence in both efficiency while being sustainable due to its ability to minimize water wastage and energy usage together with automatic labour processes. The prompt identification of leaks stops damage to constructions and leads to reduced water expenses. The system runs maintenance based on predictions thanks to which both equipment downtime and excessive service calls become unnecessary. Low operation expenses result from energy-efficient processing through the use of Raspberry Pi hardware and MQTT protocols. Approximately through its historical usage learning functionality the system optimizes resources for peak periods and drought management which helps achieve environmental targets.

The self-improving qualities of the system work to enhance its operational resilience. The framework operates beyond static systems by developing through its feedback loop while static systems experience performance decline over time. The system enables users to identify accurate or false alerts which activates detection and prediction model retraining processes. The system records both sensor measurements along with decision implementation results that it uses to enhance its following actions. A system invests numerous weeks and months into adapting to building-specific dynamics which produces improved precision and lessens false alarms while increasing overall operational performance. The system requires low structural modifications to become operational in residential buildings as well as commercial establishments and healthcare facilities and school campuses. The edge processing technology makes core features available to properties that have limited internet access including leak detection and real-time alerts.

The proposed system achieves excellence in both efficiency while being sustainable due to its ability to minimize water wastage and energy usage together with automatic labour processes. The prompt identification of leaks stops damage to constructions and leads to reduced water expenses. The system runs maintenance based on predictions thanks to which both equipment downtime and excessive service calls become unnecessary. Low operation expenses result from energy-efficient processing through the use of Raspberry Pi hardware and MQTT protocols. The system learns from historical usage behaviour through its integral data collection abilities and applies this knowledge to improve resource allocation by restricting water supply during low-demand periods as well as by switching to backup water sources during drought seasons thereby attaining larger environmental goals. The self-improving qualities of the system work to enhance its operational resilience. The framework operates beyond static systems by developing through its feedback loop while static systems experience performance decline over time. The system enables users to identify accurate or false alerts which activates detection and prediction model retraining processes. The system records both sensor measurements along with decision implementation results that it uses to enhance its following actions. A system invests numerous weeks and months into adapting to building-specific dynamics which produces improved precision and lessens false alarms while increasing overall operational performance.

7. FUTURE ENHANCEMENTS

1. The Cognitive AI Framework for Smart Water Management establishes fundamental components to develop predictive and adaptive autonomous infrastructure in building water systems. The system has several potential paths for improvement which extend its operational capability and reach. The following future enhancements represent the proposed developments:
2. The framework achieves enhanced performance when connected to a digital water network replica of the building. A dynamic digital copy of the system through simulation technologies enables predictive maintenance optimization by running different scenarios which include leak propagation and valve control strategies without interrupting actual operational activities.
3. A building-wide deployment of this framework enabled with the federated learning methodology would let sites preserve their privacy through shared knowledge acquisition. The system would achieve superior results for uncommon events such as pipe bursts and chemical contamination through its improved generalization abilities.
4. Future versions of the framework should integrate with energy management systems to jointly optimize water and energy consumption because water systems share an intricate relationship with energy usage particularly in heating or pumping applications thereby advancing sustainability objectives.
5. The system can be expanded to generate the best maintenance decisions by incorporating artificial intelligence into the scheduling process based on usage predictions and component failure probabilities and anomaly score evolution.

The system decreases the need for human staff to perform maintenance planning which prevents resource waste.

6. The development of a mobile-responsive dashboard with natural language query capabilities represents a solution for enhancing accessibility of insights and alerts to building occupants and facilities staff through mobile platforms.
7. The control module should integrate Bayesian decision theory with probabilistic models to enhance its ability to measure uncertainty factors and improve its decisions when only incomplete or unclear information exists.
8. Future releases should add explainable AI (XAI) modules to provide explanations about triggered alerts and decisions due to their mission of building trust and supporting auditability between machines and users.
9. The platform will grow to forecast municipal water demand for urban resilience by combining data from multiple smart buildings thus assisting utility companies in distributing resources sustainably.
10. Future upgrades of the framework serve two purposes by expanding its technical capabilities and establishing it as the essential base for intelligent sustainable building operations across technology advancements and expanding urban demands.

These enhancements not only broaden the technical scope of the framework but also position it as a foundational platform for sustainable, intelligent building management that can evolve with emerging technologies and growing urban needs.

8. CONCLUSION

The growing complexity of urban infrastructure and the escalating demand for sustainable resource utilization call for a transformative approach to water management within buildings. Current static threshold-based and post-failure detection systems prove inadequate for handling the multiple issues arising from leakage and contamination along with inefficient water consumption. This chapter introduces the Cognitive AI Framework which establishes a new approach toward proactive adaptive water governance that uses intelligent systems. The system merges IoT sensing with machine learning and time-series forecasting and reinforcement learning elements to form a unified framework which monitors and predicts and autonomously manages water-related situations in live environments

The analytical backbone becomes robust through a system which unites LSTM autoencoders for anomaly detection with Prophet and LSTM for usage forecasting in conjunction with XGBoost and Random Forest ensemble models for water quality prediction. The system uses two decision-making processes: rule-based logic functions together with reinforcement learning to measure context variables and select the best control choices. A closed feedback mechanism allows the system to constantly learn from user inputs as well as operational results to update threshold parameters and control protocols in line with evolving conditions.

Research connected to this topic demonstrates the major gaps in industry and academic approaches which lack integration together with contextual sensitivity and continuous learning capabilities. The proposed system framework fills gaps by offering a flexible design structure which scales across different building categories such as dense-office buildings and smart residences and institutional sites. Moreover, the system's user-friendly dashboard facilitates transparency and informed decision-making, empowering building managers and maintenance teams with real-time insights, predictive alerts, and actionable recommendations.

Finally, this research adds a visionary and pragmatic solution to one of the most urgent infrastructure problems of the 21st century—effective and smart water management. By filling gaps between data acquisition, predictive intelligence, and autonomous control, the Cognitive AI Framework not only improves operational performance but also encourages long-term sustainability, resilience, and safety in the built environment. As urbanization picks up pace and environmental issues grow more critical, such cognitive systems will be vital tools in building smarter, greener, and more responsive infrastructure ecosystems.

REFERENCES

Atanane, O., Mourhir, A., Benamar, N., & Zennaro, M. (2023). Smart buildings: Water leakage detection using tinyml. *Sensors (Basel), 23*(22), 9210. DOI: 10.3390/s23229210 PMID: 38005596

Balta, D., Kaç, S. B., Balta, M., & Eken, S. (2023). Water quality estimation and anomaly detection: A review. *EAI Endorsed Transactions on Internet of Things, 9*(4), e2. DOI: 10.4108/eetiot.v9i4.3660

Boudhaouia, A., & Wira, P. (2018). *Water consumption analysis for real-time leakage detection in a smart tertiary building*. In ICASS 2018. DOI: 10.1109/ICASS.2018.8651976

Cheng, J., Jiang, Z., Wu, H., & Zhang, X. (2025). Water Pipeline Leak Detection Method Based on Transfer Learning. *Water (Basel), 17*(3), 368. DOI: 10.3390/w17030368

Haghiabi, A. H., Nasrolahi, A. H., & Parsaie, A. (2018). Water quality prediction using machine learning methods. *Water Quality Research Journal of Canada, 53*(1), 3–13. DOI: 10.2166/wqrj.2018.025

Jha, M. K., Yogeshwari, A., Rubini, P., & Singh, M. (2023, August). Converge of iot and ai in metaverse: challenges and opportunities. In *2023 5th International Conference on Inventive Research in Computing Applications (ICIRCA)* (pp. 1462-1467). IEEE. DOI: 10.1109/ICIRCA57980.2023.10220628

Kapoor, S., Cantrell, E. M., Peng, K., Pham, T. H., Bail, C. A., Gundersen, O. E., Hofman, J. M., Hullman, J., Lones, M. A., Malik, M. M., Nanayakkara, P., Poldrack, R. A., Raji, I. D., Roberts, M., Salganik, M. J., Serra-Garcia, M., Stewart, B. M., Vandewiele, G., & Narayanan, A. (2024). REFORMS: Consensus-based Recommendations for Machine-learning-based Science. *Science Advances, 10*(18), eadk3452. DOI: 10.1126/sciadv.adk3452 PMID: 38691601

Li, W., Ma, Z., Li, J., Li, Q., Li, Y., & Yang, J. (2024). Digital twin smart water conservancy: Status, challenges, and prospects. *Water (Basel), 16*(14), 2038. DOI: 10.3390/w16142038

Liu, R., Zayed, T., & Xiao, R. (2024). Advanced acoustic leak detection in water distribution networks using integrated generative model. *Water Research, 254*, 121434. DOI: 10.1016/j.watres.2024.121434 PMID: 38484549

Mohd Yussof, N. A., & Ho, H. W. (2022). Review of water leak detection methods in smart building applications. *Buildings (Basel, Switzerland)*, *12*(10), 1535. DOI: 10.3390/buildings12101535

Patil, V. S., Shinde, S. A., Sarwade, M. P., & Dhawale, N. M. (2023). Smart robotic soil pH measurement system. In *2023 National conference on smart agriculture technologies (NCSAT). Indian Soc Tech Educ J* (Vol. 46, pp. 22-30).

Pesari, V. S., & Porika, S. (2024). A deep learning framework for leakage diagnosis and time series water consumption prediction. *AIP Conference Proceedings*, 030002. DOI: 10.1063/5.0195057

Phyn. (n.d.). *Phyn home water monitor*. https://www.phyn.com

Prasad, N., & Kulkarni, M. (2022). AI in water utilities: A review. *Journal of Water Resources Planning and Management*, *148*(6), 04022051.

Rao, A., Pang, M., Kim, J., Kamineni, M., Lie, W., Prasad, A. K., Landman, A., Dreyer, K., & Succi, M. D. (2023). Assessing the utility of ChatGPT throughout the entire clinical workflow: Development and usability study. *Journal of Medical Internet Research*, *25*, e48659. DOI: 10.2196/48659 PMID: 37606976

Sharma, V., Kumawat, P., Kumar, P., Yaminipriya, S., Chamarthi, S., & Kumar, J. (2024, December). Reinforcement Learning for Autonomous Vehicles. In *2024 1st International Conference on Advances in Computing, Communication and Networking (ICAC2N)* (pp. 1548-1553). IEEE. DOI: 10.1109/ICAC2N63387.2024.10895165

Shukla, D., & Verma, A. (2021). Modeling temporal patterns in urban water use. *Urban Infrastructure Journal*, *12*(4), 81–90.

Thakur, R., Sajid, Z., & Khan, F. (2026). Artificial intelligence (ai) safety system for safe & trustworthy autonomy. *Digital Chemical Engineering*, *19*, 100308. DOI: 10.1016/j.dche.2026.100308

Yan, M., Xu, P., Li, J., Zhang, X., Wei, L., Peng, S., Li, X., Zhou, W., Wu, X., Huang, X., & Guan, Y. (2024). A buried gas pipeline leakage model. *Journal of Pipeline Systems Engineering and Practice*, *15*(4), 04024046. DOI: 10.1061/JPSEA2.PSENG-1638

Zameer, A.. (2023). Chlorine residual estimation using IoT and regression techniques. *Water Research Advances*, *12*, 1093–1102.

Chapter 7

Data-Driven Planning for Resilient Coastal Municipalities: The Case of Machico, Madeira

Raul Costa Alves
http://orcid.org/0000-0002-3680-6626
Universidad de Extremadura, Portugal

Sérgio António Neves Lousada
http://orcid.org/0000-0002-8429-2164
University of Madeira, Portugal

ABSTRACT

This chapter presents a data-driven methodology for civil engineering to address the challenges of Short-Term Rentals (STRs) on housing and public infrastructure. Using Machico, Madeira, as a case study, the authors demonstrate how an integrated analysis of public data (Census, RNAL, mobility studies) can quantify tourist pressure. The method calculates an STR pressure ratio, which is 3.4% for Machico, a moderate value that nonetheless exceeds the 2.5% preventive threshold used in other Portuguese cities. Spatial analysis confirms the direct correlation between tourism clusters and public parking saturation. The authors conclude this replicable, low-cost model is an effective tool for planners to implement preventive and evidence-based regulations, fostering a necessary balance between tourism's economic benefits and long-term urban resilience.

DOI: 10.4018/979-8-2600-3038-7.ch007

ht © 2027, IGI Global Scientific Publishing. Copying or distributing in print or electronic forms without written permission of IGI Global Scientific Publishing is prohibited. Use of this chapter to generative artificial intelligence (AI) technologies is expressly prohibited. The publisher reserves all rights to license its use for generative AI training and machine learning model development.

1. INTRODUCTION

The growing dynamics of urban tourism and the exponential increase of digital accommodation platforms, such as Airbnb, have, in the last decade, caused a profound transformation in the housing and economic landscape of cities. This phenomenon, often referred to as "touristification", translates into the conversion of housing into tourist accommodation, resulting in rampant real estate speculation and the displacement of original residents, impacting the social identity of territories (Mendes, 2018).

This dynamic has generated new income opportunities for property owners but, on the other hand, has raised concerns that go beyond housing scarcity. Tourist concentration challenges the Tourism Carrying Capacity of destinations, a central planning concept that defines the maximum level of development an area can sustain without causing unacceptable deterioration of the physical, economic, and socio-cultural environment (Costa, 2018). The saturation of public parking, the overloading of urban services, and neighbourhood conflicts are symptoms that this capacity may be exceeded, affecting residents' quality of life (Marques & Cariço, 2023).

In this context, legislative intervention became imperative. In Portugal, Law No. 62/2018 was a fundamental milestone, amending the legal regime for Short-Term Rentals (STR) (*Diário da República*, 2014). The law granted municipalities greater autonomy in managing this phenomenon, notably through the possibility of delineating "containment zones," which function as a practical instrument for managing the carrying capacity of territories.

This study aims to apply these concepts to the reality of a medium-sized island territory. It critically and systematically analyses the main impacts of STR activity on the urban planning of Machico, assessing its pressure on the housing stock and infrastructure. The goal is to develop a replicable analysis methodology that allows for the evaluation of legislative effects and supports planning policies that promote a balance between tourism development and the safeguarding of the right to housing and quality of life.

2. THEORETICAL FRAMEWORK AND LITERATURE REVIEW

Short-Term Rentals (STR), as a global phenomenon, have been the subject of intense academic debate. This research positions itself within this discussion by mobilizing key concepts from planning, tourism, and, crucially, civil engineering, focusing on the sustainability and resilience of Machico's urban system. The framework for this study is built upon a significant body of prior research conducted by the author on the specific urban challenges of the municipality, which provides an unparalleled depth of context.

This study positions itself within a critical and emerging field for contemporary civil engineering: the application of quantitative, data-driven methodologies to address complex urban challenges that transcend traditional infrastructure design. As civil engineers, we have historically focused on the physical resilience of cities—designing structures that withstand floods, building bridges that endure maritime environments, and planning road networks that support mobility. However, the modern urban landscape is increasingly defined by invisible pressures, such as the digital transformation of the housing market, which directly impact the functional and social sustainability of our territories.

This chapter introduces an innovative and replicable methodology that integrates traditional engineering principles with new data sources and analytical tools. We argue that managing phenomena like Short-Term Rentals (STRs) is not merely a social or economic issue; it is a fundamental challenge for civil engineering and urban planning. By mobilizing publicly available data from sources like the National Statistics Institute and official tourism registries, we have created key performance indicators, such as the STR Pressure Ratio, that allow for the objective measurement and spatial visualization of urban pressure. This approach transforms abstract concepts like "touristification" into quantifiable metrics that can inform public policy and support the design of more resilient and equitable urban systems.

This methodology goes beyond a simple diagnosis. It provides a robust, evidence-based framework for municipalities to exercise their new regulatory powers—for example, by defining "containment zones" where pressure is high. It also establishes a crucial link between different sectors of urban life, demonstrating how the concentration of STRs exacerbates pre-existing infrastructural vulnerabilities, particularly in urban mobility and parking. Thus, our work serves as a practical guide for how civil engineering can move from a reactive role of building and repairing to a proactive one of integrated, strategic planning for the future.

2.1. The Dynamics of Touristification and the Challenge of Housing Sustainability

The concept of "touristification" describes the process by which a neighbourhood reorganizes to serve tourism, often to the detriment of its residential function (Mendes, 2018). This process triggers real estate pressure, where the higher profitability of short-term renting leads to the conversion of the housing stock and a rise in prices (Balthazar, 2022). The response to this challenge involves the implementation of a Local Housing Strategy (LHS), a planning instrument aimed at diagnosing shortages and defining an integrated policy to guarantee the right to housing (Alves et al., 2022). In parallel, to combat the degradation of the existing building stock, programs like the Strategic Urban Rehabilitation Program (PERU) and the delimitation of Urban Rehabilitation Areas (ARUs) emerge as essential tools for sustainability, encouraging the recovery of heritage and the improvement of living conditions (Alves et al., 2023a).

2.2. Carrying Capacity, Mobility, And Infrastructure Sustainability

To manage the impacts of touristification, planning resorts to the concept of Tourism Carrying Capacity, which defines the maximum level of visitors a destination can support without unacceptable deterioration (Cordeiro et al., 2016). This capacity is often limited by infrastructure, with mobility and parking being major bottlenecks. The analysis of mobility patterns reveals a territory's vulnerabilities, especially in areas with complex urban morphology like Machico (Alves et al., 2025). Parking saturation, exacerbated by tourist accommodation, is a clear symptom that the infrastructure's carrying capacity is being reached, with direct impacts on the quality of life of residents (Sardinha, 2018).

2.3. The Role of Civil Engineering in Building Resilient Territories

One of the greatest challenges for contemporary civil engineering is the creation of resilient territories - urban systems capable of withstanding and recovering from external shocks, particularly those related to climate change. For a coastal and port city like Machico, resilience is intrinsically linked to the robustness of its infrastructure. The author's extensive body of research on the municipality has systematically diagnosed its vulnerabilities, providing a critical foundation for any discussion on future development, including the pressure from Short-Term Rentals (STRs). This prior work underscores the need for an integrated planning approach, where new urban dynamics are assessed against the backdrop of existing environmental risks.

2.3.1. Understanding and Mitigating Fluvial Risk

Machico's location at the mouth of a major river (*ribeira*) makes it inherently vulnerable to floods, a risk that has been the focus of several key studies.

The foundational research by Alves et al. (2023d), "Flood Risk and Vulnerability: An Approach to Urban Floods in Ribeira de Machico," established a comprehensive baseline for this threat. This work was pivotal as it quantified the flood risk and identified the most vulnerable areas, providing the initial evidence that unchecked urbanization in the river's floodplain could have severe consequences.

Building on this, the study "Modelling of Flood Risk, Vulnerability and Adaptation of Machico City Center" (Alves et al., 2023b) sharpened the focus onto the city's most critical area. By modelling different flood scenarios, this research demonstrated the high vulnerability of the urban core—precisely the area with the highest concentration of housing, commerce, and, as this study shows, tourist accommodations. This work is crucial for the book chapter as it provides the direct scientific link between a known hazard and the specific urban fabric under pressure from touristification. The high vulnerability of the urban core is visually demonstrated in Figure 1, which presents the flood hazard map for the city center. Figure 1 shows five levels of risk associated with the occurrence of flooding, with red representing Very High Risk and dark green representing Very Low Risk

Figure 1. Risk of flooding in downtown Machico

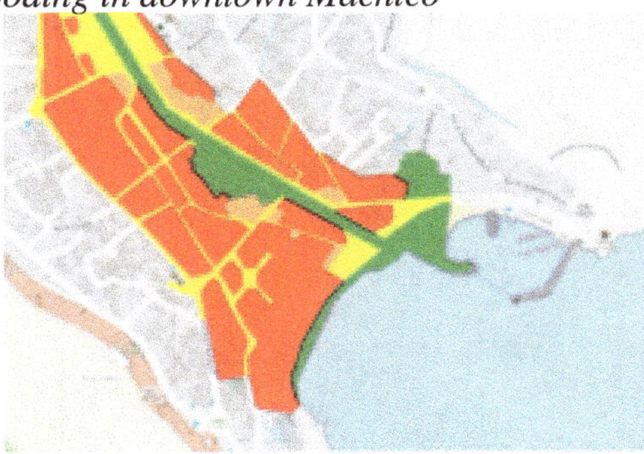

(Alves et al., 2023b)

Finally, the paper "Urban Flooding Risk in Machico and Planning Its Territory as a Form of Prevention" (Alves et al., 2023e) translates these diagnostic findings into a planning imperative. It argues that territorial planning itself must be used as a primary tool for risk prevention. This is of fundamental importance to our current analysis because it establishes that any decision regarding land use—including the authorization of new STRs—is not merely an economic or social issue, but a matter of public safety and long-term resilience.

2.3.2. Ensuring Resilience In the Maritime Interface

As a coastal city, Machico's resilience is also dependent on its maritime infrastructure. The Port of Machico is not just an economic asset but a critical line of defence for the city.

The study "Port of Machico, Madeira: Assessment of the Consequences of Overtopping Using AHP Methodology" (Alves et al., 2023c) quantified the risks associated with coastal overtopping, a phenomenon exacerbated by rising sea levels and increased storm intensity. The importance of this article lies in its assessment of consequences, showing how a failure in this maritime infrastructure could have cascading negative effects on the urban area behind it. Figure 2 illustrates the Port of Machico, a critical piece of infrastructure whose structural integrity is fundamental for the city.

Figure 2. Extension of the pier in Machico

(Alves et al., 2023c)

Furthermore, the research on "Maintenance Challenges in Maritime Environments and the Impact on Urban Mobility: Machico Stayed Bridge" (Alves et al., 2024) highlights the ongoing and complex engineering challenges required to maintain these critical structures in an aggressive maritime environment. This work is vital as it connects the resilience of coastal defences directly to urban functionality, demonstrating that the failure to maintain one piece of infrastructure (like the bridge) can severely impact the mobility and accessibility of the entire city.

In conclusion, a resilient civil engineering approach for Machico demands a planning framework that is profoundly informed by these documented risks. The decision to permit an increase in population density in vulnerable areas—whether through new construction or the intensification of use via STRs—cannot be made without considering its implications on flood risk and the integrity of coastal defences. Therefore, balancing competing land uses (housing vs. tourism) must be intrinsically linked with the analysis and mitigation of these multiple, interacting risks.

2.4. Governance and the Regulatory Response in Portugal

The response to the complex challenges posed by Short-Term Rentals (STRs) has necessitated the creation of new governance models. In Portugal, the initial legal framework was characterized by a liberal approach, promoting the growth of the sector with minimal regulatory oversight. However, the growing evidence of negative externalities, particularly in major urban centers like Lisbon and Porto, led to a significant policy shift.

The amendment to the STR regime in 2018, through Law No. 62/2018, was a crucial milestone (Diário da República, 2014). This legislation represented a move away from a "one-size-fits-all" national policy towards a model of co-responsibility and territorial differentiation, decentralizing significant regulatory powers to the municipalities. This change acknowledged that the impacts of STRs are not uniform and that local authorities are best positioned to manage the specific realities of their territories.

The law introduced several key instruments for municipal intervention. The most significant was the power to establish "containment areas" (*áreas de contenção*), defined as zones where a clear and demonstrated pressure on the housing market or urban infrastructure justifies the imposition of limits on new STR registrations (Diário da República, 2014). The creation of these areas requires a well-founded technical and social diagnosis, a role that studies like the present one aim to fulfil. Furthermore, the legislation also reinforced the rights of condominium assemblies, granting them the power to oppose the installation of STRs within their buildings, thereby addressing the widespread issue of neighbourhood conflicts.

This new legal framework has empowered municipalities to act. The case of Lisbon is paradigmatic, where the City Council commissioned an in-depth technical study to identify pressure zones and subsequently implemented strict limits on new registrations, demonstrating an initiative-taking use of its new regulatory powers (Câmara Municipal de Lisboa, 2018). It is within this broader context of evolving governance that the present study is situated. By applying a data-driven methodology, it aims to provide the municipality of Machico with the necessary evidence to exercise its own regulatory responsibilities, contributing to the creation of a more sustainable and resilient territory.

3. METHODOLOGY

This investigation adopted a mixed-method approach, combining quantitative and descriptive analysis with a spatial component, aiming for an integrated understanding of the housing and tourism dynamics in the municipality of Machico. The outlined methodology was designed to ensure the study's replicability, relying exclusively on official and publicly accessible data sources to guarantee the reliability and transparency of the results obtained.

The research process followed a structured, multi-stage approach, as summarized in the flowchart below (Figure 3). This workflow ensured that the analysis was systematically built upon a robust foundation of official and public data, leading to an integrated diagnosis and well-founded planning recommendations. The subsequent sections detail each of these stages.

Figure 3. Methodological flowchart of the investigation

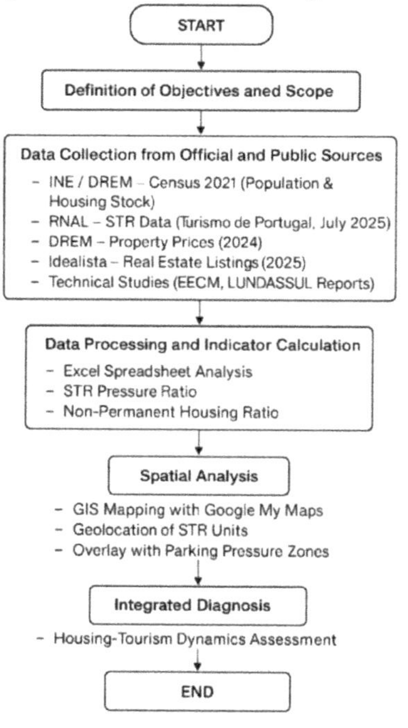

For the construction of the diagnosis, a diverse set of statistical and documentary sources was utilized. The characterization of the resident population and the housing stock was based on data from the 2021 Census, provided by the National Statistics Institute (INE), and detailed by the Regional Directorate of Statistics of Madeira (DREM, 2023). This source allowed for the collection of fundamental indicators, such as the resident population, the number of classic family dwellings, permanent dwellings, vacant dwellings, secondary residences, and the state of building conservation, with disaggregation at the parish level.

The analysis of the Short-Term Rental (STR) supply was conducted through a systematic collection of records available on the National Registry of Local Accommodation (RNAL), the official platform managed by Turismo de Portugal (Turismo de Portugal, 2025). Data updated to July 2025 were extracted, covering all municipalities in the Autonomous Region of Madeira, which made it possible to determine the number of active establishments, their respective typologies, and their accommodation capacity.

Regarding urban mobility and the impact on support infrastructure, technical studies commissioned by the Machico City Council were consulted, with special emphasis on the Parking Strategy for the City of Machico (EECM), as well as the specific report concerning the parish of Caniçal, prepared by the company LUNDASSUL (LUNDASSUL, 2023).

The analysis of the economic pressure on the housing market was supported by DREM data on median sales prices for properties (DREAM, 2024). This information was complemented by an observation of the available supply on the *Idealista* portal to capture the real-time reality of the real estate market.

All collected data were subsequently processed in a spreadsheet environment (Microsoft Excel). Among the most relevant indicators is the STR Pressure Ratio, which measures the relative weight of STRs in the total housing stock. Its formula was adopted based on reference studies, namely the one conducted for the municipality of Lisbon (Câmara Municipal de Lisboa, 2018), thus ensuring comparability. Additionally, the Non-Permanent Housing Ratio was calculated to contextualize the STR phenomenon within the universe of properties not intended for permanent housing.

The spatial component was enriched with georeferenced maps, produced using geographic information system (GIS) tools, namely Google My Maps. These maps allowed for the visualization of the territorial distribution of STR units and their overlap with areas under greater parking pressure, contributing to a more in-depth reading of the dynamics under study.

4. CASE STUDY: THE MUNICIPALITY OF MACHICO

4.1 Demographic Characterization

The municipality of Machico, a site of historical importance in the settlement of Madeira, covers 68 km^2 on the island's east coast. Its five parishes—Machico, Água de Pena, Caniçal, Porto da Cruz, and Santo António da Serra—exhibit remarkable landscape and socioeconomic diversity. With a strategic location near the airport and the capital, Funchal, the municipality has transitioned from a traditional economy (agriculture and fishing) to a more diversified model, driven by tourism and the growth of Short-Term Rentals.

With a resident population of 19.593 inhabitants, the municipality of Machico presents a demographic reality marked by strong contrasts between the more densely populated coastal parishes and the less populated inland areas (DREM, 2023). Table 1 presents the distribution of the resident population and the population density for each of the five parishes of the municipality, based on data from the 2021 Census.

Table 1. Demographic data of the municipality (DREM, 2023)

Parishes	**Inhabitants**	**Population Density (hab/km^2)**
Água de Pena	2.749	472.6
Caniçal	3.548	339.7
Machico	9.828	686.2
Porto da Cruz	2.134	111.1
Santo António da Serra	1.334	151.4
Municipality Total	**19.593**	**320.1**

From the analysis of Table 1, we can conclude that the parish of Machico is the most populous, with about 9.828 inhabitants and a population density of 686.2 hab/km^2, followed by Caniçal, with 3.548 inhabitants and a density of 339.7 hab/km^2, Água de Pena, with 2.749 inhabitants and 472.6 hab/km^2, Porto da Cruz, with 2,134 inhabitants and 111.1 hab/km^2, and finally, Santo António da Serra, the least populous, with 1,334 inhabitants and a density of 151.4 hab/km^2 (DREM, 2023).

The demographic evolution of the municipality has often been influenced by phenomena such as the progressive aging of the population, a declining birth rate, and both internal and external migration. A detailed analysis of demographic indices - such as population density, age distribution, the dependency ratio, and migration flows - allows for an understanding not only of the current population structure but also of the challenges related to social cohesion, territorial planning, and the sustainability of local public services. These indicators are particularly relevant in an insular and peripheral territory, where the effective management of human and social resources is essential to ensure the balanced development of all parishes. For a more in-depth characterization of the social structure, Table 2 details the composition of family nuclei by the number of children in each parish.

Table 2. Family nuclei in the municipality (DREM, 2023)

Parishes	No children	One child	Two children	Three children	Four or more children
Água de Pena	197	358	208	51	7
Caniçal	368	448	289	35	6
Machico	886	1.325	704	101	13
Porto da Cruz	187	246	149	26	6
Santo António da Serra	125	166	82	16	6
Municipality Total	**1.763**	**2.543**	**1432**	**229**	**38**

From the analysis of Table 2, it is evident that the municipality of Machico has a strong family-oriented demographic profile. Across all parishes, the most common family structures are those with one child (2.543 nuclei) or two children (1.432 nuclei), significantly outnumbering households with no children (1.763).

This demographic reality generates a substantial and sustained demand for specific housing typologies, particularly two and three-bedroom homes (T2 and T3), which are essential for accommodating family life. This finding is of relevance to the present study, as this segment of the housing market is often the same one targeted for conversion into Short-Term Rentals (STRs), highlighting a potential direct competition for housing resources between the resident population and the tourism sector.

4.2 Family Housing Characterization

The resident population of 19,593 inhabitants is housed in 9,965 dwellings, which represents an average of approximately 1.97 persons per dwelling, or an occupancy ratio of 196.6 inhabitants per one hundred dwellings. The parish of Machico has 5,081 dwellings, while Caniçal has 1,469, Porto da Cruz has 1,346, Água de Pena has 1,343, and Santo António da Serra has 726 [1]. Table 3 provides a detailed characterization of the municipality's housing stock.

Table 3. Housing stock of the municipality (DREM, 2023)

Parishes	Total Dwellings	Classic Family Dwellings	Permanent Residence	Secondary or Seasonal Residence	Vacant Dwellings
Água de Pena	1.343	1.338	1.012	179	147
Caniçal	1.469	1.469	1.181	165	123
Machico	5.081	5.078	3.756	579	743
Porto da Cruz	1.346	1.344	809	187	219
Santo António da Serra	726	725	525	109	91
Municipality Total	**9.965**	**9.954**	**7.283**	**1.343**	**1.323**

At a national level, housing is a central issue, marked by rising prices and tourism pressure in major urban centers, and by housing stock stagnation in less dense territories. This dual challenge creates a complex context for the management of insular territories like Machico, where the need for economic dynamism intersects with the risk of housing exclusion.

Although strengthened, public housing policies face structural challenges that fuel a national affordability crisis. In the specific case of Machico, while urban pressure is less acute than in major centers, the effects of real estate appreciation and the conversion of properties to Short-Term Rentals are felt. Factors such as the aging building stock and geographical dispersion hinder access to decent housing for permanent residents, which, as an engineer in Machico, is a visible, practical concern.

4.3 Characterization of the Building Stock's State of Conservation

In the Portuguese housing landscape, a sizeable portion of the building stock is composed of classic or older constructions, many of which show evident signs of degradation and lack structural interventions. This reality is particularly visible in peripheral or less dense areas, where market devaluation hinders private investment, perpetuating the progressive decay of the housing stock. To combat this trend, instruments such as the Strategic Urban Rehabilitation Program (PERU) have emerged as a strategic response to the growing need for intervention in old and degraded buildings in Portugal, promoting the revitalization of urban centers and the valorisation of the built heritage [1]. Through the provision of technical and financial support, PERU aims to boost the rehabilitation of priority urban areas, combat the desertification of historic centers, and promote access to dignified housing conditions.

The municipality of Machico, through its City Council, has also adopted this strategy by implementing an Urban Rehabilitation Area (ARU) in the parish of Machico, with the objective of encouraging the requalification of the building stock and the revitalization of its historic center. This measure aims not only to improve the state of conservation of existing buildings but also to promote the settlement of the resident population and enhance the local built heritage. Through the delimitation of the ARU, conditions were created for the attribution of tax benefits and technical support to property owners interested in recovering their properties, contributing to a more sustainable, cohesive, and identity-preserving urban dynamic.

This structuring project relies entirely on private initiative, with the Municipality's role being one of moral and financial incentive. Seventy-five properties were identified within the ARU as potential candidates for intervention, detected through on-site surveys, as illustrated in Figures 4 and 5.

Figure 4. Delimitation of the ARU in the parish of Machico

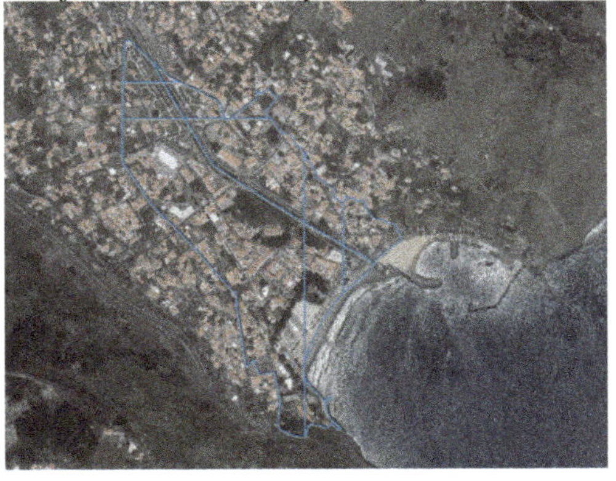

Within this priority intervention area, municipal surveys have identified the most critical points in need of rehabilitation. Figure 5 presents the precise georeferencing of these seventy-five buildings. Their distribution reveals a particular concentration in the historic core and along the seafront, highlighting the area's most in need of a regeneration boost to preserve the identity and functionality of the city center.

Figure 5. Buildings identified for recovery in the ARU

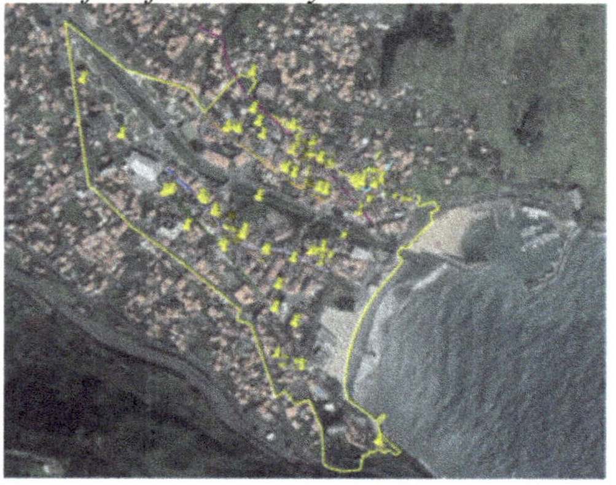

4.4 Characterization of the Housing Market

An analysis of the housing market is essential to understand the economic pressure on the territory. For this study, a mixed-method approach was employed, combining official statistical data with an analysis of the available supply on leading online platforms in Portugal.

The temporal evolution of prices in the municipality of Machico reveals significant real estate inflation. According to official data, the median price of family dwellings sold in 2023 was €1.236/m² (DREM, 2024). A complementary market analysis, based on data from the Idealista platform for the period 2021-2025, suggests a strong acceleration trend, culminating in an average asking price of €2.453/m² by mid-2025 (Idealista, 2025), as detailed in Table 4.

Table 4. Average housing price in the municipality.

Year	Average Price (€/m²)	Year-over-Year Percentage Change
2021	1.025	-
2022	1.290	+25.85%
2023	1.236	-4.19%
2024	2.343	+89.55%
2025 (July)	2.453	+4.70%

Sources: 2023 data (DREM, 2024); other years (Idealista, 2025)

This real estate inflation, disproportionate to local incomes, compromises housing access for families, accentuating residential exclusion. This appreciation highlights a pressured market and underscores the urgency for measures that balance the interests of tourism investment with the fundamental right to housing for residents.

In addition to the temporal analysis, a spatial analysis of the supply available in July 2025 reveals a remarkable heterogeneity in the municipality's housing market, as presented in Table 5.

Table 5. Housing market in the municipality in July 2025 (Idealista, 2025)

Parishes	No. of Properties for Sale	Average Gross Area (m^2)	Average Price (€/m^2)
Água de Pena	12	320	3.125
Caniçal	6	244	4.698
Machico	31	229	2.589
Porto da Cruz	10	717	1.984
Santo António da Serra	9	685	2.402
Municipality Total	**68**	-	-

The parish of Machico concentrates almost half of the total supply in the municipality (45.6% of properties), suggesting a market dominated by apartments, given its smaller average gross area. In stark contrast, Caniçal, despite its limited supply, exhibits the highest price per square meter (€4.698), indicating significant real estate valuation. Conversely, Porto da Cruz presents a different market profile, with the largest properties (717 m^2 average area) and the lowest unit price (€1.984), characteristic of a market based on single-family homes.

Overall, the data highlights very distinct local dynamics, which must be considered in any territorial analysis or real estate investment decision.

4.5 Diagnosis of the Parking Situation

This subchapter addresses the theme of parking. For this purpose, a prior study commissioned by the Machico City Council, the EECM – Parking Study for the City of Machico – authored by the technical team at LUNDASSUL and based on data from April 2022, was considered (LUNDASSUL, 2023). The objective of this study was to structure a proposal for the operation of the parking system, with special emphasis on the functioning of parking meters, including updating their locations, equipment, enforcement, and signage, and to provide guidelines for the Parking Regulation, including a proposal for parking reorganization.

With its work located in the city of Machico, the study presents a characterization of the parking supply in the city, from public and private parks to taxi ranks and informal parking. A total of 1,384 legal regular parking spaces were identified, with 666 on public roads (48%) and 718 in parking parks (52%) (LUNDASSUL, 2023). Their geographical distribution within the study area can be seen in Figure 6

Figure 6. Geographical distribution of parking places

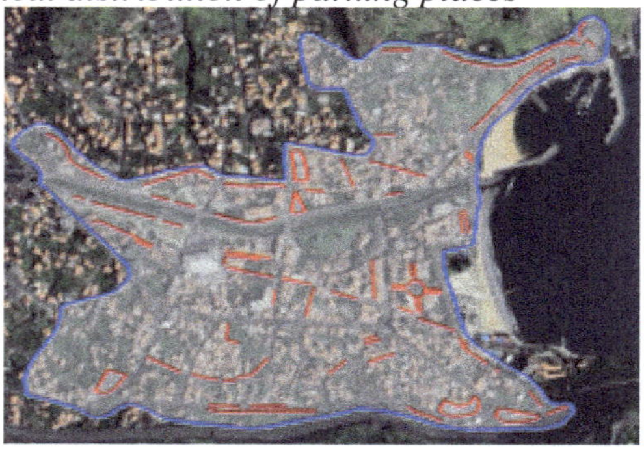

The analysis of the EECM allows for the identification of a direct correlation between parking pressure and the proximity to tourist attraction poles. Figures 5 and 6 illustrate two paradigmatic examples of this situation: the parking at the *Praceta* 25 de Abril Roundabout and the *Fórum* Machico Parking Park. Both are critical parking hubs in the city, subject to high demand from visitors due to their adjacent location to large-scale tourist accommodations, namely the White Waters and Dom Pedro hotels.

Figure 7 clearly shows the high occupancy of on-street parking spaces in a central, high-rotation area, directly impacting traffic flow and the availability of spaces for residents' short-term needs.

Figure 7. Parking at Praceta 25 de Abril roundabout

Complementing this, Figure 8 illustrates the pressure on off-street, structured parking facilities. The high demand for spaces within the Fórum Machico park demonstrates that the saturation extends beyond the public road network, affecting the dedicated infrastructure designed to absorb such demand.

Figure 8. Fórum Machico parking park

As stated in the study, "parking is directly involved with the management of land use, having a decisive impact on people's quality of life," with tourist accommodations being one of the main drivers of increased demand.

The analysis of the EECM allows for the identification of a direct link between the scarcity of parking and the growing pressure exerted by tourist accommodations. Although the report does not explicitly mention these establishments, the impacts resulting from the increased demand for vehicles in central and coastal areas, where STRs tend to concentrate, are evident. The influx of vehicles associated with tourism, both personal and collective passenger transport, intensifies not only mobility constraints in areas of high tourist demand but also the intensive occupation of public space, increasing the percentage of illegal parking spaces. This phenomenon is particularly visible near beach areas, such as *Rua do Desembarcadouro*, where a "differentiated tariff, with a low rate in winter and a high rate in summer" was even proposed, thus acknowledging the typical seasonality of tourist activity.

It is evident that the current parking situation does not meet the mobility needs of the population, making it imperative to reformulate the system. This reformulation should involve a planned expansion of the supply, with the creation of new parks and the requalification of underutilized public spaces, as well as more efficient management of the existing occupancy through measures such as pricing, controlled rotation, or even the concession of its operation.

To complement this study, the same team also analysed the parish of Caniçal in the "Analysis Report of Caniçal" of August 2022 (LUNDASSUL, 2022). This report reveals the existence of areas with high pressure on the parking system, associated with the presence of tourist and leisure attractions. The marginal area of *Rua da Pedra d'Eira*, near the restaurant complex and sea access points, stands out as a particularly critical area where the demand for parking remains high both during the day and at night.

The existence of illegal parking, especially in the vicinity of the sea access pier, constitutes a significant risk factor with potential implications in emergencies. Figure 9 illustrates precisely this improper occupation of public space at the *Varadouro of*

Rua da Pedra d'Eira, a practice that is incompatible with safety and urban mobility standards.

Figure 9. Varadouro at Rua da Pedra D'Eira, Caniçal

Furthermore, Ribeira do Natal Beach, heavily sought after by visitors and seasonal users, has two parking parks that, despite their capacity, require management adapted to seasonal fluctuations, with the recommendation of applying a differentiated tariff. Although the document does not explicitly mention short-term rentals, the combination of beach activity, restaurants, and growing tourist demand creates an urban dynamic that inevitably interferes with the mobility of residents and the availability of parking.

5. SPATIAL ANALYSIS AND PRESSURE ZONES

5.1 Framework

Spatial analysis is crucial in territories with marked geographical characteristics like Machico, where a delicate balance exists between urban, rural, and protected areas. The disparities between parishes help identify pressure zones, where housing demand competes with economic interests such as tourism. The growth of Short-Term Rentals (STRs) has been a key factor in the recent reconfiguration of housing use in the municipality.

Their proliferation in certain areas has contributed to the reconfiguration of housing use, often to the detriment of permanent housing. This reality, observed in various regions of the country, is also beginning to manifest in Machico, especially in areas with greater landscape and heritage attractiveness. Through the consultation of the RNAL, the National Registry of Local Accommodation, it was determined that there are 342 active STR units throughout the municipality's parishes (Turismo de Portugal, I.P., 2025). Table 6 quantifies the scale and territorial distribution of the Short-Term Rental (STR) phenomenon, presenting the total number of active establishments in each of the municipality's parishes.

Table 6. No. of registered STRs in the municipality in July 2025 (Turismo de Portugal, I.P., 2025a)

Parishes	No. of Registered STRs
Água de Pena	58
Caniçal	55
Machico	163
Porto da Cruz	47
Santo António da Serra	19
Municipality Total	**342**

Regarding tourist enterprises (hotels, etc.), a similar search was conducted on the RNET, the National Registry of Tourist Enterprises, yielding twelve establishments in the municipality. In addition to STRs, the formal tourist accommodation supply is detailed in Table 7, which lists the number of registered tourist enterprises (such as hotels) and their total capacity in terms of beds and units per parish.

Table 7. No. of tourist enterprises in the municipality (Turismo de Portugal, I.P., 2025b)

Parishes	No. of Tourist Enterprises	Capacity (Beds)	No. of Accommodation Units
Água de Pena	0	0	0
Caniçal	3	808	366
Machico	4	513	260
Porto da Cruz	3	64	32
Santo António da Serra	2	136	68
Municipality Total	**12**	**1.521**	**726**

Figure 10 provides a macro view of the tourism landscape, illustrating the geographical location of the main tourist enterprises within the municipality.

Figure 10. Geographical location of tourist enterprises in the municipality

To allow for a detailed spatial analysis, the georeferenced distribution of all 342 STR establishments was mapped, first for each individual parish and concluding with a consolidated map of the entire municipality. This visual analysis is crucial for understanding the specific territorial patterns of tourist pressure.

Figure 11 presents the spatial distribution of the 58 STRs in Água de Pena. The map reveals a significant concentration along the main access roads and in areas with ocean views, capitalizing on the parish's strategic proximity to the airport and Funchal. This clustering in a parish with a small housing stock is a key factor in explaining why it has one of the highest STR pressure ratios in the municipality.

Figure 11. Geographical location of STRs at Água de Pena

The distribution in Caniçal, shown in Figure 12, displays a different pattern. The 55 STR units are heavily concentrated along the seafront, near the port, the restaurant complexes, and the bathing areas. This pattern is consistent with a tourism model based on maritime activities and local gastronomy and aligns with the data showing that Caniçal has a strong presence of both STRs in houses (*moradias*) and traditional tourist enterprises.

Figure 12. Geographical location of STRs at Caniçal

Figure 13 details the situation in the parish of Machico, which, with 163 units, represents the epicentre of STR activity in the municipality. The map clearly shows a dense cluster of establishments within the city's historic and commercial center. This concentration in the urban core is a classic indicator of "touristification," where the pressure on housing and public infrastructure, particularly parking, is most acute.

Figure 13. Geographical location of STRs at Machico

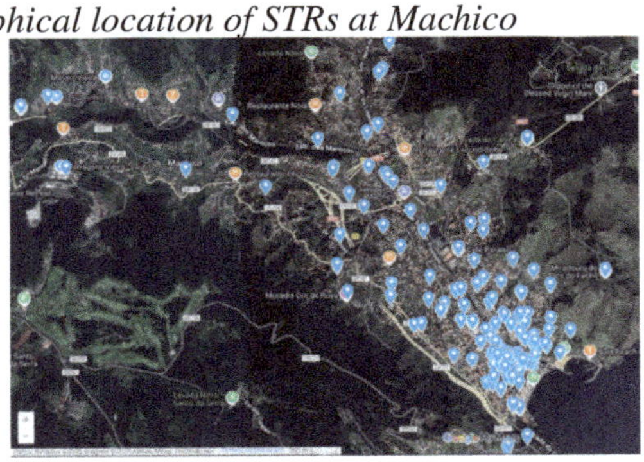

In Porto da Cruz, the distribution illustrated in Figure 14 reflects its identity as a nature and adventure tourism destination. The 47 STRs are more dispersed than in the urban parishes, clustered around the village center and along roads leading to points of interest such as surf spots and hiking trails (*levadas*). This pattern suggests a different kind of pressure, more related to the impact on landscape and rural character.

Figure 14. Geographical location of STRs at Porto da Cruz

Finally, Figure 15 shows the distribution in Santo António da Serra. With only 19 STRs, the pattern is visibly sparse and dispersed, characteristic of its rural setting. The establishments are scattered, often corresponding to larger properties or quintas, reflecting a countryside tourism model with a much lower impact on the housing stock and local infrastructure compared to the coastal parishes.

Figure 15. Geographical location of STRs at Santo António da Serra

5.2 Analysis of the Evolution and Pressure of Strs

The dynamics of STR growth are detailed in Table 8, which shows the annual evolution of new registrations by parish since 2021, illustrating the recent acceleration of the phenomenon. The growth dynamics of the STR phenomenon are detailed in Table 8, which shows the annual evolution of new registrations by parish since 2021, illustrating its recent acceleration.

Table 8. Variation of new STR registrations (Turismo de Portugal, I.P., 2025a)

Parishes	2021	2022	2023	2024	June 2025
Água de Pena	1	2	11	8	11
Caniçal	1	4	3	8	8
Machico	4	14	47	43	27
Porto da Cruz	2	2	8	15	2
Santo António da Serra	0	0	2	1	3
Municipality Total	**8**	**22**	**71**	**75**	**51**

Based on Tables 6 and 8, we can conclude that STRs have shown growing trends over the years. The observed drops at certain times can be interpreted more as a "technical pause" than as a sign of structural retraction, reflecting conjunctural factors such as the anticipation of new legal rules.

To contextualize the weight of STRs within the non-permanent housing stock, Table 9 compares the number of establishments with the stock of vacant and secondary dwellings in each parish.

Table 9. No. of STRs vs. vacant and secondary dwellings (DREM, 2023; Turismo de Portugal, I.P., 2025a)

Parishes	Vacant + Secondary Dwellings	No. of STRs	Ratio (%)
Água de Pena	326	58	17.79
Caniçal	288	55	19.10
Machico	1.322	163	12.33
Porto da Cruz	535	47	8.79
Santo António da Serra	200	19	9.50
Municipality Total	**2.671**	**342**	**12.80**

This ratio reveals how many non-primary dwellings are allocated to STRs. The parish of Caniçal shows the highest ratio (19.10%), meaning that approximately one in every five non-primary dwellings is an STR.

Finally, to assess the pressure on the housing stock, a comparative analysis was conducted for all municipalities in the Autonomous Region of Madeira.

The central analysis of this study is presented in Table 10, which provides a comparative benchmark of tourism pressure across all municipalities in the Autonomous Region of Madeira. It calculates and compares two key indicators: the ratio of STRs to Permanent Dwellings and the ratio of STRs to Classic Family Dwellings.

Table 10. STR pressure on the housing stock in the RAM (DREM, 2023; Turismo de Portugal, I.P., 2025a)

Municipality	Permanent Dwellings (2021)	Classic Family Dwellings (2021)	STRs (2025)	STR / Permanent Dwellings Ratio	STR / Classic Family Dwellings Ratio
Calheta	4.376	7.630	1.109	25.34%	14.53%
Porto Santo	2.156	4.550	453	21.01%	9.96%
Funchal	40.944	52.234	6.136	14.99%	11.75%
São Vicente	2.012	3.783	269	13.37%	7.11%
Ponta do Sol	3.194	4.888	338	10.58%	6.91%
Santa Cruz	15.637	20.373	1.056	6.75%	5.18%
Machico	**7.283**	**9.954**	**342**	**4.70%**	**3.44%**
Ribeira Brava	4.712	6.971	203	4.31%	2.91%
Porto Moniz	1.086	2.023	44	4.05%	2.17%
Santana	2.681	4.885	100	3.73%	2.05%
Câmara de Lobos	10.711	13.746	240	2.24%	1.75%
TOTAL RAM	**94.792**	**131.037**	**10.290**	**10.86%**	**7.85%**

The analysis of the ratio between STRs and Classic Family Dwellings allows for a more comprehensive reading of the tourist pressure on the total housing stock. Comparing this ratio across the municipalities of the Autonomous Region of Madeira, Machico has one of the lowest values in the region (3.44%), well below the regional average (7.85%).

Figure 16 visually summarizes the key findings from the regional benchmark analysis, offering an intuitive comparison of the STR pressure on the total housing stock for each municipality in Madeira.

Figure 16. Weight of STRs in the housing stock by municipality

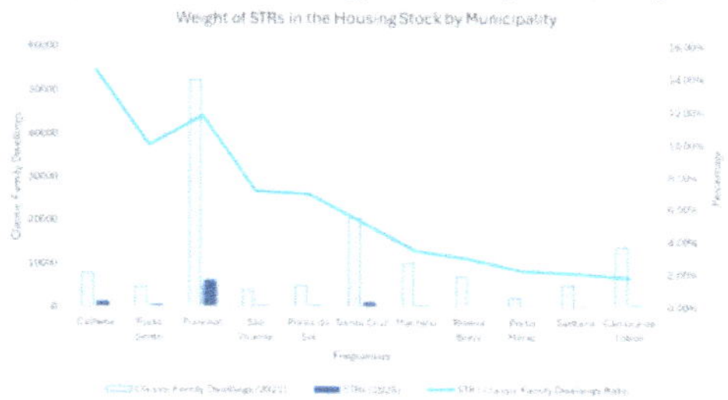

This reading suggests that the municipality still has significant room for the emergence of new STR registrations without, for now, representing an incredibly significant pressure on the existing housing stock.

5.3 The Impact of STR on Urban Parking

The activity of Short-Term Rentals (STRs), while representing an important source of economic and touristic dynamism, emerges in this context as an additional factor of pressure on the municipality's parking system. With 342 registered STR units unevenly distributed across the parishes, this type of housing occupancy attracts a considerable number of temporary visitors who frequently rent their own vehicles. This increases the pressure on parking, given that the majority of STRs do not offer private parking for their users.

In recent years, the region has witnessed a veritable "boom" in rent-a-car companies, driven by the growth of tourism and the demand for autonomy from visitors. This reality has significantly contributed to the worsening pressure on the road network and the already limited parking supply, especially in central areas and near points of tourist interest, generating notable discontent among the population.

The massive presence of vehicles associated with tourism not only intensifies congestion but also negatively affects the experience of the resident population, who are daily confronted with longer travel times, increased difficulty in parking, and constant traffic disruptions to their daily routines. Added to this is the fact that many rent-a-car drivers are unfamiliar with the steep, narrow, and winding layouts typical of the region, which often leads to blockages or improper driving, especially in places where manoeuvring space is limited. This inexperience, combined with the natural narrowing of some roads, particularly in areas of high tourist traffic, generates disturbances in the daily lives of residents, translating into frustration, insecurity, and a loss of quality of life.

Despite the lack of official data that directly correlates the number of STRs with the pressure on urban parking, various municipal and technical studies, such as those previously discussed here, point to this relationship (LUNDASSUL., 2023). The emergence of new registrations, without reserved parking spaces for their users, does cause pressure on parking. In other words, the concentration of STRs in urban areas without private parking, combined with the growing use of rental cars by visitors, constitutes an additional factor of congestion and saturation of the local road infrastructure.

6. DISCUSSION OF RESULTS

The quantitative and spatial analysis of this study reveals a complex picture of Machico's reality, where STR activity is interconnected with urban, social, and economic dynamics. This section aims to critically discuss the main results obtained, interpreting them considering the principles of urban planning and territorial sustainability.

The main indicator used, the STR pressure ratio, revealed a value of 3.4% relative to the total classic family dwellings in the municipality. This result places Machico in an ambiguous position: on one hand, the value is substantially below the saturation levels observed in municipalities with high tourist pressure, such as *Calheta* (14.5%) or Funchal (11.7%), which suggests that, at present, Machico does not face a touristification crisis comparable to that seen in those contexts. On the other hand, it is important to note that the calculated ratio already exceeds the 2.5% threshold, a benchmark adopted in the city of Lisbon's Urban Study as a reference point for applying containment measures for STR activity (Câmara Municipal de Lisboa, 2018). This finding is particularly relevant as it indicates that, although a saturation scenario is not yet present, the current situation constitutes a strategic moment for implementing preventive planning policies. The experience accumulated in other territories proves that, in the absence of timely regulation, moderate ratios can rapidly evolve into situations that are difficult to reverse.

The investigation also highlighted a structural aspect often overlooked in the public debate: the existence of a significantly higher number of vacant or secondary-use properties (2.671) compared to the number of registered STR units (342). This reality dismantles the simplistic narrative that attributes to STRs the exclusive responsibility for the housing shortage. The data suggests that the problem of housing affordability in Machico is deeply rooted in factors such as real estate disinvestment, inheritance issues, and existing tax regimes. However, it is important to stress that this finding does not exempt STRs from their implications: their higher profitability, when compared to long-term renting, constitutes an economic incentive that diverts unoccupied properties to the tourist market, to the detriment of their reintroduction into the permanent housing market. The discussion, therefore, should not focus on prohibiting the activity, but rather on designing regulatory and fiscal mechanisms that encourage a balance between both forms of occupancy, ensuring that the right to housing is safeguarded.

Another particularly relevant contribution of this study lies in the connection between the spatial analysis of STRs and the data from the Parking Strategy for the City of Machico (EECM). The geographical coincidence between the areas with the highest concentration of STR units and those with the greatest pressure on urban parking cannot be seen as mere chance, but rather as evidence of a negative synergistic effect. Although STRs are not the origin of the parking problem, they unequivocally function as an amplifying factor of a pre-existing vulnerability in the urban infrastructure. Each new tourist accommodation unit that does not provide its own parking solution transfers an additional cost to the public space, a cost that is collectively absorbed by residents and visitors. This negative externality alone legitimizes the need to integrate tourism policy with urban mobility policy, reinforcing a systemic and intersectoral approach in municipal planning.

In this context, the case study of Machico provides valuable contributions to the reflection on resilient cities. A resilient city is one that can maintain its essential functions even in the face of external shocks, which requires economic diversification, housing affordability, and adaptability. Excessive dependence on tourism and the inability to ensure decent housing for its active population seriously compromise this resilience. Likewise, territorial sustainability requires a permanent balance between economic viability, social justice, and environmental protection. The methodology adopted in this work - based on open data, objective indicators, and spatial analysis tools - represents a promising path for the actions of engineering and urban planning in the construction of strategies that reconcile development with quality of life.

7. CONCLUSIONS AND FUTURE PATHWAYS

Throughout this study, an effort was made to develop and apply a data analysis methodology to assess the impact of Short-Term Rentals (STRs) in the municipality of Machico, framing it within the broader context of the tourism and housing reality of the Autonomous Region of Madeira. The adopted approach allowed not only for the construction of an updated and rigorous snapshot of the current situation but also laid the groundwork for a prospective reflection on sustainable urban planning and territorial resilience, which are central challenges for contemporary civil engineering.

In the final phase of the investigation, it was confirmed that Machico is in an intermediate stage of its touristification process, evidenced by a 3.4% pressure ratio of STRs on the classic housing stock. Although this value remains below that of markedly tourist-heavy regions of Madeira, it exceeds thresholds such as Lisbon's (2.5%), signalling an emerging tension that requires strategic action. It is important to note that this pressure manifests unevenly across the territory, being more intense, in relative terms, in the parish of Água de Pena and, in absolute terms, in the municipal seat, Machico. This geographical asymmetry reinforces the need for differentiated and territorially sensitive policy responses.

Concurrently, the study highlighted the overlap between STR pressure and pre-existing structural frailties—such as the saturation of infrastructure and parking, and the significant volume of vacant or secondary properties, which outnumbers active STRs by a factor of 7.8. This evidence points to an urban ecosystem where STRs function as a catalyst for already latent imbalances.

In this context, the implications for urban planning are profound and multidimensional. Firstly, the opportunity for Machico to transition from a reactive response model, common in many urban centers, to a preventive model anchored in data and objective criteria stands out. This change in basic assumptions would allow for the anticipation and mitigation of impacts before they become irreversible, safeguarding the territory's carrying capacity and the quality of life of the resident population. Secondly, it becomes clear that an effective housing policy cannot be limited to controlling STRs. The abundance of unoccupied properties constitutes a strategic opportunity for the municipality to develop incentive mechanisms for long-term renting, thus promoting greater equity in access to housing.

Finally, this analysis demonstrates that civil engineering, as an integrating discipline, must play a significant role in reconciling economic development with infrastructural sustainability. The pressure on parking clearly illustrates the risks of ignoring the articulation between new economic activities and the physical and functional limits of the territory.

Considering these findings, several future pathways are outlined to enhance the resilience of Machico and other municipalities in similar contexts. Firstly, the development of an adaptive management model is proposed, supported by a dynamic municipal regulation sensitive to the pressure indicators presented here. Secondly, the necessary integration of tourism, housing, and mobility planning axes is advocated. The revision of the Municipal Master Plan (PDM) and the issuance of new licenses must be based on a systemic logic, requiring, for example, traffic impact studies and parking solutions as prerequisites for the approval of tourist developments. Lastly, the relevance of the methodology itself as a tool for local governance is emphasized. The use of open data to produce objective indicators represents a replicable, low-cost, high-impact innovation capable of strengthening the transparency, technical foundation, and robustness of municipal decisions.

In summary, this study constitutes not only an in-depth diagnosis of the current impact of Short-Term Rentals in Machico but also a concrete proposal for transforming the urban planning model, grounded in data science, civil engineering, and the pursuit of more just, functional, and sustainable territories.

The relevance and urgency of a preventive regulatory approach, such as the one proposed for Machico, are corroborated by recent actions in other municipalities. A case in point is the city of Lisbon, which, even in the absence of a new regulation, suspended approximately four hundred new Short-Term Rental registrations between October 2024 and July 2025, based on the same housing pressure criteria analysed herein (Público, 2025). This action demonstrates that continuous monitoring and the application of containment thresholds are active and essential planning tools in contemporary territorial management.

8. STRATEGIC RECOMMENDATIONS

Based on the conclusions and future pathways outlined, it is essential to present a set of recommendations that can guide the strategic actions of the municipality of Machico towards a more sustainable, integrated, and initiative-taking management of Short-Term Rentals (STRs). The recommendations proposed here aim not only to respond to the identified challenges but also to anticipate future dynamics, promoting urban resilience, equity in access to housing, and infrastructural efficiency. To this end, it is fundamental that decision-making be based on a systemic, multidisciplinary, and data-driven vision.

8.1 Establishment of an Adaptive and Evidence-Based Municipal Regulation

The cornerstone of a sustainable strategy lies in updating the municipal regulatory framework. The current Municipal Master Plan (PDM) predates the STR phenomenon, creating a regulatory vacuum. It is therefore recommended to establish an integrated framework that is both evidence-based and adaptive.

This can be achieved by developing a municipal regulation for STRs that is linked to the monitoring methodology created in this study. The regulation should define maximum pressure ratio thresholds per parish, allowing for the automatic suspension of new registrations in critical zones and their eventual reopening if pressure subsides. Furthermore, this new regulatory thinking must be integrated into the broader revision of the PDM. This process should overcome sectoral fragmentation by requiring that the approval of new tourism projects (both STRs and hotels) be conditional on a holistic assessment of their impacts on housing supply and mobility, for instance, by requiring traffic impact studies and the inclusion of private parking solutions in high-pressure zones.

8.2 Sectoral Integration in Urban Planning: Tourism, Housing, and Mobility

Beyond regulation, effective management requires initiative-taking interventions in the urban fabric. One of the key strategic opportunities identified is the large stock of unoccupied properties. It is recommended that the municipality create active policies to mobilize the vacant housing stock for the long-term rental market. These measures could include tax incentives (e.g., property tax reductions), targeted urban rehabilitation programs, or direct mediation between owners and potential tenants. This strategy would rebalance the housing market by addressing the needs of the resident population without completely halting tourism dynamics.

Simultaneously, the pressure observed on the parking system must be addressed with a vision for sustainability. As a Civil Engineer in Machico, you know that simply expanding car capacity can exacerbate problems. Instead, interventions should focus on sustainable and innovative solutions, such as developing peripheral parking hubs with public transport links, creating soft mobility zones in historic centers, and implementing intelligent traffic management systems. Any intervention must be aligned with the principles of sustainable development to enhance, rather than degrade, urban resilience.

8.3 Mobilization of the Vacant Housing Stock for Long-Term Rental

Finally, to ensure that these actions are not a one-off response but part of a continuous improvement cycle, it is recommended that the municipality institutionalize the monitoring methodology developed in this study. The systematic use of open data to produce indicators on STR pressure, spatial distribution, and their overlap with critical variables (housing, parking) should become a standard administrative practice. This will foster a more transparent, responsive, and evidence-based governance model. By embedding this data-driven approach into its regular planning instruments, Machico can facilitate communication with citizens, legitimize public policies, and position itself as an example of good local governance in managing the challenges of contemporary tourism.

FINAL CONSIDERATIONS

The recommendations presented here are intended to be more than mere technical proposals: they are invitations to build a strategic and sustained vision for Machico. The balance between economic development, the preservation of social cohesion, and urban functionality is a collective goal that requires political courage, technical rigor, and community involvement. By implementing these guidelines, Machico can navigate the transformations of the insular territory and set a benchmark for resilient and sustainable planning.

REFERENCES

Alves, R., Lousada, S., & Cabezas, J. (2025). Understanding Mobility Patterns in Machico's Low-Density Rural Areas Through Stated Preference Surveys. In *Sustainable Civil Engineering for a More Resilient Society* (pp. 235–257). IGI Global., DOI: 10.4018/979-8-3693-2448-6.ch012

Alves, R., Lousada, S., Cabezas, J., & Gómez, J. M. N. (2022). The Importance of the Strategic Urban Rehabilitation Plan in the Sustainable Development of the Municipality of Machico. *Sustainability (Basel)*, *14*(24), 16816. DOI: 10.3390/su142416816

Alves, R., Lousada, S., Cabezas, J., & Gómez, J. M. N. (2023a). Local Housing Strategy: Analysis of Importance and Implementation in Machico Municipality, Madeira. *Land (Basel)*, *12*(9), 1778. DOI: 10.3390/land12091778

Alves, R., Lousada, S., Gómez, J. M. N., & Cabezas, J. (2024). Maintenance Challenges in Maritime Environments and the Impact on Urban Mobility: Machico Stayed Bridge. *Infrastructures*, *9*(10), 180. DOI: 10.3390/infrastructures9100180

Alves, R. M. C., Lousada, S. A. N., Cabezas, J., & Gómez, J. M. N. (2023b). Modelling of Flood Risk, Vulnerability and Adaptation of Machico City Center. In *Sustainable Development of Urban Tourism Systems* (pp. 19-41). IGI Global. DOI: DOI: 10.4018/978-1-6684-6449-6.ch002

Alves, R. M. C., Lousada, S. A. N., Cabezas, J., & Gómez, J. M. N. (2023c). Port of Machico, Madeira: Assessment of the Consequences of Overtopping Using AHP Methodology. In *Sustainable Development of Urban Tourism Systems* (pp. 61-80). IGI Global. DOI: DOI: 10.4018/978-1-6684-6449-6.ch004

Alves, R. M. C., Lousada, S. A. N., Cabezas, J., & Gómez, J. M. N. (2023d). Flood Risk and Vulnerability: An Approach to Urban Floods in Ribeira de Machico. In *Sustainable Development of Urban Tourism Systems* (pp. 81-103). IGI Global. DOI: DOI: 10.4018/978-1-6684-6449-6.ch005

Alves, R. M. C., Lousada, S. A. N., Cabezas, J., & Gómez, J. M. N. (2023e). Urban Flooding Risk in Machico and Planning Its Territory as a Form of Prevention. In *Sustainable Development of Urban Tourism Systems* (pp. 42–60). IGI Global., DOI: 10.4018/978-1-6684-6449-6.ch003

Balthazar, B. S. (2022). *Gentrificação, Turismo e Conflitos: Análise comparada entre o Bonfim (Porto, Portugal) e Koukaki (Atenas, Grécia)* [Gentrification, Tourism and Conflicts: A comparative analysis between Bonfim (Porto, Portugal) and Koukaki (Athens, Greece)]. Master's Thesis, University of Minho.

Câmara Municipal de Lisboa. (2018). *Estudo Urbanístico do Turismo em Lisboa* [Urban Study of Tourism in Lisbon].

Cordeiro, I. D., Korossy, N., & Selva, V. (2016). Determinação da capacidade de carga turística a partir do método Cifuentes et al. (1992): Aplicação à Praia dos Carneiros (Tamandaré/PE) [Determination of tourism carrying capacity using the Cifuentes et al. (1992) method: Application to Carneiros Beach (Tamandaré/PE)]. *Revista Brasileira de Ecoturismo*, *9*(5), 991-1011.

Costa, C. (2018). A Saturação do Turismo e a Governança: Desafios para as Comunidades Locais. O caso do Alojamento Local em Portugal [Tourism Saturation and Governance: Challenges for Local Communities. The case of Short-Term Rentals in Portugal]. *Proceedings of the XI Portuguese Congress of Sociology*, Lisbon, (p. 132).

Diário da República. (2014). *Decreto-Lei n.º 128/2014, de 29 de agosto* [Decree-Law No. 128/2014, of August 29]. Diário da República, 1st series, no. 166. Consolidated Version.

Direção Regional de Estatística da Madeira (DREM). (2023). *Censos 2021: Resultados Definitivos - Região Autónoma da Madeira* [2021 Census: Definitive Results - Autonomous Region of Madeira].

Direção Regional de Estatística da Madeira (DREM). (2024). *Estatísticas da Construção e Habitação da Região Autónoma da Madeira - 2023* [Construction and Housing Statistics of the Autonomous Region of Madeira - 2023].

Idealista. (2025). *Analysis of the property sales market in the municipality of Machico*. [Data collected in July 2025].

LUNDASSUL. (2022). *Relatório de Análise do Caniçal* [Analysis Report of Caniçal].

LUNDASSUL. (2023). *EECM – Estudo de Estacionamento da Cidade de Machico*. Relatório Final. [EECM – Parking Study for the City of Machico (Final Report)]

Marques, J. A., & Cariço, C. (2023). Capacidade de Carga Turística: Uma Análise sob a Ótica do Planejamento Urbano e Regional [Tourism Carrying Capacity: An Analysis from the Perspective of Urban and Regional Planning]. *Revista Brasileira de Pesquisa em Turismo*.

Mendes, L. (2018). *Gentrificação turística em Lisboa: Impactos do alojamento local na resiliência e sustentabilidade social do centro histórico* [Touristic gentrification in Lisbon: Impacts of short-term rentals on the resilience and social sustainability of the historic center]. Poder Local. *Revista de Administração Democrática*, *155*, 58–73.

Público. (2025, July 26). *Ainda sem regulamento, Lisboa já travou 400 novos Alojamentos Locais desde Outubro* [Even without a regulation, Lisbon has already blocked 400 new Short-Term Rentals since October].

Sardinha, L. F. S., Fernandes, D. N., & Moura, J. (2018). Alojamento Local na Região Autónoma da Madeira - Tipologia e distribuição do Imóvel. O Perfil do Investidor [Short-Term Rentals in the Autonomous Region of Madeira - Typology and distribution of the Property. The Investor Profile]. *Revista de Economia. Empresas e Empreendedores na CPLP*, (3), 11–29.

Turismo de Portugal, I. P. (2025). *Registo Nacional de Alojamento Local*. RNAL. [National Registry of Local Accommodation (RNAL)]. [Data extracted in July 2025]

Turismo de Portugal, I. P. (2025). *Registo Nacional de Empreendimentos Turísticos*. RNET. [National Registry of Tourist Enterprises (RNET)]. [Data extracted in July 2025]

KEY TERMS AND DEFINITIONS

Classic Family Dwelling (*Alojamento Familiar Clássico*)**:** An official statistical category from the Portuguese Census (Censos 2021) that represents the total housing stock of a territory, including permanent residences, secondary homes, and vacant dwellings. It is used as the denominator in the STR Pressure Ratio to provide a comprehensive measure of pressure on the entire housing market.

Containment Zone (*Zona de Contenção*)**:** A legal instrument provided by Portuguese law that allows municipalities to suspend or limit the registration of new STRs in specific geographical areas. These zones are typically established in areas where tourism pressure is deemed to have reached a critical threshold, threatening the social and urban balance.

Georeferencing: The process of assigning geographic coordinates to data, in this case, the addresses of STR establishments. This technique allows for the creation of maps that visualize the spatial distribution and concentration of a phenomenon, enabling a more detailed and localized analysis.

Short-Term Rental (STR) / Alojamento Local (AL): Furnished accommodation units (such as apartments, houses, or rooms) that are rented out to tourists or temporary visitors for short periods, typically less than 30 days. In the Portuguese context, this activity is legally defined and regulated by the RNAL (National Registry of Local Accommodation).

STR Pressure Ratio: A key performance indicator developed in this study to measure the intensity of the STR phenomenon. It is calculated by dividing the total number of registered STR units by the total number of Classic Family Dwellings in each area, expressed as a percentage. It serves as a tool to compare tourism pressure across different territories.

Tourism Carrying Capacity: The maximum level of visitor use and development an area can accommodate without causing unacceptable deterioration of the physical environment, the socio-cultural fabric, and the quality of the visitor experience. In this study, it is used to assess the limits of tourism growth in relation to housing and public infrastructure.

Touristification: The process by which a residential area or city is progressively transformed to cater primarily to the needs and consumption patterns of tourists. This often leads to the displacement of the local population, the alteration of the traditional commercial fabric, and a loss of local identity.

Urban Resilience: The capacity of an urban system, community, or society to maintain its essential functions, identity, and structure in the face of shocks and stresses. In the context of this study, it refers to the ability of Machico to absorb tourism growth without compromising its housing affordability and quality of life for residents.

Chapter 8
AI and Data-Driven Approaches for Sustainable and Inclusive Transport Networks

Sitesh Kumar Singh
http://orcid.org/0000-0002-7108-0808
College of Engineering, National University of Science and Technology, Oman

Jaishri Gothania
http://orcid.org/0000-0002-2656-642X
Lingaya's Vidyapeeth, Faridabad, India

Abby Balkis Gorayah
http://orcid.org/0009-0006-7326-2249
College of Engineering, National University of Science and Technology, Oman

H. R. Shyam Agasthya
http://orcid.org/0009-0001-9382-9971
College of Engineering, National University of Science and Technology, Oman

ABSTRACT

This chapter examines how Artificial Intelligence (AI) and information data-driven methods can be transformative in the realization of sustainable and inclusive urban transport systems. It studies some of the core principles of fair mobility and compares the conventional approaches to transport planning with AI-enhanced solutions. Some of the main uses are the surveillance of traffic, the anticipation of demands, cross-modality, low-carbon and accessibility improvements for the elderly, the disabled, and underprivileged groups. The Singapore, Amsterdam, Copenhagen, and Bengaluru

DOI: 10.4018/979-8-2600-3038-7.ch008

ht © 2027, IGI Global Scientific Publishing. Copying or distributing in print or electronic forms without written permission of IGI Global Scientific Publishing is prohibited. Use of this chapter to generative artificial intelligence (AI) technologies is expressly prohibited. The publisher reserves all rights to license its use for generative AI training and machine learning model development.

case studies give a real-world influence and shed light on efficiency improvement, environmental issues, and social inclusion. Lessons focus on adaptive learning, data integration, and implementation in context. Future considerations are also identified in the chapter, detailing how an AI-based system for transporting the population will work in the future, including on a more sustainable and inclusive scale.

1. INTRODUCTION

1.1 Urban Mobility Challenges

The urbanization trend is changing transport at the global level where more than half a billion of world population is already living in the cities, and it is predicted to ascend to nearly 70 percent in 2050 (United Nations, 2019). The upcoming urbanization has intensified already existing transport issues in the city and changed to a greater number of road congestion, air pollution, road safety, and unfair access to mobility services. Developed and developing economies must deal with cities which experience a high amount of traffic compared to the available infrastructure capability consequently colossal delays, losses and environmental damages. In fact, the estimates provided by International Transport Forum (2021) research put the traffic generated by the cities alone at nearly 2-4% of GDP per year due to congestion.

To this it is added that the urban transport systems are increased as far as the environment is concerned. Almost one-fourth of the total CO 2 emissions are comprised of transportation by road, and a significant portion of the emissions is emitted by personal vehicles (International Energy Agency, 2022). The dependency on the means of transportation which is mostly dependent on fossil fuels aggravates climate change and that is why sustainable mobility is burning policy issue. It is further aggravated by the fact that in most of the developing countries, the transport infrastructure is in poor state, and the informal transport system has emerged thus forcing people to adopt a more connected yet dangerous and inefficient mode of transportation which has made transportation in most countries to an aspect of disjointedness and precariousness.

Of importance is also the issue of transport inequality. The systematic barriers to the provision of high quality, comfortable, and safe services of transport services are likely to affect weaker audiences (those with very low incomes, the elderly, disabled people). This kind of disparity prevents their mobility and incomes, employment, education, and medical care. Singh et al. (2025) think that radical shifts in the organization of urban transport need to be leveraged because the existing strategy to expand the infrastructure is not effective enough to manage the increasing complexity, and artificial neural networks and other AI may find application in solving the problem.

Hence, the current issues of urban transport are not only in the number of facilities that have been provided to support the functions of the transport infrastructure, but in the reconfiguration of mobility to make it more inclusive, resilient and environment friendly.

1.2 Need for Sustainable and Inclusive Networks

The trend under consideration of sustainable and inclusive transporting nets is not a technological shift but rather a social, financial and ecological need. Green mobility frameworks consider the significance of minimization of carbon emissions, modal shifting towards the utilization of public and non-motorized transport, and maximization of the utilization of resources. Enhanced access to sustainable transport to everyone particularly the vulnerable to the city is an important element that is noted in the Sustainable Development Goals (SDG 11: Sustainable Cities and Communities) of the United Nations (Martin, 2025).

The mobility systems are responding to the mobility needs of all these masses of people in terms of age, income, sex and physical conditions to access safe and affordable mobility through the concept of inclusiveness in mobility systems. It was also found that transport exclusion often leads to social exclusion, which limits the individuals in their ability to have full capability of economic and social interaction activities (Lucas, 2019). The concept of inclusivity should therefore be regarded as the long overdue aspect to transport planning, which should be taken care of alongside sustainability.

Traditional transport structures that are concerned with the flow of motorized modes do not optimize these mutually related goals. Car-centric planning models have been changed by the provision of new road capacity and infrastructure that promotes the notion of inequality founded on the investment of the long-term use of available modes of transportation and pedestrian amenities. With this imbalance there is the reinforcement of inequities, and the cities are locked in unsustainable mobility patterns.

In this respect the strategy of sustainable transport targets three related priorities:

- **Decarbonization:** The reduction of the reliance on the transportation of fossil fuels.
- **Integration:** The interconnection of multiple networks of mobility.
- **Accessibility:** Taking care of the mobility of every group of individuals.

The climate pressure, the growing traffic jams, and the social inequality are just a few of the things that make the necessity of developing the inclusive and intelligent transport solutions that will bring urgency to the situation. A solution should be data-driven and AI based to provide a revolutionized course to fulfill these dreams by supporting smarter, dynamic, and user-friendly mobility infrastructures.

1.3 Role of AI and Data-Driven Methods

The recent tendency in the reconstruction of urban mobility has been the use of Artificial Intelligence (AI) and data analytics as disruptive facilitators. Transport systems generate massive amounts of real-time data in the form of sensors, GPS systems, mobile applications and linked infrastructure. Applied to AI techniques, i.e. machine learning, deep learning, and neural networks, this data can be used to provide powerful incentives to optimize traffic flows, predict travel demand, and tailor mobility services.

AI can be used to manage traffic: the current state of traffic will be monitored, the adjustment of the timing of traffic lights may be carried out in real-time, vehicles may be redirected, and traffic jams may be anticipated. Similarly, demand forecasting with the use of artificial intelligence will allow transport planners to adjust the resources to the actual travel needs, therefore, increasing the efficiency of the services rendered. Jain et al. (2025) note that the combination of Internet of Things (IoT) and AI in ensuring sustainable and safe transportation could be helpful to predictions of any emerging ethical and societal concerns and guarantee greener and safer transportation.

Moreover, evidence-based tools are the key factor in the development of welcoming mobility models. The AI can inform and even distribution of the available public transportation resources through planning the impoverished areas and through socio economic data to determine whether the system is accessible to the poor. The predictive modeling can also be utilized in assessing the infrastructure stability regarding climate stress to make sustainable investments.

The use of AI (a) fleet optimization and (b) route planning to take climate action helps to reduce emissions significantly. Using the example of electrified fleets through AI-powered energy management systems, the carbon intensity can be lowered to obtain the growing number of urban mobility (Wang and Yang, 2023).

In such a way, the introduction of AI and data-driven approaches to transportation can no longer be seen as a matter of operational efficiency alone, but instead, as a strategic necessity, to achieve sustainable, comprehensive and durable urban transportation systems.

2. FOUNDATIONS OF SUSTAINABLE AND INCLUSIVE MOBILITY

2.1 Principles and Goals

The idea behind inclusive and sustainable mobility systems is that mobility systems should be both environmentally, socially equitable and economically efficient simultaneously. The concept of transport to facilitate economic growth has grown to be understood as broader awareness of resources as a method of advancing social inclusion or ecological outcomes and urban sustainability (Gebresselassie & Sanchez, 2019).

The main principles of sustainable mobility are:

- **Environmental sustainability:** It is a highly sensitive issue, but we will minimize the emission of greenhouse gases and will not be so reliant on fossil fuels and will promote the use of energy-efficient transport technology.
- **Equity and inclusion:** Enabling accessibility of transport systems to all and marginalized groups in the population, women, children and elderly.
- **Adaptability and resilience:** Developing transport networks which are resilient to shock-related to climate change, pandemics or economic shocks.
- **Transport modes integration:** The facilitation of multimodal means of transport that include transportation in the field of public transport, bicycle and walking mobility and the integration of new mobility services into single networks.

Efficiency and innovation: With the help of digitalization, data analytics and AI are used to make mobility services and infrastructure use more efficient.

The purpose of inclusive mobility is not only physical connection. They are oriented to reduce access inequality, offer equal opportunities to jobs, education and health and integrate social integration. Sustainable mobility must also improve overall accessibility as an inherent, rather than an added value aspect by addressing the well-being of citizens (Elvas & Ferreira, 2021).

This multi-dimensional approach, which has become identified in the agendas of the world and is reflected in the Paris Agreement, the New Urban Agenda, UN Sustainable Development Goals (SDGs) and other international agendas, is in line with the overall world agendas.

2.2 Traditional Transport Planning Approaches

Transport planning, in the past, has always been laden with infrastructure-based and automobile-oriented solutions. The traditional measures that were aimed at increasing the issues of urban mobility were the conventional strategies that focused on overbuilding roads, traffic engineering, and capacity maximization as the primary measures. Even though these actions resulted in alleviation of the congestion in the short term, in most instances, they resulted in creation of other demands, a self-dependence cycle of privately owned cars persists.

These are the strategies that have not considered imperative elements of inclusiveness and sustainability. The immense emphasis on car travel marginalized the pedestrians, cyclists, and those who used bowers, and this left the traveling systems disjointed and maleficent (Cervero, 2014). The upper-income groups were targeted as they own a private vehicle to the detriment of the low-income and vulnerable groups in terms of investments.

The other shortcoming of traditional planning is that it consumes outdated information and forecasting mechanisms. The old surveys and aggregate data that contain travel demand models could not capture dynamic travel behavior variation particularly in a highly urbanizing environment. These models barely incorporated socio-economic disparities, gender, or disability, a factor that helped the solutions in transport to generate inequalities.

In addition, the conventional approaches did not consider the environmental externalities in many cases. Road enlargement processes led to urban sprawl, energy consumption, and emission. This contributed to the fact that transport systems occupied their niche in the climate change process and air pollution exacerbating the issues of sustainability in urban areas (Babb, 2016).

Though traditional planning was the foundation of the modern system of transportation, it is increasingly becoming clear that it is not sufficient in addressing the complex relationship of sustainability, equity and resilience of the modern cities. This has facilitated the switch to data-driven and AI-based mobility planning that would be able to adapt itself to the dynamically shifting demands.

2.3 Shift Towards AI and Data-Driven Solutions

The flaws of traditional approaches are the ones that have contributed to the shift to AI and data-based transport planning which puts more emphasis on real time responses, predictive analytics and user centered design. The development of connected devices and sensors is turning transport systems into large data systems like never before, and transport data has never been as rich as it has the potential to be used to have new visibility into mobility trends.

AI-based means:

- **Dynamic traffic control:** Using machine learning to control signals, traffic diversion, avoiding traffic jams.
- **Predictive demand forecasting:** It operates based on past information and streams information to predict the number of passengers and offers the maximum number of services.
- **Fair planning:** Planning based on socio-spatial data to comprehend the unmet needs in communities and targeting to aid the equitable policy action.
- **Enhancement of sustainability:** Use of optimization techniques to reduce emissions, encourage electrification and convert renewable energy to transport systems.

The process of AI integration will also allow making the transition to the Mobility-as-a-Service (MaaS) ecosystems where the several modes would be interconnected via the digital platform seamlessly. The new orientation is on the issue of flexible and personalized and service based solutions to mobility issues as opposed to structural solutions-based issues.

Speaking of the ways in which AI and IoT can be utilized in transportation, I would like to note that the two can not only help in making the transportation service more efficient but also address the ethical and societal issues of equal opportunities and use of resources. Still on the same note, Sietchiping et al. (2012) indicate that big data analytics and AI have become a necessity to ensure that the twin goals of sustainability and inclusion in urban mobility system can be realized.

This change may be regarded as a subset of a broader paradigm shift: it was grounded on supply-based, infrastructure-centered planning, and now it is data-based and people-centered and responsive transport systems. The application of AI in mobility planning can transform transportation in cities into an inclusive and resilient arrangement due to its new fusion of technological advancement and sustainability, inclusion and accessibility aspirations.

3. AI APPLICATIONS IN TRANSPORT SYSTEMS

Artificial intelligence (AI) is quickly transforming the way transport within the city is occurring, enabling cities to reduce the complexity of their movement in a customized way. The congestion of cities has led to congested traffic jams, high commuting speed and environmental waste and this has been a challenge that the traditional practice in the transport planning has not been able to balance well. The AI offers an opportunity to operate with a significant volume of real-time data, detects certain meaningful patterns and an optimization of the decisions that will provide the maximum throughout the transport network.

The AI also can manipulate and process non-homogenous data, which includes traffic sensors, GPS signals, video feeds, social media updates, and weather reports, the possibility of which is one of the considerable advantages of the AI (Zhang, 2025). By using machine learning and deep learning methods, transportation authorities can predict traffic jams, use more efficient route planning and deploy tactics to manage traffic lights including the mitigation of delays. Besides traffic, AI is used to organize multimodal transportation, whereby it is possible to use buses, train, bikes, and sharing mobility services as a single convenient lower-carbon travel experience.

In addition, sustainability and inclusiveness increase because of the AI-based transport systems. Integrated AI within the low-carbon plans, like electric vehicles (EV) or freight/public transportation routes planning, is made possible through the modeling of network energy consumption and emissions. It also enables a justifiable use by localizing the underserved accommodations and informing the resource on the provision of sound visual mobility to the aged, disabled and the marginalized people.

Overall, the strategy to combine predictive analytics, real-time monitoring, and intelligent decision-making that AI displays is revolutionizing the transport network in urban areas. It is not only offering efficiencies in operations, but also a way of making mobility networks sustainable, diverse and resilient that would be flexible to the evolving needs of the modern cities.

3.1 Traffic Monitoring and Management

Traffic monitoring systems which are AI-based guarantee the optimal traffic flow and reduce congestion through advanced algorithms and real-time data analysis. Computer vision algorithms, such as Convolutional Neural Networks (CNNs), are applied to stream of video images of the cameras mounted on roads to give the identification of vehicle types, traffic congestion and vehicle accidents (C. Wang et al., 2025). This data is subsequently fed into the adjustment of the timing of the traffic lights in such a manner that is optimal to the overall traffic. Table 1 depicts the most significant Artificial intelligence application in the urban transport systems, the objective, sample technologies and worth of application in the purpose areas. The table is a summary of the application of AI in traffic monitoring and management, demand forecasting, multimodal integration, low-carbon planning and improvement of accessibility. The extensive view of the potential of AI to increase efficiency, along with sustainability and inclusiveness of the contemporary transport networks, as entailed in Table 1, is achieved by connecting some of the technologies to operation-related objectives.

Table 1. Key AI applications in urban transport systems

AI Application Area	Purpose	Example Technologies	Benefits
Traffic Monitoring & Management	Reduce congestion, optimize traffic flow	Computer vision, IoT sensors, traffic cameras	Shorter travel times, reduced emissions
Demand Forecasting & Optimization	Predict transport demand and allocate resources	Machine learning models, neural networks	Efficient vehicle deployment, cost savings
Multimodal Integration	Coordinate public transport, cycling, walking, micro-mobility	Data analytics platforms, mobility apps	Seamless travel, reduced private vehicle dependency
Low-Carbon & Energy-Efficient Planning	Promote sustainable transport	AI-based energy optimization, route planning algorithms	Lower carbon footprint, improved sustainability
Accessibility & Inclusion	Enhance mobility for all, including vulnerable populations	AI-assisted navigation, predictive services	Equity, social inclusion, better access to services

As Figure 1 indicates, partnerships in urban transportation systems pivot around AI technologies that connect the usages of traffic monitoring, the forecast of demand, low carbon initiatives, and the multimodal scheduling and the enhancement of new accessibility.

Figure 1. AI integration in urban transport systems

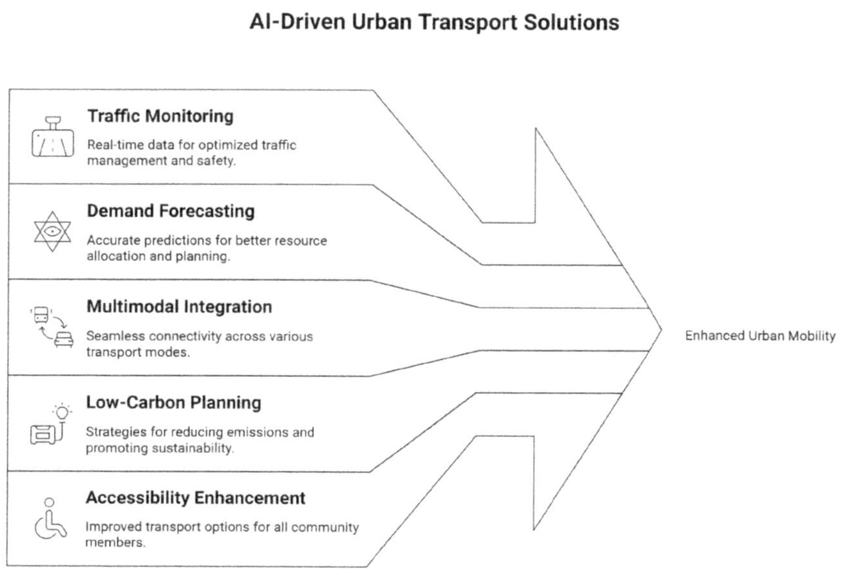

Indicatively, AI-driven adaptive traffic signs have been deployed in Chennai, India, and already indicate a tremendous change in traffic control. These systems can alter the signal phase in real-time basing on the traffic, and therefore, the drivers are not waiting as much, and the congestion of large intersections is reduced (India, 2025). These examples demonstrate the potential that AI can bring to the mobility of the cities by making them more sensitive and efficient to traffic control.

Moreover, Artificial intelligence serves to predictive maintain traffic infrastructure. Based on past data and sensors, AI models can predict when traffic lights or road surfaces may require repairs and can enable taking proactive measures before the damage is done, reducing the downtimes and prolonging the service life of infrastructure.

3.2 Optimization and Demand Forecasting

Transportation planning and resources investment require proper forecasting on demand. Particularly, machine learning algorithms in the form of long short-term memory (LSTM) networks can be used to handle the time-related trends of transportation data to predict future demand. These models can handle lots of data including historical ridership, weather condition, special events and with great precision forecast the number of passengers.

The AI-assisted demand forecasting tools were applied in Mobility-as-a-Service (MaaS) platforms in Singapore where the dynamically planned and routed transportation services of a service provider, namely public transportation, are scheduled (Huiling and Goh, 2017). These applications optimize the service frequencies and routes on real-time basis as per the estimated demand in such a way that the resources can be put into use and the passengers also need not wait too long. The case of such applications highlights the importance of AI in the design of adaptive transportation systems that can react to dynamically changing needs.

Also, AI can be utilized in optimization of multi-mode transportation networks. By processing information in various forms of transport such as buses, trains, and ride-sharing companies, AI systems will guide individuals on how to travel best, or the least time and cost as well as the environmental effect. This broad approach to thinking of transportation prompts individuals to travel with sustainable means to bring the use of personal cars to near extinction.

3.3 Multimodal Integration

The integration of various modes of transport into an adequate ecosystem is the clue to the sustainability of urban mobility. AI has assisted in integration, which provides users with an opportunity to unite data of various transportation services and plan the movement and pay easily.

Its examples include multimodal integration enabled by AI, i.e. Whim in Helsinki or SimplyGo in Singapore (Hub, 2020). Through such platforms, users can manage travel among multiple modes of transport such as buses, trams, trains, and bicycles and pay for amenities with a single interface. The AI algorithms analyze live data to suggest the most possible and sustainable travel paths, considering the traffic condition, availability and preferences of users.

Secondly, the AI can facilitate the optimization of shared mobility services such as biking and carpooling. The AI systems will allow the operators to shorten the waiting time to their users by predicting the location and time when such services will be needed based on the analyzed pattern in user behavior and demand and what services will be needed to place the vehicles strategically. The efficiency of shared mobility services is increased in this dynamic assignment, and it stimulates the use of services instead of the use of personal cars, and the implementation of sustainability solutions.

3.4 Low-Carbon and energy-efficient planning

The solution to the planning of low-carbon and energy-efficient transportation is AI. With AI models, the reduction of emissions is possible because routes could be devised using traffic sensors, weather forecasts among other data sources to generate fuel-efficient routes.

Case of making transportation system smart by using AI and Internet of Things (IoT) technologies where minimal energy will be consumed and carbon emission reduced is interesting. In a study published in Multimodal Technologies and Interaction, researchers developed an AI-based framework that supports traffic control and energy efficiency, grounded on the synthesized information on variables such as traffic density and consumer preference. The system demonstrated energy savings of 15 percent and CO2 reduction of 10 percent compared to the traditional methods (Reis, 2025).

Moreover, AI can be applied to enable the transition to electric vehicles (EVs), simplifying the fast-charging infrastructure and integrating EV into the current transportation system. The charging demand forecasting by AI-supplied algorithm enables optimization of the standard of charging the EVs and efficiency of the charging infrastructure utilization since they can be charged in accordance with the parameters of traffic flows and availability of charging stations.

Another area of application of AI can be freight transportation whereby it will help make the process sustainable in terms of optimal routes and schedules. Traffic situations, delivery schedules, capacities of vehicles, etc. are some of the data sources that machine learning algorithms use to compute routes with the lowest fuel consumption and emission levels. The solution does not only ensure that freight transport becomes more environmentally friendly but also assists in ensuring that transportation is made more efficient.

4. AI FOR ACCESSIBILITY AND INCLUSION

Inclusive mobility is aimed at ensuring that the transport networks are accommodative to all people living in the city, including the aged and physically challenged among others who are ordinarily part of the marginalized groups. Traditional transportation planning does not consider the mobility needs of all these groups and introduces a gap in service coverage, challenges in access, and social differences. In such challenges, Artificial Intelligence (AI) has emerged as a revolutionary tool to handle them and enables cities to plan their transportation systems to accommodate them because it is flexible, equitable, and efficient (Singh, Alasali, and Belbase, 2025).

The AI technologies permit the collection and analysis of big real-time information, including travelling patterns, ridership, demographics, etc. Such data-driven approach to work will help locate the spheres of underserved areas in which children may rely on the use of public transportation or safe walking paths. ML has the ability of distributing services in the most stable manner, whereby resources are found where they can be exploited in the most advantageous manner and the resources that require distribution to the vulnerable groups travel less distance.

Also, AI-based systems, such as smart navigation routes, intelligent time scheduling of paratransit, and smart maintenance of available infrastructure enhance the entire experience of the different people with special needs who would use mobility services. The usage of AI in planning the urban transport will enable the city to create systems that will not only optimize their level of operational efficiency but will also create the social inclusiveness and minimize the level of inequalities and can help to achieve sustainable urban development as well.

4.1 Mapping Underserved Areas

Taking advantage of or evading a shortage because of transportation is a fundamental card in designing fair and inclusive mobility systems. The urban residential neighborhoods, particularly in the peri-urban regions or economically disadvantaged areas, lack transport infrastructure and by virtue of this they take longer time to travel and inaccessibility to jobs, healthcare and education, which graphically translates to them being outcasts in society. Artificial Intelligence (AI) is among such tools that offer delicate tools to plan to detect and analyze such gaps (Esri, n.d.). An amalgamation of geo-spatial information, census data, land use surveys and real time travel activities enables AI to develop finer maps of transportation desert and underserved areas.

By comparing residency histories, machine learning can detect the problematic spatial mobility patterns (i.e. untimely, crowded, or unreliable service) (Site et al., 2022). These data will help plan the transportation in the city and will give us the opportunity to make the best decision about the implementation of the services, such as bus service, bike-sharing, or micro-mobility. Additionally, predictive analytics would enable the future mobility demands to be realized based on population dynamics, urban transformation and socio-economic changes, and emphasize proactive rather than reactive transport measures. The conceptual mapping of underserved areas, as shown in Figure 2, provides an idea of how AI can be used to identify neighborhoods with poor access to transport and determine the mobility gaps so that they can be used during the inclusive planning process.

Figure 2. AI-driven accessibility mapping

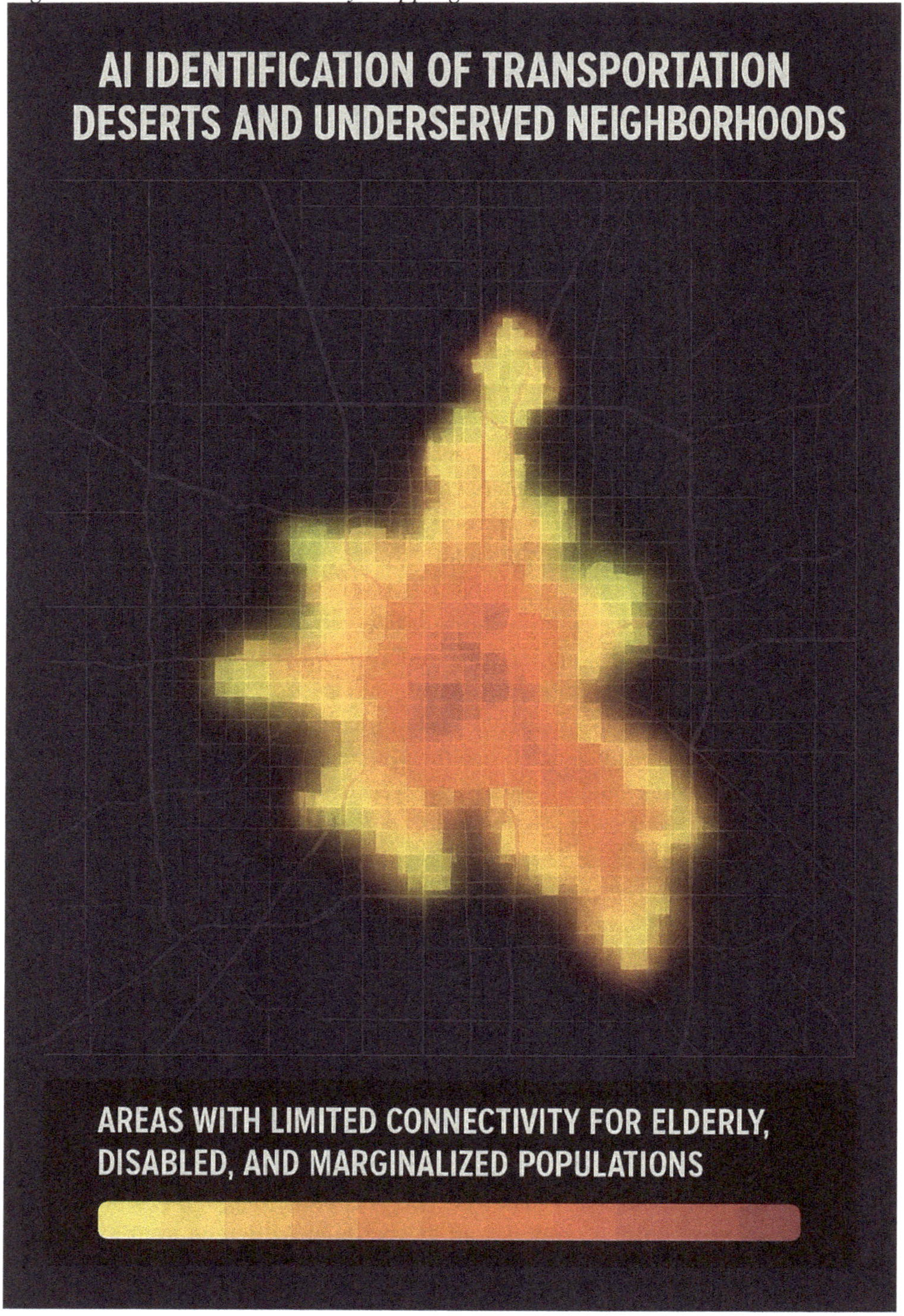

Due to the assistance of the AI technology in mapping under-served places, there is an opportunity to make the data-driven decisions made by the cities to achieve the ability to get access improved, reduce travel inequalities, as well as improve the overall social inclusion. It does not only support vulnerable populations in such ways but also creates more sustainable and efficient urban transport systems (Cina et al., 2025).

4.2 Inclusive Mobility for the Elderly, Disability, and Marginalized

Ensuring fair access to the urban transport by non-working adult generations, the disabled and disadvantaged classes of the society is also an important step. The traditional transport systems fail to meet the special needs of these customers entirely, which in turn causes several problems with employment and education, healthcare and social engagement. Adaptive solutions can eliminate such problems with the aid of Artificial Intelligence (AI) and become the means of providing individual, responsive, and efficient transport services (Zemmouchi-Ghomari, 2025).

Personally customized path directions in real-time, including that of an AI-managed navigation application, could be offered to those with visual, hearing, or mobility impairments, to mention but a few. Such tools may indicate such readily available paths or indicate an available opening that lacks steps or simply provide sound to locate which is the safest and most convenient path that disabled users could use, which might significantly improve the freedom and safety of disabled users (UNRIC, 2024). Besides navigation, AI can also assist in simplifying paratransit as well as ride-sharing agencies to reduce waiting time, match riders with cars that can accommodate their mobility requirements and make their movement seem like a door-to-door experience.

More than that, the AI systems, always based on prior travel patterns and reaction on the side of the user, it is possible to enhance the type of service provided over time. Another illustration is that predictive analytics can anticipate the time when travelers are most likely to be using the mode of transport (the aged and disabled) and make the frequency of service in a manner that will reduce congestion and queuing to the smallest possible. It is also AI that enables integrated systems to integrate different types of transport (e.g., buses, shared vehicles, other forms of micro-mobility) into one network, which is concerned with issues of accessibility and vulnerable populations. Having this potential, urban transportation systems can become inclusive, responsive, and socially equitable in providing services (Sustainability Directory, 2025).

4.3 Equitable Allocation of resources based on data

Equitable allocation of transportation means is quite essential to the creation of sustainable and inclusive mobility networks. AI plays an essential role in the process since it enables decision-making on the data and traffic flows, ridership and demographic patterns analysis. Advanced analytics helps the planners to learn where the people require it most and when people will require its services and consequently utilize vehicles, infrastructure and services efficiently (Jain, Upreti, et al., 2025).

To illustrate, AI will be able to adjust the frequency of buses, capacity of vehicles and the locations of the station to serve demand in underserved neighborhoods. This will ensure that those communities that were underprivileged to receive quality transport services (previously) are fairly served, as well as spatial and social disparities are reduced to the minimum. Under predictive modeling, the planners can explore theoretical interventions (e.g. new transit lines or shared mobility hubs) and learn about the consequences of their impact on accessibility, environmental sustainability and social inclusion long before implementing the interventions. Figure 3 is the AI-based decision-making approach to the fair distribution of resources.

Figure 3. AI-based decision-making for equitable resource allocation

The allocation of resources with the help of data also promotes the efficient utilization of new mobility technologies, including electric buses, bike sharing systems and autonomous vehicles. The analysis of the trends in travelling and demand can make routes and schedules smoother and use less energy, enhance the work of low-carbon transports projects using AI. In addition, the services should also enhance their efficiency and performance with the assistance of AI, which will provide the opportunity to regularly monitor the situation and receive comprehensive insights that will contribute to the increased maintenance, safety, and satisfaction of users.

Combined, the applications provided reveal that the integration of AI into the realm of accessibility planning is not only effective in the context of making operations more effective but can also increase social equity. Three primary ways cities can create inclusive, sustainable, and resilient transport systems include mapping underserved areas, changing the mobility services to reach vulnerable populations, and redistributing resources based on the knowledge informed by the data. AI can therefore be the central part of urban mobility, as the way to meet the intricate needs of every citizen, reduce the distance and make cities greener, fair and more transparent.

5. CASE STUDIES AND GLOBAL PRACTICES

A discussion into the existing applications of Artificial Intelligence (AI) in transportation would come in handy by informing the cities on how to build smarter, more efficient and sustainable transportation systems. The practical conditions and possibilities of integrating AI in the transport networks can be learned when urban planners, policymakers and researchers attempt to analyze global case studies. The examples of successful cities where solutions utilizing AI are actively used have demonstrated real gains in control over traffic, functioning of the network of mass transport services and ensuring the environmental safety and made the transportation network more approachable to the population of other backgrounds.

One such use of AI technologies in transportation is the application of real-time information that can be accessed on sensors, GPS, smart cameras and user applications, to optimize processes and predict traffic jams and improve service quality (Zemmouchi-Ghomari, 2025). Besides efficiency in its operations, AI in designing an inclusive mobility solution is meant to access underserved regions and vulnerable population demand predicting and facilitating equal distribution of resources. Further, the presence of adaptive learning systems that would become more competent with subsequent assessment due to the data on the historical and real-time processes is also demonstrated in such case studies, thus, rendering the transport networks sensitive to the new needs within the city.

The comparison of different settings such as crowded urban centers and even mid-tier towns indicates that it is possible to adapt AI to the local infrastructure and population, social-economic situations. Such analyses propose lessons that are helpful regarding replica, scale, and the implementation of transport systems, which are not only smart but also socially and environmentally sustainable.

5.1 Examples from Cities

5.1.1 Singapore: Pioneering AI-Driven Traffic Management (LTA | Intelligent Transport Systems, n.d.)

Singapore is a city and urban transportation pioneer in the application of Artificial Intelligence (AI). This Intelligent Transport System (ITS) in the city-state is seen as a system of hundreds of sensors and cameras with an effective road tunnel and expressway network (over 160 kilometers) to measure and monitor the traffic flow, travel times and road demand. This is an empirical plan that enables practical real-time optimization of traffic and reduces congestion and enhances safety of commuters.

Of great significance is Cooperative and Unified Smart Traffic System (CRUISE) that is being developed by the Agency of Science, Technology and Research (ASTAR). CRUISE involves the use of smart sensors and communication technologies, and artificial intelligence predictive potential to create an island-wide smart traffic-light network that will control the traffic of vehicles and pedestrians on the island in a smart manner.

In addition, dynamism in the toll setting in Singapore Electronic Road Pricing (ERP) has the capability to align the adjustment of the toll fees to the real traffic conditions, hence minimizing congestions. These projects indicate that Singapore is one of the nations that embrace the successful and sustainable mobility of urban areas with the application of AI.

5.1.2 Amsterdam: Multimodal Transport with AI

The example of Amsterdam gives an idea of how AI was incorporated into multimodal transport systems, and the emphasis was put on efficiency and sustainability. This bike-sharing initiative has turned out to be one of the iconic programs in the transportation network in the city and would highly benefit through AI-enhanced demand prediction models. The weather, the time of the day, and historicity in the history of usage in determining the ideal availability of bikes in the stations are among factors that these models take into consideration (Yin et al., 2025).

The use of AI algorithms can also assist in scheduling the public transport according to the demand prediction, which offers employment in an energy-saving mode and reduces the time of waiting. With the analysis of mobility trends, Amsterdam prioritizes the paths of bicycle users and the public transportation users, reducing the dependence and the amount of carbon released on the streets. Such a humanistic theme reveals how AI can be easily integrated into a collaborative life in a city mobility environment.

5.1.3 Copenhagen: AI to improve accessibility

Copenhagen has also employed AI to enhance access and inclusivity towards the urban transport system. AI applications map unserved and underserved areas and predict calls of old people and individuals with disabilities. These considerations assist the city in the process of modeling transport systems that will support all urban dwellers despite their age, disability, and marginalization (Vujadinovic et al., 2024).

Moreover, Copenhagen demonstrates its interest in sustainability, which is evidenced by its extensive cycling network and reliable 24/7 transportation that is behind the enormous rating of this city in urban mobility preparedness. Such efforts testify to the inclusion by Copenhagen towards coming up with an inclusive transport system that is both green and environmentally friendly.

5.1.4 Bengaluru: Intelligent Transportation Management System with the help of AI

AI technologies are now being used in Bengaluru to become one of the pioneers in controlling the complexity and demand of traffic within the cities. The Bangalore Traffic Improvement Project (B-TRAC) is an urban project deployed to the city to implement adaptive traffic signal control using AI to make the city traffic of over 300 intersections decrease peak-hour congestion and improve flow in the traffic. According to TrafficInfraTech - Editor, (2025), the preliminary findings show that the travel time is reduced by approximately 15 percent, which can be attributed to the more-or-less real time changes in the signals based on the increased or decreased density as well as changes in the travel patterns. This is done with the connection of sensors, cameras and IoT devices to the AI learning processes that identify and adjust to the changing traffic patterns.

What is even greater, the authorities of Bengaluru, jointly with C-DAC have developed extremely advanced models which combine data that is accumulated in GPS and traffic cameras and compare rides-hailing services. These models can forecast congestion days in advance to occur and optimize the dynamic route of both the emergency and the public transport. One such pilot AI system developed by NEC is one that provides real time information based on automatic number recognition and detection of red-light violations and where it is already proposed to incorporate such intelligence into a city-wide Intelligent Traffic Management System (ITMS).

In summary, the Bengaluru instances of AI-driven traffic management solutions through the magnitude and complexity of the traffic issues in the rapidly growing cities serve as an example of a practical and the expanding solutions. They offer very much as learning opportunities in the undertaking of predictive analytics, supportive regulation, inclusive and demand receptive planning of infrastructures, and all these procedures are crucial towards reaching reliable and sustainable urban movement systems.

5.2 Conclusions and Discoveries

The case studies of Singapore, Amsterdam, and Copenhagen have proven the radical ability of AI in transport structure of cities. Technologies of artificial intelligence give cities a challenge to simplify traffic movement and reduce traffic jams and maximize efficiency in activities with the help of predictive analytics, real-time monitoring, and optimization of resources. Such data-based solutions to AI advantageous include CRUISE system and Electronic Road Pricing in Singapore to preemptively manage traffic to reduce congestion, and Amsterdam adopt AI-enhanced multimodal transport networks to manage cycling, public transport and ride-sharing networks. Similarly, there is an AI-driven accessibility planning in Copenhagen that generates equilibrium between the elderly, disabled, and underserved groups and represents the benefits of an inclusive transportation policy in society (OECD et al., 2020; Vujadinovic et al., 2024). Table 2 gives examples of AI applications in the field of urban transport, a global sample of proved results, which contains the names of the cities, technologies of AI use, and desired results and the attained results. The table provides real-life examples of Singapore, Amsterdam, Copenhagen, and one city in India in which AI has made traffic more efficient, accessible, sustainable and convenient to use. Table 2 will enable the comparison between the contextual differences and strategies and provide an insight into the future transport planning and policies of AI interventions and lessons that ought to be learned.

Table 2. Global case studies of AI in transport

City	AI Applications	Outcomes	Key Lessons
Singapore	Predictive traffic management, mobility-on-demand	Reduced congestion, optimized public transport schedules	Importance of integrated data, real-time monitoring
Amsterdam	AI-driven cycling infrastructure, demand prediction	Lower private vehicle use, energy-efficient transport	Multimodal focus, environmental sustainability
Copenhagen	Accessibility mapping, AI-based micro-transit	Equitable access for elderly and disabled, low emissions	Inclusive planning, adaptive services
Bengaluru (India)	Traffic signal optimization, predictive congestion, ride-sharing platforms	Improved traffic flow, reduced commute times	Localized AI solutions, context-specific implementation

These are some of the key lessons learned in these global practices.

- The first requirement is the need to unify different sources of data, i.e., GPS and IoT sensors, demographic information and mobility trends to offer adequate model and decision-making.
- Second, AI-related policies and algorithms must be transparent, so that the population can trust the policy and algorithms, particularly when it comes to sensitive issues related to accessibility and prices.
- Third, community engagement and participatory planning offer the opportunity that the AI solutions may ground on the needs and priorities on a local level.

Further, this learning may be based on historical and real-time facts and thus transport systems can be continuously modified according to the changes in demand and anticipate future changes in the city. The Environmental impact of the measures can be assessed by cities, with the decreasing use of personal vehicles, the maximum level of water and energy consumption of the cases of public transport, and the encouragement of the low-emission mobility utilities. Overall, these findings re-establish that AI-based transport planning is not only useful in the light of augmenting operational efficiencies but also enables sustainability, inclusivity and better urban mobility network.

5.3 Potential for Replication in Other Contexts

The experience gained due to the successful example of AI implementation in the Singapore, Amsterdam and Copenhagen transport system can be educative to other cities worldwide that want to enhance mobility, accessibility and sustainability. However, any efforts to implement said AI-based solutions locally require a significant focus on the local conditions, consisting of such factors as the level of infrastructural development, the density of the population, the restrictions put on the terrain, the social and economic inequalities, the shapes of the existing transportation patterns. An example of a highly congested city with limited access to mass transit might be interested in AI-assisted traffic control and mobility-on-demand, but a cycling-centric area in the region may be interested in AI assisted multimodal integration and route optimization.

Developing sound data collection and management systems should be one of the most significant processes to take place to replicate. GPS, IoT sensors, and citizen reporting systems give high-quality real-time information on which AI-based analytics is based on predictive modeling and dynamic resource allocation (Saki and Soori, 2025). The interaction with governmental institutions is planned to be structured by the means of the public-private approach to the investments in AI technologies, expertise sharing, and innovative mobility services. Transport authority, technology providers and community members will ensure that AI solutions are technically possible, socially acceptable and will address long term objectives about urban planning.

The pilot programs are the cheapest approach of experimenting AI interventions on small and fixed scale before expanding it to the city. It is advantageous since these pilots can empower the planners to evaluate the efficacy of the system and identify any operational challenges and modify the solution in accordance with the local contexts. Research into the tenets of inclusivity would be useful in ensuring that the weak sectors of the population, the old, the infirm or the inhabitants of the underserved regions will not lose out in the process of enjoying the mobility advantages that AI provides. Focusing on the scalable, context-based, and sustainable model, cities of various size and resource supply will be able to implement AI to make their transport system more efficient, reduce its environmental impact, and improve the quality of the entire urban mobility in general.

6. CONCLUSION AND FUTURE DIRECTIONS

6.1 Conclusion

The transportation systems have been transformed using artificial Intelligence (AI) and data-driven technologies to enhance urban commuting and present a new solution to old problems of time-of-day traffic/mobility congestion, inefficient operations, environmental damage, and bad access and accessibility. All AI technologies are machine learning, predictive analytics, and real-time monitoring that enable the optimization of traffic patterns by cities, enhance the work of public transport, and aerodynamics the travels of citizens.

When examining the practices globally, it might be observed that AI could be utilized to create more sustainable and inclusive transport systems that can be realized by ensuring that the multi-modal integration is facilitated, low-carbon mobility is promoted, and ensuring that the underserved populations have equal access to the transportation services. Singapore, Amsterdam, and Copenhagen are among such cities that develop a competitive edge due to data-driven knowledge, smart algorithms, which inform strategic planning, dynamically allocate resources and address real-time needs of the city. All these illustrations indicate that AI has the potential to not only guarantee efficiency in working but also promote the sustainability of the environment and social equity within the transport systems within cities.

AI applications are also prioritized on inclusivity. Personalized urban mobility services to be run to the aged, handicapped, and the marginal groups will ensure that the urban mobility is covered to all. With the help of demographic and geospatial data, AI can find underserved areas, allocate services in the most efficient manner, and make safety and convenience of access more comfortable among vulnerable groups. The AI-based solutions in transport planning, in its turn, will contribute to the creation of more resilient, flexible, and people-oriented transport mobility systems.

6.2 Future Directions

The application of AI in the transport sector will only increase in the future due to technological advancement, increasing population in cities, and the increasing demand to achieve sustainability. Two areas that require attention in the future research and implementation work are as follows. On the one hand, AI combined with other new topics, such as the Internet of Things (IoT), connected vehicles, and autonomous mobility systems will enable a more advanced real-time decision-making process, including the ability to foresee. This merger will also be able to reduce congestion and enhance energy efficiency and create smoother multimodal travel.

Second, the cities will benefit by keeping in mind systems of incessant learning and adapting. A system performing AI-based research on historical trends over real-time data can respond to the trends of cities and respond to the shifting travel patterns, infrastructure constraints, and environmental conditions prior to the occurrence. Adaptive learning processes will be included as a critical component to the overall efficiency and stability of such urban settings and where the city grows at a quick pace or in the densely populated cities.

Third, an increasing role of data governance and ethical requirements will be present. To have trust and acceptance of AI-driven mobility systems by the people, transparency, accountability, protection on privacy of information processes should be implemented. Planners and policymakers must establish regulations regarding the use of their data and equity in algorithms and also make an inclusive service design, since unexpected social disparities can arise.

Finally, this collaboration between the community, the public, and the private will be instrumental in the appropriations of AI to greater extents. The distinct strategy can enable cities endowed with technological savvy, policy support and citizen engagement at a governmental stage to apply country-specific solutions, which are cheap and could address concerns of local mobility and those of the global mobility crisis. The key factors to be addressed in the next generation of transport structures are their sustainability, accessibility, and resilience, and AI can be one of the strategic resources to achieve the objectives.

REFERENCES

Babb, C. (2016). The End of Automobile Dependence: How Cities are Moving Beyond Car-based Planning. *Urban Policy and Research, 34*(1), 90–92. DOI: 10.1080/08111146.2016.1155194

Cervero, R. (2014). Transport Infrastructure and the Environment in the Global South: Sustainable Mobility and Urbanism. *Jurnal Perencanaan Wilayah Dan Kota, 25*(3), 174–191. DOI: 10.5614/jpwk.2015.25.3.1

Cina, E., Elbasi, E., Elmazi, G., & AlArnaout, Z. (2025). The role of AI in Predictive Modelling for Sustainable Urban Development: Challenges and opportunities. *Sustainability (Basel), 17*(11), 5148. DOI: 10.3390/su17115148

Elvas, L. B., & Ferreira, J. C. (2021). Intelligent transportation systems for electric vehicles. *Energies, 14*(17), 5550. DOI: 10.3390/en14175550

Esri. (n.d.). *Transportation Equity | Improving Mobility for Underserved Communities.* Retrieved from https://www.esri.com/en-us/industries/roads-highways/transportation-equity

Gebresselassie, M., & Sanchez, T. W. (2019). Banister: Inequality in transport. *Journal of the American Planning Association, 85*(4), 593–594. DOI: 10.1080/01944363.2019.1641385

Hub, G. I. (2020, November 4). Mobility as a service. https://www.gihub.org/infrastructure-technology-use-cases/case-studies/mobility-as-a-service/#:~:text=Cost:%20Whim%20offers%20three%20tiers,Vienna%2C%20Antwerp%20and%20the%20US.&text=Implementation:%20Created%20by%20the%20Transport,Timeframe:%20Launched%20in%20September%202018.&text=Implementation:%20Pilot%20in%20Scotland's%20Dundee,month%20pilot%20held%202017%2D2018

Huiling, E., & Goh, B. (2017, December 31). AI, Robotics and Mobility as a Service: the Case of Singapore. https://journals.openedition.org/factsreports/4411

India, T. O. (2025, August 4). AI-driven adaptive signals to cut wait times at 165 junctions. The Times of India. https://timesofindia.indiatimes.com/city/chennai/ai-driven-adaptive-signals-to-cut-wait-times-at-165-junctions/articleshow/123103169.cms

International Energy Agency. (2022). *CO_2 Emissions from Fuel Combustion.* IEA Publications. https://www.iea.org/reports/co2-emissions-in-2022

International Transport Forum (ITF). (2021). *Reforming Urban Transport: Lessons from Global Practices*. OECD Publishing. https://www.oecd.org/en/publications/2021/05/itf-transport-outlook-2021_fff2ec9f.html

Jain, R., Singh, S. K., Palaniappan, D., Parmar, K., & T, P. (2025). Data-Driven Civil Engineering: Applications of artificial intelligence, machine learning, and deep learning. *Turkish Journal of Engineering*, *9*(2), 354–377. DOI: 10.31127/tuje.1581564

Jain, R., Upreti, K., Ekatpure, S. R., Gupta, V. K., Rai, A. K., & Gothania, J. (2025). The convergence of IoT, ML, and big data in Self-Service innovation. In IGI Global eBooks (pp. 241–270). https://doi.org/DOI: 10.4018/979-8-3373-1295-8.ch009

LTA | Intelligent Transport Systems. (n.d.). https://www.lta.gov.sg/content/ltagov/en/getting_around/driving_in_singapore/intelligent_transport_systems.html

Lucas, K. (2019). A new evolution for transport-related social exclusion research? *Journal of Transport Geography*, *81*, 102529. DOI: 10.1016/j.jtrangeo.2019.102529

Martin. (2025, July 30). Cities - United Nations Sustainable Development Action 2015. United Nations Sustainable Development. https://www.un.org/sustainabledevelopment/cities/#:~:text=Goal%2011%3A%20Make%20cities%20inclusive%2C%20safe%2C%20resilient%20and%20sustainable&text=Cities%20are%20centers%20of%20innovation,%2C%20safe%2C%20and%20resilient%20cities

OECD. Kamal-Chaoui, L., Matta, A., Fritz, K., Kim, B., Kim, S.-J., Akhmouch, A., Ministry of Land, Infrastructure and Transport of Korea (MOLIT), Permanent Delegation of Korea to the OECD, & Philip, P. (2020). Smart cities and inclusive growth. https://www.oecd.org/content/dam/oecd/en/publications/reports/2020/08/smart-cities-and-inclusive-growth_332850c0/8a4ce475-en.pdf

Reis, M. J. C. S. (2025). Internet of Things and Artificial Intelligence for Secure and Sustainable green mobility: A multimodal data fusion approach to enhance efficiency and security. *Multimodal Technologies and Interaction*, *9*(5), 39. DOI: 10.3390/mti9050039

Saki, S., & Soori, M. (2025). Artificial intelligence, machine learning and deep learning in advanced transportation systems, a review. *Multimodal Transportation, 100242*. Advance online publication. DOI: 10.1016/j.multra.2025.100242

Sietchiping, R., Permezel, M. J., & Ngomsi, C. (2012). Transport and mobility in sub-Saharan African cities: An overview of practices, lessons and options for improvements. *Cities (London, England)*, *29*(3), 183–189. DOI: 10.1016/j.cities.2011.11.005

Singh, S. K., Alasali, B. E., & Belbase, Y. (2025). Urban Transportation Planning: Artificial Neural Network Applications. In Syam Sundar, L., Balakrishnan, D., & Sousa, A. C. M. (Eds.), *Expert Artificial Neural Network Applications for Science and Engineering* (pp. 469–504). IGI Global., DOI: 10.4018/979-8-3693-7250-0.ch017

Site, A., Vasudevan, S., Afolaranmi, S. O., Lastra, J. L. M., Nurmi, J., & Lohan, E. S. (2022). A Machine-Learning-Based Analysis of the Relationships between Loneliness Metrics and Mobility Patterns for Elderly. *Sensors (Basel)*, *22*(13), 4946. DOI: 10.3390/s22134946 PMID: 35808440

Sustainability Directory. (2025). *AI In Transportation Equity*. Retrieved from https://sustainability-directory.com/area/ai-in-transportation-equity/

TrafficInfraTech - Editor. (2025, June 26). Artificial intelligence in India's intelligent transportation systems: driving the next mobility revolution - TrafficInfraTech magazine. TrafficInfraTech Magazine. https://trafficinfratech.com/artificial-intelligence-in-indias-intelligent-transportation-systems-driving-the-next-mobility-revolution/

United Nations. (2019). Shifting demographics | United Nations. https://www.un.org/en/un75/shifting-demographics

UNRIC. (2024). *Building an Accessible Future for All: AI and the Inclusion of Persons with Disabilities*. Retrieved from https://unric.org/en/building-an-accessible-future-for-all-ai-and-the-inclusion-of-persons-with-disabilities/

Vujadinovic, V. L., Damnjanovic, A., Cakic, A., Petkovic, D. R., Prelevic, M., Pantovic, V., Stojanovic, M., Vidojevic, D., Vranjes, D., & Bodolo, I. (2024). AI-Driven Approach for Enhancing Sustainability in urban public Transportation. *Sustainability (Basel)*, *16*(17), 7763. DOI: 10.3390/su16177763

Wang, C., Shang, Q., Liu, K., & Zhang, W. (2025). Traffic congestion recognition based on convolutional neural networks in different scenarios. *Engineering Applications of Artificial Intelligence*, *148*, 110372. DOI: 10.1016/j.engappai.2025.110372

Wang, J., & Yang, H. (2023). Low carbon future of vehicle sharing, automation, and electrification: A review of modeling mobility behavior and demand. *Renewable & Sustainable Energy Reviews*, *177*, 113212. DOI: 10.1016/j.rser.2023.113212

Yin, A., Chen, X., Haitao, H., Morris, A., Yuan, Q., Ma, X., & Yang, Z. (2025). Shared e-bikes demand in urban mobility: Temporal heterogeneity, driving factors, and strategic implications. *Travel Behaviour & Society*, *41*, 101075. DOI: 10.1016/j.tbs.2025.101075

Zemmouchi-Ghomari, L. (2025). Artificial intelligence in intelligent transportation systems. Journal of Intelligent Manufacturing and Special Equipment. https://doi.org/DOI: 10.1108/JIMSE-11-2024-0035

Zhang, X. (2025). Artificial intelligence in intelligent traffic signal control. *Applied and Computational Engineering, 118*(1), 113–120. DOI: 10.54254/2755-2721/2025.20846

Chapter 9

AI in Integrated Transport Systems: Pathways to Sustainable and Equitable Mobility

Jaishri Gothania
http://orcid.org/0000-0002-2656-642X
Lingaya's Vidyapeeth, Faridabad, India

Sitesh Kumar Singh
http://orcid.org/0000-0002-7108-0808
College of Engineering, National University of Science and Technology, Oman

M. K. Harikeerthan
http://orcid.org/0000-0001-9738-2495
Dayananda Sagar Academy of Technology and Management, India

H. R. Shyam Agasthya
http://orcid.org/0009-0001-9382-9971
College of Engineering, National University of Science and Technology, Oman

ABSTRACT

This chapter discusses how Artificial Intelligence (AI) can be used to develop unified transport systems that are sustainable and socially fair. It emphasizes how AI technologies, including machine learning, computer vision, predictive analytics, and natural language processing, can be used in real-time optimization, demand prediction, and individualized mobility. Based on international case studies, it demonstrates such applications as optimization of public transport schemes, safety in active mobility, and rural accessibility. The challenges such as data governance, algorithmic bias, and interoperability are also present within the chapter, and gov-

DOI: 10.4018/979-8-2600-3038-7.ch009

ght © 2027, IGI Global Scientific Publishing. Copying or distributing in print or electronic forms without written permission of IGI Global Scientific Publishing is prohibited Use of this chapter to generative artificial intelligence (AI) technologies is expressly prohibited. The publisher reserves all rights to license its use for generative AI training and machine learning model development.

ernance models, collaboration frameworks, and open data standards are offered to reduce risks. Through its AI innovation to inclusive design and strong policy pathways, it provides the solutions to scale low-carbon, accessible, and resilient mobility systems with an emphasis that future AI-enabled transport must be people-focused and environmentally sustainable.

1. INTRODUCTION TO AI AND INTEGRATED TRANSPORT SYSTEMS

The paradigm shift in the modern transport systems has been brought on by the conditions of urgent need to accelerate the rapid urbanization process, environmental degradation, and worsening social inequality. The cities around the world must face a twofold task of the creation of efficient and sustainable networks on the one hand and, on the other hand, networks that must be socially accommodating to all the groups of the population and to all the social-economic layers. All these should be done within a greater context of reducing environmental footprint in line with the vision of climate change mitigation and the global sustainability pledge.

Traditionally, mode-specific, siloed approach to transportation planning has been the staple of transportation planning where all modes of transportation i.e. buses, rail, cycling, and freight are planned and operated in relative silos. Though the strategy may have been effective in terms of functional efficacies at a per-mode level, it has been associated with a poorly smooth user experience, lack of connectivity, and poor management of the various resources. This is due to lack of full-scale integration and hence the travelers tend to find it difficult to switch between modes and there is also increased travel time and absence of systematic performance among the transport agencies.

Artificial Intelligence (AI) works in contrast to that and offers a paradigm shift to overcome these disadvantages. Machine, deep, and natural language processing and computer vision AI have the capacity to extrapolate large and heterogenous data and cut across a variety of transport and infrastructure systems. This analytical capability enables it to develop real-time and user-controlled, dynamic, and data-driven mobility systems with adaptive infractions, which respond to dynamic travel demand and operational variables over an environment (Singh, Alasali, and Belbase, 2025).

1.1 Definition of Integrated Transport Systems

An Integrated Transport System (ITS) is merely the harmony of the diverse transport modes, e.g, rail, bike, walk, shared mode, and freight logistics, and a transport system that acts as a single convenient transport system. This is to enable the passengers to travel between origin and destination without necessarily operating on different schedules, ticketing systems and information channels. Higher-dense integration, such as Transportation 5.0, is neither technical integration nor human-centered priorities but is inclined towards inclusiveness, safety, and sustainability (Rong et al., 2025).

The main characteristics of a coordinated transport system are the following:

- Extracurricular cross-mode communication (e.g. bus-train-bike).
- Information systems that are integrated with real-time information on travelling.
- Efficiency in utilization of the transport resources in terms of passenger and freight.
- Active and green traffic assistance.

1.2 Artificial Intelligence in Integrated Transport

An Integrated Transport System (ITS) is concentrated on the free flow of mobility where diverse modes of transport would be incorporated in a single network. These include scheduling, fare and information integration causing the passengers to easily transition between modes. The vision is augmented into the application of AI that helps to optimize the schedule in real-time, predicts the maintenance of the infrastructure, predicts the demand, and makes individualized travel recommendations.

Giving a practical example, the algorithms that the AI offers can include the introduction of live traffic information, weather forecast, and passenger traffic prediction that will make sure the planning of the infrastructure of the public transportation is altered dynamically and the concession and reliability of the services are enhanced. The computer vision systems can monitor traffic flows and incidents real time and ensure quick responses with minimum disruption and delay. Alternatively, accessibility can be enhanced by relying on the concept of natural language processing in applications that interact with the passengers by providing passengers with multilingual and voice-powered journey-planning apps that cater to the needs of passengers with varying requirements.

With Mobility-as-a-Service (MaaS) systems, AI can serve as the decision-making engine, which will subsequently be able to propose the most optimal travel choices based on time, cost, environmental footprint, and accessibility preferences of the journey. This makes the experience of the user easier and enhances modal alternatives over the need to drive a personal vehicle in less responsible ways (e.g.: public transit, biking and sharing mobility) (Servou et al., 2023).

One can refer to Artificial Intelligence as a vast area of computing techniques like machine learning, artificial neural networks, computer vision, and natural language processing where the apparatus can be taught new information by the latest information, identify patterns, and make predictions or judgements with less human intervention. AI can:

- Predict the demand of passengers based on the new and previous data.
- Optimize buses, trains and ride-share services routes and schedules.
- Enhance the safety of cyclists and pedestrians by hazard detection under perception of the computer.
- Enhance access by tailoring the mobility services to the needy.

An example of this is that the dependent traffic flow models, which rely on neural networks, have proved to be far more effective in managing congestion levels and dependability of the transportation medium of the populace than the conventional models (Leonardi et al., 2025). It is the ability of AI to handle complexity of integrated networks that is unique since it is the only system that can handle large, multimodal data in real-time.

1.3 Linking AI with Sustainability, and Social Inclusion

The benefits of using AI-based integrated transport systems on the environment are enormous. The applications of AI can be used to mitigate greenhouse gas (GHG) emissions, improve air quality, and energy savings through the optimization of the operations and provoke modal shift to less-carbon modes. The analytics of prediction can be utilized to determine the periods when the congestion is the most problematic and traditional methods such as congestion pricing, route optimization, and adaptive switching of the signal can decrease the consumption of fuel and emissions. The only area that AI in transport is useful is in terms of operational efficiency as it will help in the process of sustainability. The possibility of having the modal shift to working with the public or active modes of transport is achieved through predictive analytics and leads to a decrease in greenhouse gas emissions. The AI-based demand forecasting also adheres to allow them to deploy their fleets in the most efficient ways to ensure that they do not have to waste their fuel and idle time (Cohen, 2024).

Equity-wise, AI would be able to identify transport deserts that are underserved by public transport and make investments more informed to serve it better. It is equally important that there is a chance of AI that enables improving social inclusion. The vulnerable populations that are often faced with incapacitation of affordable and convenient forms of transport are the elderly, people with disabilities, and low-income populations. These differences could be resolved with the aid of AI through the application of personalized mobility services, which would be suitable to the needs of an individual. One such example is the AI-enabled wayfinding solutions that can provide step-free paths to passengers with mobility impairments and the demand-based transport that is activated by user-request may be capable of accessing the underserved locations. Assistive technologies like an AI-powered wayfinding system must be accessible to individuals with impairments in their vision, and dynamic pricing algorithms must be developed to maintain the prices as affordable to the low-income users. However, unless it operates in a manner of clear governance, AI can entrench inequalities even further, since biased choices can be made, and with poor governance these choices can even reinforce the existing inequalities (Mirindi, 2024). Figure 1 below gives a conceptual framework of how AI is used as an intermediary between integration, sustainability, and equity of transport systems.

Figure 1. Conceptual relationship between AI, integrated transport systems, sustainability, and equity

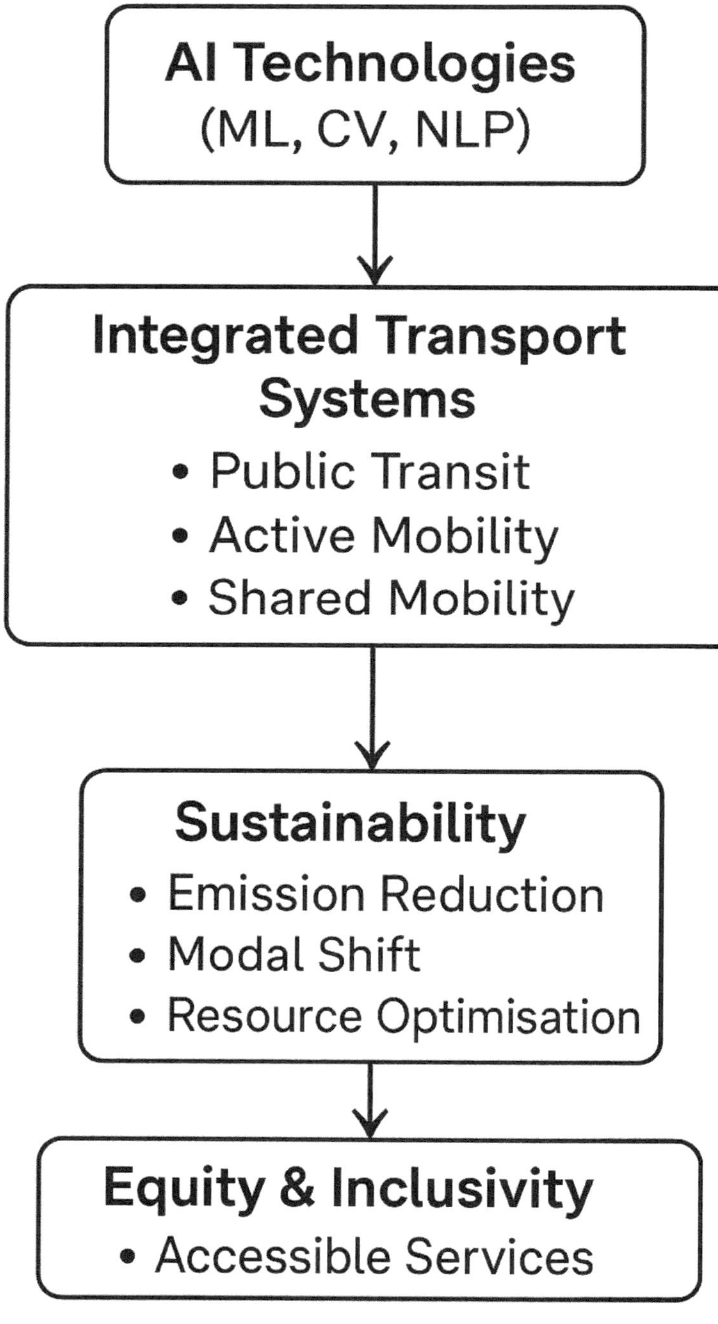

Computing public transport deserts Policymakers will also be able to identify transport deserts in a district by processing spatial and demographic data with the help of AI. These studies can be applied to take a wise decision and invest in infrastructure in an equity-based approach whereby all communities will not be left behind in transitioning to smart mobility.

1.4 Global Momentum and Research Relevance

Other cities around the globe are Singapore, Helsinki, and Barcelona, where in specific cities are implementing AI in Mobile-as-a-Service (MaaS) where they are centralizing payment, journey planner, and real-time communications. Pilots have demonstrated a significant decrease in congestion and travel time with assigners of smarter signals and congestion charges via AI. The way AI is being marketed as a sustainable mobility policy pillar can be seen in the AI4Cities program implemented by the European Commission, and Ventures by the World Bank in their Sustainable Mobility for All (SuM4All).

The combination of AI and integrated transport systems as a research field is yet to become a baby step. Recent research somehow focuses on one of the policy implications or technological possibilities without focusing on the other. The chapter attempts to address that gap by connecting the technological enablers to sustainability and equity objectives and is thus a more comprehensive answer.

2. AI FOR SUSTAINABLE MOBILITY

Sustainable mobility is a highly significant part of global efforts to reduce the volume of greenhouse gas (GHG) emissions and improve air pollution in the cities, not to mention increased accessibility to transportation of equitable services. As the cities face pressure due to the pressure of rapid urbanization and climate change, Artificial Intelligence (AI) has come into the scene as a game-changing enabler that may increase operational effectiveness, bring a modal shift into low-carbon transportation and maximize congestion management (Singh and Jain, 2025; Singh and Jain, 2025). The other way AI can reduce environmental impact of the transport network is through high-quality services that are achieved by making decisions based on real-time data. In Figure 2, the conceptual pathways demonstrate the process through which AI facilitates sustainable mobility results.

Figure 2. Conceptual pathways showing how AI supports sustainable mobility outcomes

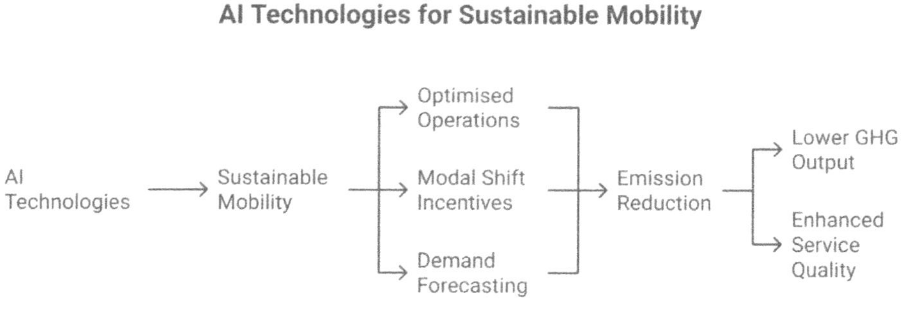

2.1 Emission Reduction through Optimized Operations

The process of optimizing transport with references to applying the Artificial Intelligence (AI) methods plays a significant role in facilitating the process of minimizing the greenhouse gas (GHG) emissions and in increasing the mobility systems in overall efficiency. Transportation agencies and operators can restructure its schedules, routes and deployments of its fleet dynamically in response to the shifting demand patterns through machine learning, predictive analytics, and processing of real-time data. This dynamic solution is a way of reducing congestion, reducing waste of power, and enhancing service reliability.

The one of the brightest ones is the implementation of the adaptive signal controls systems following which a sequence of traffic lights can be altered based on the real-time traffic situation. The systems can reduce idling by as much as 20 percent by the reduction of unnecessary vehicle idling in intersections, particularly in busy urban corridors; thus, contributing to the reduction of fuel consumption and vehicle emissions to a considerable extent (Saki and Soori, 2025). The benefits do not only result in the targets of climate action but also enhance the quality of the air particularly in the urban centers.

In the provision of public transport, AI has also been influential in synchronization of the schedule. The arrangement of buses and trains to shape up the schedules of the conveyances based on the demand data will assist the operators to not just reduce the delays in the transportation of people, but also the unnecessary loss of energy that may be experienced when the vehicles are not loaded. Indicatively, real-time-estimating deployed by the Land Transport Authority of Singapore through

AI-enabled scheduling algorithms that minimize train headways and concurrently decide on the number of passengers that are expected has been implemented. This has already resulted in measurable energy savings across the transport system demonstrating the environmental and operational benefits of AI-based schedule optimization (Jin et al., 2014).

Combined with the work on the system level, AI technologies are helping people and autonomous drivers drive more eco. Individual AI-based eco-driving systems provide real-time feedback and optimal route instructions to promote the driving habits of fuel efficiency, such as smooth acceleration, optimum gearing, and regenerative breaking where available. These measures directly reduce the amount of fuel consumed and vehicles wear and thus reduce emissions.

These tactics of operations are also supported by the inclusion of AI within predictive maintenance, which implies that the condition of transport fleets is maintained in the state of the most efficient mechanical performance. Predictive models can be used to analyze the sensor and performance data to identify the maintenance requirements to avoid the breakdowns that will cause congested traffic movement between offices and pollute the environment with too much gas emissions leading to stalled or detoured traffic.

One of the actual and pyrophobia impacts of such AI interventions is the decrease in fuel consumption. It has been demonstrated that 5-15 percent of fuel efficiencies can be achieved through AI-enhanced eco-driving and optimizing the fleet (Ma et al., 2017). It is a direct contribution toward the achievement of targets stipulated to cut GHG emission as per the international treaties such as Paris Agreements and contribute to economic benefits e.g. reduced costs of operation of transport providers.

In conclusion, the idea of optimization of transport activities through AI is not only a technological advancement, but also a strategic means of introducing sustainable and low-carbon mobility. The combination of adaptive schedule, eco-driving, and predictive analysis enables cities to make their services more efficient and provide their passengers with improved experience and take a great leap forward in the direction toward the environmental sustainability goals.

2.2 Modal Shift Strategies using AI

The best sustainability action is, perhaps, the redirection of those who travel by car with trains, bikes, and walking activities, which are low-carbon modes of transport. Enhancing the service that sustainable transport offers with AI is one of the ways this modal transition can be facilitated.

Mobility-as-a-Service (MaaS) platforms under AI management will consist of planning, ticketing, and real-time information in the various modes of travel and will offer the end-users a seamless experience of travelling. There are Whim and other

AI-powered solutions in Helsinki, including personalization of travel recommendations about sustainable first-choice modes, altering user behavior, and rendering people less reliant on cars (Cruz & Sarmento, 2020).

Moreover, the dynamic pricing models used through the AI can reward off-peak trips, manage the network balance, and decrease the price of the transport provided on a public basis. To illustrate that, some of the European cities are experimenting with the idea of lowering fares on trips with low carbon emissions that are measured with the assistance of mobile technology, and AI analytics can be used to promote the practice (Sariyer et al., 2024).

AI is also useful in terms of active mobility particularly through the computer vision systems that monitor the cycling and pedestrian infrastructure. Urban planners can get localized information on the usage trends using such systems, such that special investments can be undertaken to make walking and cycling more appealing and safer.

2.3 Demand and Congestion Management Predictive analytics

Predictive analytics is one of the main AI-related applications in sustainable transport. AI will be able to identify the demand of passengers, and the state of the traffic with high accuracy through the analysis of previous data, weather, event times and streamed information presented through sensors.

This kind of forecasting is also valuable in ensuring the transport agencies in the management of their resources; with extra buses being put in during the cues time or buses being deployed to areas where they are in high demand unpredictably. These aggressive managers reduce unnecessary wastage of miles and associated emissions. Otherwise, as Transport for London (TfL) in London uses AI models to predict future traffic congestion up to 24 hours in advance to enable pre-emptive traffic diversion or the setting of the dynamic signal control time (Mutambik, 2025).

On a more advanced level, AI-based future congestion management systems are a combination of environmental impact (predicting) and the traffic flow models. This will allow the cities to consider the GHG impact of the different traffic control options and compare those that can offer the greatest incentives in sustainability.

3. AI FOR INCLUSIVE AND EQUITABLE MOBILITY

Equitable access to mobility plays a vital role in an actual sustainable transport system. Equity in mobility outlines the concept that all people (income, age, gender, physical ability and geographic location) ought to have equity and consistency in having access to safe, reliable and affordable means of transportation. It is not a question of operations efficiency and meeting the average demands, but it is concerned with systematic inequalities to make sure that certain groups have not been underserved in the past.

Under the conventional transport planning, the standards have often been concerned with attempts to achieve efficiency and meet the demands of most of the population even by compromising the demands of the sidelined people. This has led to inequalities in mobility of certain groups of people, particularly the low-income groups, in the countryside or in places where there is minimal transportation. The social-economic segregation is further aggravated by these inequalities which pose a hindrance to access to basic services, employment, education, and social inclusion and participation.

Artificial Intelligence (AI) and the application of it to transport planning represent a paradigm shift. The fact that an AI can process large amounts of data and make informed decisions based on them implies that it is possible to apply the AI as a tool to intervene on the needs of underserved populations in relation to mobility in the most specific ways (Singh and Saraswat, 2019). Having both demographic, spatial, and behavioral data, AI can identify so-called mobility deserts (areas with a low level of transport coverage) and propose improvements that are not only technically but also socially inclusive in nature.

An illustration is that, in demand-responsive transport (DRT) systems, AI-driven systems can be able to dynamically change routes and schedules to serve low population density regions, or human mobility patterns typically not considered by traditional fixed-route transport. Continuing with the same topic, it is also possible to use AI in the building of inclusive infrastructure, such as the location of bus stops so that elderly individuals would be able to walk a shorter distance or accessibility guidelines that are available in real-time to those unable to walk.

Equity contribution is not limited to the service provision in AI. Predictive analytics also give a chance to the policymakers to find out future transportation demands of the growing or ageing population based on this choice. The AI tools can also be used to simulate the equity impacts of proposed policies, which, consequently, allows the planners to take into consideration the change more than vulnerable populations in advance before implementing it.

In addition, AI will promote diversity in decision-making by integrating the community engagement systems feedback in the models of planning. One can apply techniques like natural language processing e.g. to retrieve the outcomes of a survey of crowded opinions on matters regarding accessibility and/or holes in services one can offer (individuals might not be willing or even give negative feedback, so one will have to use aggregate statistics). These lessons can be considered to make planning decisions sensitive to the experiences of the various groups within the community.

The outcome of the equity ideals in the AI-run transport systems is the sealing of the access inequity in the past. Whereas the traditional system has, in some incidences, been instrumental in marginalization, AI may be employed with the aim of leveling the playing field so that the mobility technology advancements can positively transform the whole society.

Figure 3. Conceptual framework of AI tools translates into equitable mobility

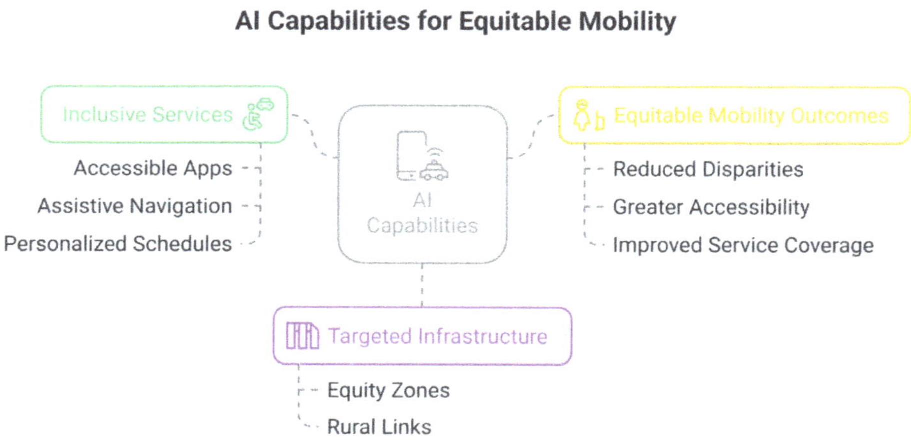

Figure 3 shows the theoretical framework as to how AI tools will be translated into equal mobility results. As a result of adopting data-based diagnostics, participatory design, and inclusive service design, AI will not only facilitate efficient transport but social justice, either, and equitable mobility will become a tangible, quantifiable and attainable concern in the context of sustainable urban development.

3.1 Accessibility Solutions to Vulnerable Populations

The artificial intelligence technologies can substantially improve the availability to disabled people, the elderly, and immobile ones. Computer vision and natural language processing tools can be used to develop assistive navigation systems such that the visually impaired passengers can be provided with a voice track in the public transport environment in real-time.

As an example, Transport for London (TfL) has conducted pilot projects on AI-assisted mobile applications that can locate their position using a survey without a station and calculate step-free routes (Iyer, 2021). Similarly, AI speech recognition front-end may allow passengers to interactively use ticketing and information kiosks without requiring a manual raised the frequency of the inaccessible device without the need to touch conventional input devices and can be useful in meeting the requirements of people with limited literacy or dexterousness

Besides physical access, AI can also act in enhancing the temporal accessibility-that is, by ensuring that the services are active at the time that is convenient to the shift workers, care givers and students. Predictive modelling would be capable of revealing at what time the unsafe users' groups would be most required to travel, and the best way of developing schedules would be to accommodate them and relieve them of being marginalized.

3.2 AI in Fair Infrastructure Planning

Traditional methods of infrastructure planning often operate on aggregated information, in which such differences in the supply of service can be concealed. According to AI, inequities can be thought of on a neighborhood level using fine-grained space and temporal analysis. This assists planners in determining the areas where investments can be made in transport to be given priority in a bid to deal with the greatest transport deficiencies when compared to economic and social participation.

The city of Los Angeles, as an example, has employed the use of AI-based spatial analysis to identify the so-called mobility equity zones, in which there is a lack of public transport and congestion among low-income households and minority populations (W. Chen et al., 2020). This information is applied in the identification of areas where new bus rapid transit corridors will be constructed or where the networks of pedestrian and cycle paths can be updated.

In addition, AI can be used in the context of scenario modelling since it will help evaluate how any infrastructure investment will affect different demographics. This will enable the policymakers to choose the option of paying in cash and not to interfere with the optimization of either efficiency or equity outcomes through simulation of the outcomes of the different planning choices.

3.3 Addressing Transport Deserts through Data-Driven Insights

A transport desert is a neighborhood that has an imbalance between transport demands and the supply of services. They are more likely to be in peri-urban, rural or low-income urban areas where there is low demand to the extent that investing in the traditional fixed-route services is not viable. The answers to the questions that are given by AI are demand-reactive TRANS (DRT) model that involves the dynamic real-time deployment of vehicles, which relies on demand.

The AI-driven DRT systems have been piloted in rural counties in Kansas in the United States due to its capability to predict clusters of demand by depending upon the information concerning previous bookings, weather conditions, and local event schedules. These systems enhance efficiency of the services and reduce the operating costs to ensure that an operator can offer services to more passengers with the working resources being limited (Kimery, 2025).

On mobile phone data, satellite images, and socio-economic factors, AI can be used in the developing world to identify areas where transport has been continuously under-provided. This information can then be incorporated to guide the government and the commercial sector to come up with certain mobility solutions e.g. community-based ridesharing of shared mobility provision of electric shuttles.

3.4 Global Case Studies

London, UK: TfL Go app was enhanced to increase route planning of an accessible non-step route and the addition of accessibility tools AI supports such designs positively influence wheelchair users and passengers with mobility problems (Iyer, 2021).
Bengaluru, India: AI-based bus route planning software is one that takes into consideration the demand of the low-income populations and then they increase the frequency of services offered in the early mornings and late nights (Desai and Dawda, 2025).
Finland: MaaS schemes, including Whim, offer subsidized package transport services to low-income families, and minds are artificial intelligence-based that design the most effective pricing models to encourage uptake (Hsieh, 2025).
United States (Kansas): Unfixed demand forecasting of rural transport services based on AI broke the market barriers in those locations, which might be regarded as transport deserts previously (Kimery, 2025).

4. TECHNOLOGICAL ENABLERS

The high level of technology is also a condition of effective implementation of AI in combined transport. Internet of Things (IoT) sensor networks, machine learning (ML), Mobility-as- a- Service (MaaS) platforms are the other critical enablers that offer the behests and are related to interoperability data frameworks and open standards. Technology acts as the support of the operations process with the possibility to make real-time decisions, predictive analytics, and ensure the efficient working of multimodal (Jain et al., n.d.). Figure 4 illustrates technological layers that facilitate the introduction of AI systems into transport systems, including IoT data capture as well as user-interfaces MaaS.

Figure 4. Technological layers enabling AI-integrated transport systems, from IoT data capture to user-facing MaaS applications

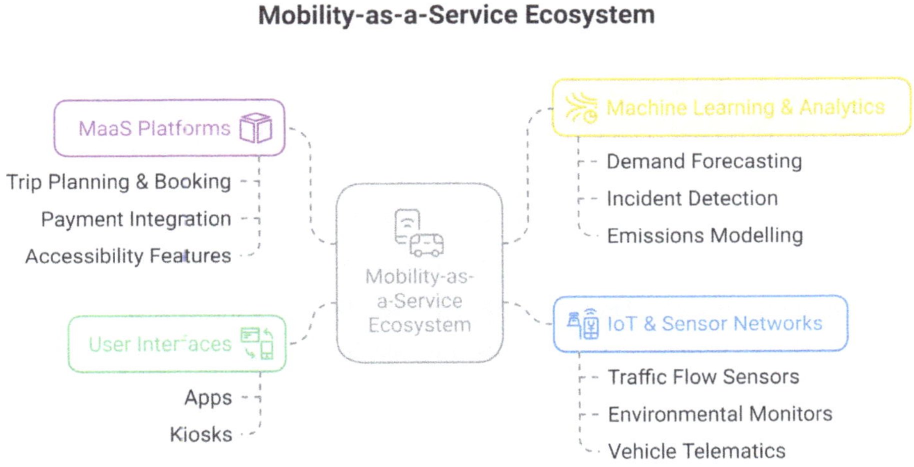

4.1 IoT and Sensor Networks for Real-Time Data Collection

The first link in the chain of AI-powered transport is IoT devices that are fed with large amounts of real-time data information about vehicles, infrastructure, and human beings. Traffic can be monitored; incidents of traffic can be notification and parameters that determine the condition of the environment such as air quality and noise pollution can be traced using sensor networks.

The Smart Mobility 2030 project being done in Singapore is another such example that uses over 5,000 IoT traffic sensors to integrate the feeds of CCTV, inductive loops, and connected vehicles into a centralized operations platform. Real time such data may be used to dynamically regulate the operation of traffic lights and reduce traffic jams and shorten travel time by up to 15 percent (Orissa International & Intralink Limited, 2025).

IoT can also facilitate predictive maintenance. Vibration and temperature sensors of the Shinkansen rail system in Japan feed the AI models that forecast when the components will be worn out and the operator can be notified ahead of time to fix and maintain the component before the failure and the resulting service disruption and wastage of resources is encountered (Turner, 2019).

General Transit Feed Specification (GTFS) and Sensor Things API are some examples of interoperability that enable the data of different IoT devices to be interconnected into an AI system, and, therefore, the data sharing between agencies/platforms becomes a reality.

4.2 Predictive Modelling with Machine Learning Algorithms

Machine learning is the central AI part of the AI-enabled transport, and it converts IoT raw data into insightful insights. Modelling supervised learning can be used to predict the demands of customers and identify the patterns of behavior of travel which is unsupervised learning.

ML models that utilize historic ticketing history and GPS tracking predict on a stop level (to minimize resource wastage on the bus routes) the demand, which is consumed by the bus routes. This has reduced the operation cost by 12 percent keeping the service cover (Cruz & Sarmento, 2020).

In computer vision, incident detection is also being done using a subcategory of ML, deep learning. One can mention such applications as convolutional neural networks (CNNs) that can identify hazards on the road, unauthorized parking, or overcrowding platforms with the assistance of live video streaming and call timely interventions.

The application of environmental modelling to ML workflows can also ensure sustainable mobility since cities would be capable of estimating the opportunities of the influence of transport management on the level of emissions and planning it (Ma et al., 2017).

4.3. Integration with Mobility-as-a-Service (MaaS) Platforms

MaaS is based on the idea of connecting transport modes via a single interface, whereby it is possible to plan and book trips, pay and support customers via one application.

To exemplify, the Whim platform is a Helsinki system that integrates buses, trams, taxis, as well as bike-sharing using AI to plan the most efficient and sustainable ride. It also implements dynamic pricing incentives to low emission and off-peak traveling and means and modes, which contributes to the decrease of congestion and decarbonization (Cruz & Sarmento, 2020).

The UK has piloted projects to investigate MaaS platforms with an AI-based access-focus, the Transport for West Midlands (TfWM) has been experimenting with voice-assisted transport and notifying passengers that they need to take a step-free journey. Such inclusive features are supported by the availability of lift and escalator data availed in real-time IoT.

The MaaS scalability requires the API standardization that provides the possibility to standardize communications between the third-party app developers and the private operators and the public agencies without any proprietary concerns. This will ensure that MaaS systems can expand to accommodate new types of transport e-scooters or autonomous shuttles and will not need the system to be developed extensively.

4.4 Best Practices

Lessons on the sustainability of integrated transport systems based on real life experiences of Artificial Intelligence (AI) are other lessons that will provide insights into how sustainable and equitable mobility solutions could be scaled up. This section makes comparative international case studies on the three areas of optimizing public transit, active mobility safety and planning, and rural/peri-urban mobility solutions and demonstrates the methods employed, their outcomes and best practices that may be adapted. The overall overview of all the case studies that are presented in this section is stated in Table 1 and the summary of the best practices and applications of AI globally is introduced as Figure 5.

Figure 5. Global best practices in AI-enabled integrated transport

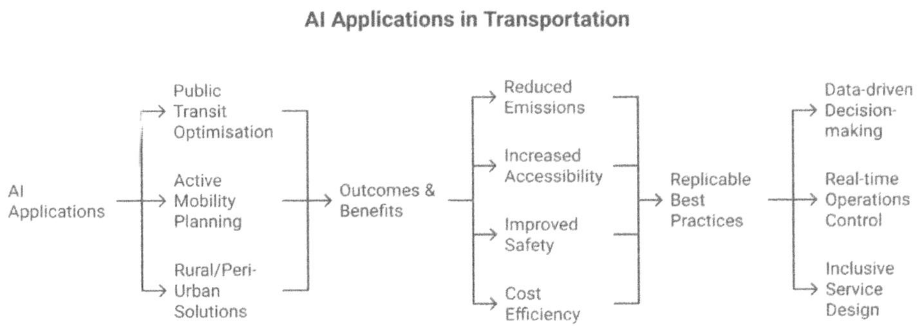

4.5 AI in Public Transit Optimization

AI technologies have interfered with transportation in most cities. The agencies will have the capability to maximize resource use and generate real-time data (IoT, predictive analytics, dynamic scheduling algorithms) to improve operations, national security, or passenger experience, and lower emissions.

Singapore: The Land Transport Authority implements AI-based predictive models to optimize the headway between trains and bus frequencies. The past ticketing information and instant sensor control are the basis of respective forecasts of the passenger loads which result in a 15 percent boost in punctuality and 9 percent reduction in energy expenditures (Orissa International & Intralink Limited, 2025).

Barcelo, Spain: AI-controlled traffic lights organization reduced bus commuting by 12 percentage during rush hour conditions. The city also uses machine vision systems to detect the illegal use of the bus lanes as parking spots in case the opportunity to enforce suddenly can be implemented and the continuity of the services delivered (Miller, 2025).

Bengaluru, India: Bengaluru Metropolitan Transport Corporation (BMTC) takes advantage of AI-powered fleet management software and provides dynamically re-routed buses based on the predictive congestion forecasts in this method, increasing the on-time performance rates by 18 percent and reducing the operation costs by 10 percent (Roy, 2025).

4.6 AI in Active Mobility Safety and Planning

Active mobility, i.e. walking and bicycling, is a vital component of sustainable networks of transportation. By examining the utilization of the infrastructures, predicting the needs, and detecting the risks, AI makes the infrastructure safer and more planned.

Helsinki, Finland: They possess computer vision systems which with the power of AI to watch over the movement of cyclists and pedestrians at the crossroads and therefore provide high-resolution data to the planners which reveal their conflicts. This fact informed a redesign that reduced the number of cyclist-vehicle crashes by 22 percent over two years (Cruz & Sarmento, 2020).

London, UK: TfL embarked on the government-sponsored initiative of Vision Zero, based on AI that interprets the video footage of the safety cameras and retrieves the near miss instances among pedestrians and cyclists. The acquired knowledge has led to the application of certain measures to address the problem, such as signaling phasing and having protective biking infrastructure (Mutambik, 2025).

Melbourne, Australia: Bike-share applications need to be AI-guided hazard warnings now that it runs in real-time using crowd-sourced alerts and camera feeds, which help to make routes safer (Abduljabbar et al., 2019).

4.7 Artificial Intelligence in Rural and Peri-Urban Movement Solutions

The rural and peri-urban areas are marred by the occurrence of transport deserts where services are either insignificant or non-existent. Here is where the power of AI-based demand-responsive transport (DRT) systems, which close this gap, comes in.

Kansas, USA: Rural counties predict demand hotspots with the help of AI based DRT based on previous booking data, events schedule and weather forecast. This way, the operating cost has been lowered by 14 percent, and the ridership is increased by 20 percent (Kimery, 2025).
Oromia, Ethiopia: Mobile phone data and satellite images are used along with AI algorithms to determine under-served regions and map out special minibus routes. The services have improved the access of the market and healthcare to over 50,000 of its residents (Kenea et al., 2021).
Japan: In the peripheral areas of Aichi Prefecture, an AI ride-sharing match-making system that paired ride-sharing riders with their destinations that matched maximized the vehicle occupancy rate, and reduced emissions per passenger by 25% (Shimamoto & Kitazono, 2025).

Table 1. Comparative summary of case studies

City/Region	Application	Key Outcomes	AI Technologies Used
Singapore	Transit schedule optimization	+15% punctuality, –9% energy use	Predictive modelling, IoT analytics
Barcelona, Spain	Bus lane enforcement & traffic signal sync	–12% travel time	Machine vision, dynamic signal control
Bengaluru, India	Fleet rerouting & congestion prediction	+18% on-time performance	ML-based traffic forecasting
Helsinki, Finland	Active mobility safety planning	–22% cyclist-vehicle collisions	Computer vision, data analytics
Kansas, USA	Rural demand-responsive transport	+20% ridership, –14% costs	Predictive demand clustering
Oromia, Ethiopia	Rural service design using mobile data	+50,000 people served	Geospatial AI, socio-economic mapping

5. CHALLENGES AND RISKS

Despite AI huge potential in changing the integrated transport system, it is not without challenges in implementing the transport system. These risks are categorized into the subgroups of technical, ethical, and institutional and refer to the topic of privacy, fairness, interoperability, and trust of the people. The challenges must be identified and addressed to make AI-based mobility systems sustainable, inclusionary, and resilient (Singh and Jain, 2025). Risks, mitigation measures and impacts are summarized in Table 2 and the AI risk landscape and mitigation measures in Integrated Transport Systems are provided in Figure 6.

5.1 Data Governance and Privacy

AI in transportation is reliant on the prevalence of the information gathered with the assistance of IoT devices, GPS-based applications, CCTVs, and ticket systems. This leads to the problem of privacy of data, loss, and unwarranted monitoring.

In another instance, European MaaS, which has been proposed by the number of cities in 2022, is being rescinded because of the outcry against the disclosure of the location information of the passengers to third-party advertisers without their express approval (Garroussi et al., 2025). They lead to loss of trust and can prevent the use of AI.

The best practice standards include:

Institute Privacy by Design.
Observation by anonymizing using data and differential privacy.
Open data-sharing relations between agencies and operators.

Presence of extremely high standards in legal regulations such as the General Data Protection Regulation (GDPR) in the EU provides a regulatory basis and most rising economies do not have this protection and thus can be abused.

5.2 Fairness and mitigation of bias in Algorithms

Applied carelessly, though, AI algorithms can become victims of the recreation/reproduction of the existing inequalities. Bias may be ascribed to:

Discriminative trends have been shown in the past.
The sample bias is due to underrepresentation of certain groups in the data.
The efficiency-based optimization objectives (as opposed to equity).

Using the example of a study in the United States in 2021, the low-income neighborhoods have very low probability of using AI-powered ride-hailing services due to optimization that tended to place the demand-based services in these neighborhoods (Akter et al., 2022).

Mitigation measures are:

- Algorithms audits to detect bias periodically.
- Photo-perform AI models with equity-based measures.
- Using different and representative data sets in training.

These measures align with the approach of Ethics by Design since it does not do anything unfair initially but instead it does it as an add-on (Bezrukov, 2025).

5.3 Technical and institutional Interoperability

Among the operational issues is the necessity to ensure that the systems that are capable of functioning in different modes and jurisdictions can communicate in the system that is facilitated by AI. Interoperability requires:

- Technical compatibility (data formats, APIs).
- Aligned policies (common policies, joint operations).

This is the reason that MaaS pilots failed in most cities, as the information across proprietary systems in one operator was not accessible to other operators, and as such, the integrated journey planning was not achievable (Eze, 2025).

Solutions include:

- Adopting an open industry standard such as GTFS and Mobility Data Specification (MDS).
- The creation of shared types of governance including the formation of common trusts of the public data.
- Standardization of system architecture through the development of national or regional interoperability requirements.

Table 2. Summary: Risks, impacts, and mitigation measures

Risk	Potential Impact	Mitigation Measures
Data Governance & Privacy	Loss of public trust, regulatory breaches	Privacy by design, anonymisation, GDPR compliance
Algorithmic Bias	Exacerbation of inequality, service exclusion	Audits, equity metrics, diverse training datasets
Interoperability Gaps	Fragmented user experience, failed MaaS integration	Open standards, joint governance, interoperability laws

Figure 6. Risk landscape for AI in integrated transport systems

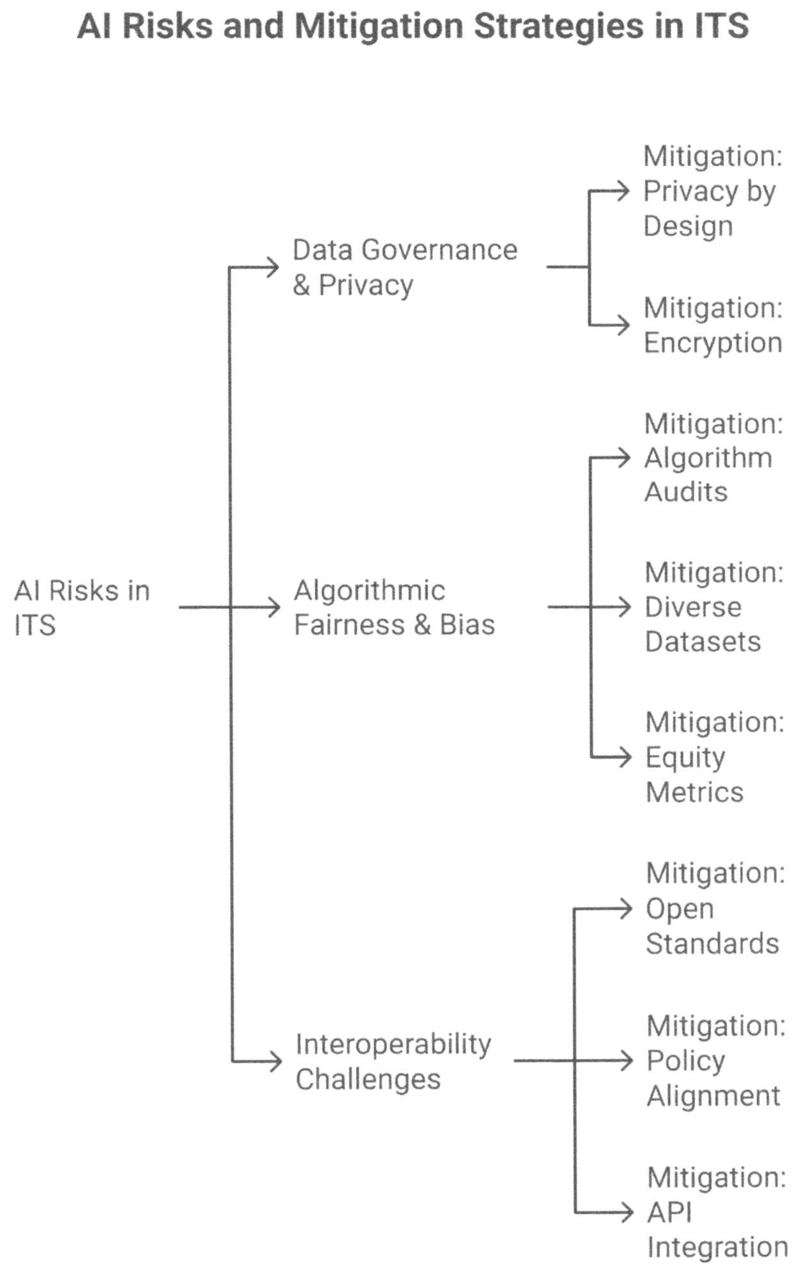

6. POLICY PATHWAYS AND RECOMMENDATIONS

Numerous advantages of AI in the transport system would be unable to achieve the transformational potential without effective governance models, effective public-private collaboration systems, and scalable means of its implementation. The policies should create a balance between innovation and equity, privacy, and interoperability since AI-based mobility can be used to achieve the aim of sustainable and inclusive urban futures (Singh et al., 2017). Table 3 is the policy matrix of enabling AI in integrated transport.

6.1 Governance Models for AI-enabled Transport

There must also be effective governance systems in place to regulate the uses of data, responsibility for the algorithms and trust of the populace. Three forms of governance are emerging in the world; they are:

a. **Regulation-based Models:** Indicative countries, i.e., the EU member states are already integrating AI into transport but under the regulation which has already been adopted by the EUAI Act that presupposes transparency, risk analysis, and human control (Cancela-Outeda, 2024).
b. **Innovation-driven Models:** The Singapore model has regulatory sandboxes as well: AI solutions are tested in a controlled environment before being scaled.
c. **Hybrid Models:** The UK puts duality between national recommendations and flexibility of the city level policy with a balance between standardization and localization.

The costs of best practice governance are:

- Pre-implementation of algorithmic Prototypes impact assessment.
- Assessing AI ethics committees in transport.
- The distribution of the people in decision-making to achieve buy-in of the society.

6.2 Sustainable and Equitable AI Solutions

Scaling is associated with the policies of technical, financial and institutional readiness. The next method to prove to be effective:

- **Technical Preparedness:** The necessity of interoperable standards (e.g. GTFS, MDS) not only makes investors in renders future proof but also enables service to be developed on the same platform: A major contribution to the successful development of other transport services and networks (e.g. open standards in the airline and rail sector).

- **Financial Preparedness:** This should be done through subsidies or grants to adopt AI in neglected areas to ensure even distribution.
- **Institutional Readiness:** It involves the training of transport planners and operators in AI literacy so that they can become more adopters and operators.

The pilot-to-scale models should have the evaluation measures of not only efficiency but also the outcomes on the equity parameters, e.g., access to services by the low-income users and disabled users.

Table 3. Policy matrix: Enabling AI in integrated transport

Policy Area	Key Actions	Expected Outcome
Governance	Algorithmic impact assessments, ethics boards	Transparency, reduced bias
Data & Privacy	GDPR-like protections, secure data-sharing protocols	Public trust, compliance
Interoperability	Adoption of open data standards	Seamless multimodal integration
Equity	Subsidies for underserved regions, inclusive design	Wider accessibility, reduced disparities
Capacity Building	AI literacy programs for planners and operators	Improved adoption and informed oversight

7. CONCLUSION

Utilizing Artificial Intelligence (AI) in modern-day transportation offers an unprecedented opportunity to introduce sustainability and fairness in the urban transportation system. Cities will become transformed in real-time using AI transportation planning and operations and expansion, predictive analytics, personal mobility services and smart demand management. It has been demonstrated by the best practices across the world, and case studies that have been considered by this chapter that AI is no longer a conceptual creation, but rather a practical and real tool that can be employed to create measurable efficiencies, increases in accessibility as well as environmental benefits in numerous scenarios.

Nonetheless, the possible power of AI is deadly complicated. Questions of data governance, system interoperability and algorithmic fairness demand vigorous and forceful policy efforts. These protections are essential since without such AI it would be more likely to increase the existing inequality and inefficiencies of the system rather than address them. Ethical deployment, transparency, and accountability should therefore be used to frame all the AI-based mobility programs to ensure that technological advancements do not overtake the social responsibility in technology.

Innovation should now be supported with an inclusive design and collaborative governance to deliver a sustainable and equitable transportation future. The partnership between policymakers, transport agencies, urban planners, and the private operator is required to ensure that AI-empowered systems would not be people-centered, rigid, or unclear. These include the way of making sure that the necessary solutions of mobility will meet the needs of many users including the differently abled people to the underserved communities and the ecological effects of these solutions.

The connections between the technical capacity, social responsibility and environmental stewardship, which AI can produce in transportation to make it more than smart, but fair and green, require future research. The AI can provide sustainable transformations in the mobility in the city whereby technological progress is associated with the conscious policy frameworks and citizen participation in its discourse to ensure the stipulation of accessibility, reduced levels of carbon emissions, and urban resilience in the cities worldwide.

REFERENCES

Abduljabbar, R., Dia, H., Liyanage, S., & Bagloee, S. A. (2019). Applications of Artificial Intelligence in Transport: An Overview. *Sustainability (Basel)*, *11*(1), 189. DOI: 10.3390/su11010189

Akter, S., Dwivedi, Y. K., Sajib, S., Biswas, K., Bandara, R. J., & Michael, K. (2022). Algorithmic bias in machine learning-based marketing models. *Journal of Business Research*, *144*, 201–216. DOI: 10.1016/j.jbusres.2022.01.083

Bezrukov, O. (2025, May 29). AI in Transportation: Ethical Balance between Safety and Privacy. Blog. https://tech-stack.com/blog/ai-in-transportation/

Cancela-Outeda, C. (2024). The EU's AI act: A framework for collaborative governance. *Internet of Things : Engineering Cyber Physical Human Systems*, *27*, 101291. DOI: 10.1016/j.iot.2024.101291

Chen, W., Chen, L., Xie, Y., Cao, W., Gao, Y., & Feng, X. (2020). Multi-Range attentive bicomponent Graph convolutional network for traffic forecasting. *Proceedings of the AAAI Conference on Artificial Intelligence*, *34*(04), 3529–3536. DOI: 10.1609/aaai.v34i04.5758

Cohen, A. P. (2024). *Artificial Intelligence in transportation: Use cases and potential.* Mineta Transportation Institute. https://transweb.sjsu.edu/sites/default/files/2430-Cohen-Artificial-Intelligence-Transportation-Use-Cases.pdf

Cruz, C. O., & Sarmento, J. M. (2020). "Mobility as a Service" Platforms: A Critical Path towards Increasing the Sustainability of Transportation Systems. *Sustainability (Basel)*, *12*(16), 6368. DOI: 10.3390/su12166368

Desai, D., & Dawda, N. H. (2025). New approaches for integrated multimodal transport systems. SSRN Electronic Journal. DOI: 10.2139/ssrn.5080740

Eze, C. C. (2025). Integration of AI-Driven multimodal transport systems for optimizing Real-Time urban and intercity mobility solutions. *International Journal of Research Publication and Reviews*, *6*(4), 9288–9304. DOI: 10.55248/gengpi.6.0425.1574

Garroussi, Z., Legrain, A., Gambs, S., Gautrais, V., & Sansò, B. (2025). A systematic review of data privacy in Mobility as a Service (MaaS). *Transportation Research Interdisciplinary Perspectives*, *31*, 101254. DOI: 10.1016/j.trip.2024.101254

Hsieh, F. (2025). Emerging Research Issues and Directions on MaaS, Sustainability and Shared Mobility in Smart Cities with Multi-Modal Transport Systems. *Applied Sciences (Basel, Switzerland)*, *15*(10), 5709. DOI: 10.3390/app15105709

Iyer, L. S. (2021). AI enabled applications towards intelligent transportation. *Transportation Engineering*, *5*, 100083. DOI: 10.1016/j.treng.2021.100083

Jain, R., Singh, S. K., Palaniappan, D., Parmar, K., & T, P. (2025, June 30). Data-driven civil engineering: Applications of artificial intelligence, machine learning, and deep learning. *Turkish Journal of Engineering*, *9*(2), 354–377. DOI: 10.31127/tuje.1581564

Jin, J. G., Tang, L. C., Sun, L., & Lee, D. (2014). Enhancing metro network resilience via localized integration with bus services. *Transportation Research Part E, Logistics and Transportation Review*, *63*, 17–30. DOI: 10.1016/j.tre.2014.01.002

Kenea, L. M., Kumar, L., Singh, S. K., Chala, G., & Chimdi, J. (2021). Assessment of asphalt road pavement rehabilitation and maintenance practices: The case of West Showa Zone in Oromia, Ethiopia. *IOP Conference Series. Materials Science and Engineering*, *1187*(1), 012033. DOI: 10.1088/1757-899X/1187/1/012033

Kimery, A. (2025, August 8). Kansas City equips buses with AI-powered FRT ahead of World Cup. Biometric Update | Biometrics News, Companies and Explainers. https://www.biometricupdate.com/202508/kansas-city-equips-buses-with-ai-powered-frt-ahead-of-world-cup

Leonardi, S., Distefano, N., & Gruden, C. (2025). Advancing Sustainable Mobility: Artificial intelligence approaches for autonomous vehicle trajectories in roundabouts. *Sustainability (Basel)*, *17*(7), 2988. DOI: 10.3390/su17072988

Ma, H., Xie, H., & Brown, D. (2017). Eco-Driving assistance system for a manual transmission bus based on machine learning. *IEEE Transactions on Intelligent Transportation Systems*, *19*(2), 572–581. DOI: 10.1109/TITS.2017.2775633

Miller, J. (2025, June 28). Barcelona implements AI technology to monitor traffic violations in bus lanes. Rentacar Trends, News and Special Offers. https://blog.getrentacar.com/sr/news/barcelona-ai-urban-transportation/

Mirindi, D. (2024). A review of the advances in artificial intelligence in transportation system development. *Journal of Civil Construction and Environmental Engineering*, *9*(3), 72–83. DOI: 10.11648/j.jccee.20240903.13

Mutambik, I. (2025). IoT-Enabled Adaptive Traffic Management: A multiagent framework for urban mobility optimisation. *Sensors (Basel)*, *25*(13), 4126. DOI: 10.3390/s25134126 PMID: 40648380

Orissa International & Intralink Limited. (2025). Opportunities in Singapore. In Trade Report for the Ministry of Foreign Affairs of Republic of Estonia [Report]. https://www.vm.ee/sites/default/files/documents/2025-03/Singapore%20Strategy_Smart%20City.pdf

Rong, R., Ma, S., Ren, N., Lin, Q., & Jia, N. (2025). Generative artificial intelligence in intelligent transportation systems: A systematic review of applications. *Frontiers of Engineering Management*, *12*(4), 1020–1036. Advance online publication. DOI: 10.1007/s42524-025-4241-9

Roy, P. (2025). AI-DRIVEN FLEET ANALYTICS: REVOLUTIONIZING MODERN FLEET MANAGEMENT. International Research Journal of Modernization in Engineering Technology and Science, 07. https://www.irjmets.com/uploadedfiles/paper//issue_3_march_2025/70285/final/fin_irjmets1745248405.pdf

Saki, S., & Soori, M. (2025). Artificial intelligence, machine learning and deep learning in advanced transportation systems, a review. *Multimodal Transportation*, *100242*. Advance online publication. DOI: 10.1016/j.multra.2025.100242

Sariyer, G., Mangla, S. K., Sozen, M. E., Li, G., & Kazancoglu, Y. (2024). Leveraging explainable artificial intelligence in understanding public transportation usage rates for sustainable development. *Omega*, *127*, 103105. DOI: 10.1016/j.omega.2024.103105

Servou, E., Behrendt, F., & Horst, M. (2023). Data, AI and governance in MaaS – Leading to sustainable mobility? *Transportation Research Interdisciplinary Perspectives*, *19*, 100806. DOI: 10.1016/j.trip.2023.100806

Shimamoto, H., & Kitazono, M. (2025). Feasibility assessment of ride-sharing system in mountainous area in Japan using a matching model. *Transportation Research Procedia*, *82*, 11–23. DOI: 10.1016/j.trpro.2024.12.014

Singh, S. K. Dr. Nikhil Agrawal, & Dr. Surendra Yadav. (2017). Review on key elements of traffic & transportation planning approaches for a mid-sized city. *International Conference on Recent Development in Engineering Science, Humanities and Management*, 86–95. Singh, S. K., & Jain, R. (2025). Future of mobility: Potential challenges and ethical considerations. In *AIoT for Smart Transportation: Transforming the future of mobility* (pp. 235–253). IET. DOI: 10.1049/PBTR047E_ch12

Singh, S. K., Alasali, B. E., & Belbase, Y. (2025). Urban transportation planning: Artificial neural network applications. In Syam Sundar, L., Balakrishnan, D., & Sousa, A. C. M. (Eds.), *Expert artificial neural network applications for science and engineering* (pp. 469–504). IGI Global., DOI: 10.4018/979-8-3693-7250-0.ch017

Singh, S. K., & Saraswat, A. (2019). Design service volume, capacity, level of service calculation and forecasting for a semi-urban city. *Revue d'Intelligence Artificielle, 33*(2), 139–143. DOI: 10.18280/ria.330209

Turner, J. (2019, July 2). Lines of data: using IIoT and AI to improve predictive rail maintenance. Railway Technology. https://www.railway-technology.com/features/ai-and-iot-in-rail/

Chapter 10
The Current Development in the Real Estate Sector With the Application of Artificial Intelligence

Aditya Singh
http://orcid.org/0000-0001-9347-5627
Amrita Vishwa Vidyapeetham, India

ABSTRACT

Artificial Intelligence including ChatGPT has attracted the attention of people quickly in the recent years. It could be easily noticed that the day to day lives of people have been drastically affected by it and people's reliance over it for almost everything has grown tremendously. The same can be said for various sectors, and it could be noticed that nowadays real estate is also being driven by Artificial Intelligence. This chapter will discuss the ways AI is affecting real estates and how people are relying on it to find suitable properties for them. Further, the way real estate sector using AI to improve their customer services will be talked about. Then, the way Artificial Intelligence could help in the management of properties will be explained in the book chapter. Some graphical analyses will be performed to support the study and to understand the current trends.

INTRODUCTION

Real Estate could be understood as any structure which is permanently attached along with the land, which includes man made or natural enhancements to a given land, houses, growing crops like timber, roads, utility systems, buildings, structures,

DOI: 10.4018/979-8-2600-3038-7.ch010

Copyright © 2027, IGI Global Scientific Publishing. Copying or distributing in print or electronic forms without written permission of IGI Global Scientific Publishing is prohibited. Use of this chapter to train generative artificial intelligence (AI) technologies is expressly prohibited. The publisher reserves all rights to license its use for generative AI training and machine learning model development.

or fixtures. In other words, it is a property which includes land as well as its attachments which are tangible in nature. The said land comprises of earth's real surface as well as any natural object which might be permanent ones like rock, dirt or water, any particulars which might be present under the surface of that land or minerals. Basically, it is a form of real property which the individual or group owns and is stuck to a portion of the land. As per the Legal Information Institute, Law School of Cornell University, Real Estate can be described as the portion of land which comprises natural or artificial property permanently stuck to that land, including things which are beneath or above the said land, trees, buildings, etc. Plus, the term Real Estate as per the law, the first word symbolizes real property like land, whereas the second word relates to a person's or group's interest in the said land. Here, it is necessary to understand that the real property is not the same as personal property, as the latter is not permanently stuck or simply put temporarily stuck to the said land property, like jewellery, vehicles, tools, boats, furniture, farm animals, farm equipment, as well as farm's rolling stock. In the context of business, Real Estate indicates managing, producing, selling, in addition to purchasing physical properties to earn profits. For people, normally Real Estate could be counted as one of the most precious assets which they could possess. The value of Real Estate changes with time, it could either increase or decrease with the span of time, depending on the several factors. Therefore, from an economic perspective, if someone wants to know the health of an economy, then value of Real Estate will definitely help in that matter. The market of Real Estate directly affects jobs in millions, including lending, improvement of homes, insurance, development, in addition to business. Reflection of its value could also be seen in the development rates of properties, rental, as well as homeownerships. Normally, a Real Estate Agent, aids in arranging matching buyers, transactions, acting representatives, sellers, in the case of negotiations, as a licensed professional, but it is important to know more about Real Estate.

Real Estate Types

This subsection will cover the various kinds of Real Estate present in the world, to give a better understanding of real properties, which are as follows:

- Land
- Residential
- Commercial
- Industrial
- Special Purpose

Land

For all kinds of real property, the starting point is the land. It generally indicates to vacant land as well as property which is underdeveloped. To increase the value of property, typically developers purchase land and then merge the land with properties known as assembly, in order to rezone the said land, which further raises the density. This also includes agricultural lands like orchards, timberland, ranches as well as farms. Normally there are two options for underdeveloped land, either income generation by using them timber, grazing, agriculture or other similar uses, or to let the said underdeveloped land be vacant to be developed in the future. Apart from the functional uses of such land, the value of land can fluctuate with time, but in normal circumstances the possibility of increasing the value of land is high, which makes it a nice regular investment strategically.

Figure 1. Real estate land present in Salem, Tamil Nadu, India

Source: Surya Prakash.S.A.

Here, the underdeveloped land shown in Figure 1, is an example of Real Estate Land, which is a major type of Real Estate.

Residential

This kind of real property is known as Residential Real Estate, which includes houses for not only individuals, but also groups of people including families. This is the famous kind of estate which is considered as asset class by many people. There are subtypes in this particular category of real property, which are single-family or multifamily homes. In the case of former, there is one dwelling, and in the case of latter, on a plot of land, there are more than one dwelling at least. Here, either centuries old homes could also be there or property which are accounted as pristine developments, in addition resale homes. This category comprises of condominiums, homes, duplex, triplex or triple-deckers, quadplex, town homes, cooperatives, vacation homes, multi-generational homes, high value homes, as well as portable homes such as houseboats. In essence, the real property must be utilized for residential purposes only, where people could live. it is essential to understand that Residential Real Estate could develop considerable equity with the passage of time, even though occupying in addition to owning individual homes would not be able to help in income generation for individuals. When the value of home of individual or group raises, and considering the mortgage payment is done by the said individual or group, then the equity in the mentioned home increases, giving the individual or group an important asset with time, which could be not only borrowed but also leveraged to additional objectives which are financial ones. However, when considering multifamily estate, then not only they could help in considerable rental income generation but also gives the individual or group their own respective residence as well as their value increases with time.

Figure 2. Collage of duplexes

Source: Sightline Institute: Missing Middle Homes Photo Library, Wikimedia

Here, the duplexes shown in Figure 2, are an example of Residential Real Estate.

Commercial

This kind of real property is called as Commercial Real Estate, which is utilized perform professional activities or business in order to use commercial means for income generation. It can be explained in other words, that owner of the estate permits other businesses to lease property on his/her land, that gives revenue to the owner. Here, it is also possible that the owner themselves owns a business on the same estate. This comprises nail salons, shopping centres, restaurants, medical offices, offices, stores, hotels, theatres, strip malls, parking facilities, retail spaces, shopping mall, as well as large scale apartment buildings having five or more units owned by a company or investor, which are rented out for many people to reside, for income generation.

Figure 3. Square one shopping mall present in Batu Pahat, Malaysia

Source: Wee Hong, Wikimedia

Here, Square One Shopping Mall shown in Figure 3, which is located in Malaysian town Batu Pahat, is an example of Commercial Real Estate.

Industrial

This kind of real property is known as Industrial Real Estate, which is utilized for industrial as well as manufacturing purposes. Industrial Real Estate just like Commercial Real Estate is meant to make profit. This includes goods distribution centres, research and development particularly lab based as well as hardware testing ones here which excludes software or theoretical work, mines, warehouses, manufacturing, heavy processing, steel plants, power plants, cement plants, food processing plants, factories, storage, in addition to logistics hubs. The distinctive categorization of this type of real property affects practices in the sales, regulations in the case of construction activities, as well as zoning of the area.

Figure 4. Ravenswood generating station in Long Island City, present in Queens, New York City, USA

Source: King of Hearts, Wikimedia

Here, the Ravenswood Generating Station in Figure 4, is an example of Industrial Real Estate.

Special Purpose

This kind of real property is known as Special Purpose Real Estate, which is normally ignored because of above mentioned types by the masses. So, they are often counted as a part of Commercial Real Estate, but they are fundamentally different as they include real properties which are designed for particular uses and are extremely hard to repurpose for other functions apart from their original respective particular functions. They comprise libraries, amusement parks, government buildings, schools, colleges, universities, stadiums and arenas, hospitals, museums, airports, cemetery, in addition to religious buildings like churches, mosques, and temples.

Figure 5. Naturhistorisches Museum present in Vienna, Austria

Source: Pudelek, Wikimedia", Figure 6 "Source: Soica2001 (talk), Wikimedia

Here, Nataurhistorisches Museum shown in Figure 5, is an example of Special Purpose Real Estate.

After understanding the types of Real Estate, it is time to realize the pillars of Real Estate, which can't be overlooked.

Pillars of Real Estate

This subsection will focus on the pillars of Real Estate which are needed to make it smoothly functional or people about to be involved in Real Estate, can't ignore them. They are its investment which are considered as 4 aspects of investing in Real Estate which aids in wealth creation. When these pillars are combined together, then they look like a grand efficient machine which is able to generate revenue, so the mentioned pillars are discussed as follows:

- Leverage & amortization concepts
- Appreciation
- Cash flow
- Tax benefits

Leverage & Amortization Concepts

With the debt, it is possible to create wealth while utilizing the investment in Real Estate which acts as precious tool in this process. When there is a significant purchase, then the amount spent on the said cost could be spread out with the passage of time, which is a process known as Amortization, that could be attained by paying on a monthly basis on a loan for the same. This consequent in making the payments completed on a monthly basis to be lower that could be favourable in the case of cash flow. Further, if mortgage or loan is added, then investors could be leveraged to convert the inflation into higher rate of return on the said property, which decreases the associated risk on it.

Appreciation

With the passage of time, normally the value of Real Estate increases, which further raises the success of investing in Real Estate. The strategies like flipping, fixing in addition to renting, are used by the investors to enhance the values of the bought and occupied properties. The holding period as well as economic climate of the said properties decide the increase in the value of those properties, which leads the path of possible sales in addition to rent charges being higher than normal. The profit could be made from this process apart from the appreciation of the property by the refined and experienced investors in Real Estate.

Cash Flow

Here in the case of property, the expenses in addition to income generated from the said property is called as Cash Flow. With the flow of time, profit generation could be aided with cash flow which is considered essential in the investment of Real Estate. If investors want returns to be good on their investments in Real Estate, then it is crucial to maintain the said cash flow to be positive as well as the management needs to be proper.

Tax Benefits

Considerable benefits in tax could be obtained while investing in Real Estate, which comprise of deductions in the interest of mortgage, deductions in the case of the wear and tear of the given property to be depreciated, in addition to tax rates on capital gains. In the case of property tax privileged investors particularly, major tax savings could be done with the assistance of depreciation deductions. While investment properties are being sold, then tax rates on capital gains could lead to considerable savings, on the other hand taxable income could be decreased with deductions in the interest of mortgage.

Since the earlier subsection covered the pillars of Real Estate, so it is time to concentrate on its characteristics which will be covered in the next subsection.

Real Estate Characteristics

This subsection will cover several characteristics of Real Estate, which must be paid attention to while proceeding further, they are commonly categorized into two broad categories, which are:

Economic Characteristics

Economic characteristics include four subcategories which are improvements, location, scarcity, as well as permeance of investment, which could be explained as:

Permanence of Investment: When a property goes through any kind of improvement, then the capital invested on that improvement is normally an investment which is fixed and huge in amount. So, the return on that fixed investment is normally not only reasonably secure and constant but also long-term, which is why it leads to permanent investment.

Improvements: In case there is a piece of land, where improvement of any kind is made, then the use in addition to value of the land could be affected, preferably positive in nature. This might affect the community where the land is located after the improvements, in addition to neighbours.

Scarcity: The planet Earth has vast uninhabited or unused land, but there is a limit to the quality as well as location of the land, making it scarce in nature. For instance, in the case of downtown Chicago, the possible number of office buildings which are high-rise ones are finite in number.

Location: Among the economic characteristics, the location of Real Estate is the most crucial one. The reason behind is that humans give higher preference to properties in particular geographic areas in comparison to others, which consequence in properties having difference in values in various locations. For instance, one thousand square feet land value will be different in Tokyo metro area compared to Makkari village in Hokkaido, Japan. Similarly, the same can be said for San Francisco and Oklahoma City in USA or Mumbai and Coimbatore cities in India will have different worth for the same one thousand square feet land even if two places are located in the same country.

Figure 6. Makkari village in Hokkaido, Japan

Here, Makkari which is a village in Hokkaido in Japan as shown in Figure 6, will have lower value of one thousand square feet land in comparison to Tokyo metro are in Figure 7.

Figure 7. Satellite map of Tokyo metro area

Source: Google Earth

Here, one thousand square feet land will have higher value in Tokyo metro area in Figure 7 compared to the same area of land shown in Figure 6.

Social Characteristics

Social characteristics include three subcategories which are uniqueness, indestructibility, as well as immobility, which are as follows:

Uniqueness: Even if the homes have same layout or design and land could be similar, the said pieces of land will always be different, making each of them unique in nature.

Indestructibility: Land is considered not only permanent but indestructible in normal circumstances, as well as its location would not change ever.

Immobility: The surface of the land, or soil could be changed as per the requirements of the owner, but the said parcel of land's location remains the same always.

As the above subsection gave an understanding of the characteristics of Real Estate, it is essential to know that either commercial or private lender could help in financing with a mortgage or the people having cash to purchase Real Estate, are the normally considered the best ways to invest in Real Estate. Then, when Real Estate development is talked about, then it simply means the development of properties, which comprise of activities from the sale of parcels or developed land to other interested parties, purchasing of raw land, in addition to renovation of buildings which already exist. Now, it is time to get a rough idea of Real Estate industry, which the next subsection will cover.

Real Estate Industry

This subsection will discuss the working of the Real Estate industry, along with significant careers as well as jobs involved which completes the said industry. It can be said that this industry is broadly divided into numerous areas which are:

- Development
- Sales and marketing
- Brokerage
- Property management
- Lending
- Professional services

Development

The development of Real Estate can be explained as a process which requires not only the raw land to be purchased, rezoning of that land, then construction of buildings or renovation of existing buildings, putting the finished building on lease or sale to the interested people. Here the Developers are the ones who often finance a project which involves high risk, in order to earn profit later on by increasing the value of the said land, which could be done through rezoning, improvements in addition to constructing buildings on the said land. A novel product is created by Development Firms, which could be considered primary market, which will help novel inventory to be generated.

Figure 8. Real estate developer in Noida Extension, (Delhi NCR), U.P., India

Source: Keltechgroup0121, Wikimedia

Sales and Marketing

The firms which are for sales and marketing, often help in selling the units as well as buildings constructed by developers, making such firms to be close to developers. The Sales and Marketing Firms not only utilize their own sales agents to complete the selling of the buildings or apartments, etc., but also marketing materials are created by them, through where they earn their commission. Normally, these companies focus on novel units for the above process, as the charm of new units to the end users is higher than old ones.

Figure 9. Satellite map of TREMGroup registered address in Miami, USA

Source: Google Earth

Brokerage

A team of realtors or real estate agents are employed by companies like Real Estate Brokerage, who assist in enabling a property transaction between two parties where one is seller and other is buyer. They might be acting as the representative of either of the two mentioned parties, and then they aid the concerned party in attaining the sale or purchase of the property with the terms which could be the best for their clients.

Figure 10. Satellite map of Charles Schwab Headquarters in Westlake, Texas, USA

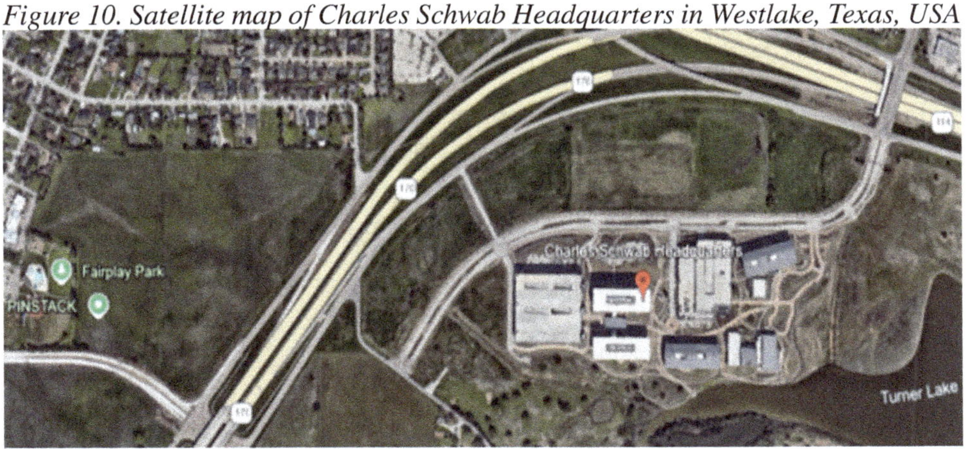

Source: Google Earth

Property Management

Firms which are involved in the management of properties, aid in the owners of real estate to rent out the particular properties, which might be an apartment or unit in a building, and so on. The Property Management Firms have the job to not only show the said units, repair those units, collect rent from the tenants, to manage the tenants in those units, as well as to fix any kind of deficiency in those units. The owners of those properties are supposed to give them a fixed percentage of the collected rent as the fee for the performed jobs of these firms.

Figure 11. Satellite map of Cushman & Wakefield headquarters near Chicago, USA

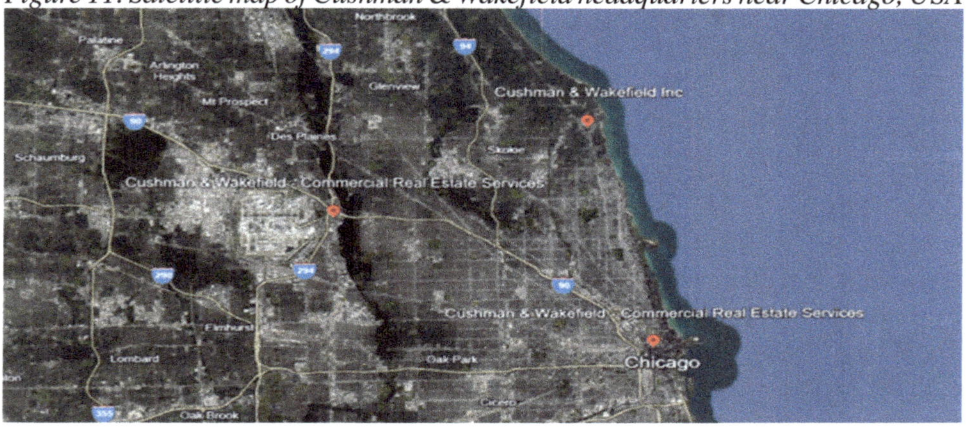

Source: Google Earth

Real Estate Lending

In the real estate industry, the developments as well as properties require debt or leverage, which are provided by Lenders, in order to finance the concerned projects normally. So, these Lenders are crucial in this industry, and they include credit unions, government institutions, banks, as well as private lenders.

Figure 12. KeyBank headquarters in Key Tower in Cleveland, Ohio, USA

Source: Cards84664, Wikimedia

Professional Services

To make the real estate industry functional, there are various professionals offering a wide range of services to the users. These professional services include interior designers, accountants, construction workers, lawyers, tradespeople, as well as general contractors.

Since the above subsection covered the overview of the real estate industry, so question comes in the mind about the investment in real estate is good or bad. Normally, many people think the business of real estate as not only an easy option of investment but also a considerably profitable one, which is the case in most situations. People who invest in real estate are able enjoy numerous benefits which include regular cash flow, leveraging the investment, appreciation, in addition to benefits in tax. This helps the investors in obtaining another way of acquiring passive income. However, it is also noteworthy that the market of real estate could definitely crash in case many negative factors continue with the passing time. The reasons behind the market crash of real estate, could be mortgage offerings being larger than normal, interest rates being flexible, risky investment movements, overbuilding, economic recession and so on. Now, it's time to see the consider the condition of real estate in India, which will be covered by next subsection.

Indian Scenario

In this subsection, Indian real estate market situation will be given focus to have a rough idea about the where the country is heading in regard to real estate sector.

The REIT platform also known as Real Estate Investment Trust was given permit by SEBI, which is the Securities and Exchange Board of India, due to which all types of investors will be permitted to invest in the country's market of real estate. This will help in building an opportunity of around $19.65 billion USD in the forthcoming years in India in the case of real estate sector. The developers are now accelerating the speed to meet the current rate of globalization, as well as the growing knowledgeable base of the users, to accept new challenges. Recently the trend of businesses which were family-owned ones has been changed to the ones which could be managed professionally, to indicate the above-mentioned change in the Indian real estate market. To manage numerous projects in different Indian cities, with the increasing necessity, the centralized processes are being invested by the developers, in order to not only hire professionals who are qualified in areas such as engineering, architecture, as well as project management, but also in organizing manpower, and to source materials in the projects. As per the earlier plans of the Central Government of the country, urban areas in the country were supposed to have construction of twenty million houses at reasonable prices by the targeted year 2022. This means the residential real estate sector was showing drastic growth in the past several years under Union Ministry of Housing and Urban Affairs supported PMAY scheme, which is Pradhan Mantri Awas Yojana, even though the practical construction growth of houses missed the planned deadline, but still significant growth has been noticed. However, this also means that the demand for retail offices and commercial buildings will increase further, because of increased houses in the urban areas in the country. There has been shortage of around 10.1 million houses in the urban areas of India, which is expected to increase in the future, so the IBEF says that further twenty-five millions of houses at reasonable prices will be required in the urban areas in India by the year 2030, because of the growing population. The Indian real estate sector received increasing flow of foreign direct investment, which is motivating the sector to be more transparent. Developers in India has been restored their management as well as accounting systems to attract funding for the projects in order to meet the industry standards. The earlier expectations were that by the fiscal year 2022, there will be considerable amount of capital infusion of foreign direct investment of around $ 8 billion USD in the real estate sector of India.

Figure 13. National review or consultation on PMAY in New Delhi, India in 2017

Source: Ministry of Housing and Urban Poverty Alleviation (GODL-India), Wikimedia

Till now this section covered the details of Real Estate in a conventional way, but with the advancement of technologies over the decades, it is natural to see how the Real Estate sector is being influenced by such technologies. This primarily include Artificial Intelligence, which has been changing a number of sectors and disciplines over the years, and of course the Real Estate sector wasn't ignored by its effect. There has been a significant transformation of the Real Estate sector with the application of AI in the past few years, which will be discussed briefly in the next subsection and later on in the coming sections in detail in this chapter.

Artificial Intelligence in the Real Estate Sector

This subsection will give a glimpse of Artificial Intelligence emergence in the area of Real Estate and the possible transformations occurring in this sector. When machine learning, sophisticated algorithms, in addition to analytics of the data to not only optimize but also improve as well as automate different processes in the Real Estate sector, then it can be said that AI technology is being utilized in the said sector. Earlier when AI technology was not introduced in the Real Estate industry, many things were manual, repetitive and time consuming, which made the processes slow and inefficient. However, with AI technology, market trends could be easily predicted, able to get tailored experiences whether tenant, sellers or buyers, in addition to property management to be modernized and automated, along with large datasets being analysed quickly. Targeting the right client, marketing efficiency being improved drastically, and aid in correct investment decisions all thanks to AI technology, increases the revenue generation considerably for people involved in the Real Estate industry. Communication could be done with virtual assistants, smart building management, valuation tools, as well as predictive analytics, are being highly helpful for agents, owners, sellers, buyers, and tenants.

Objectives

This book chapter has several objectives associated with the current study, which are as follows:

- To understand the concept of Real Estate and the major types involved in the Real Estate sector.
- To get an idea of the important pillars of Real Estate as well as the characteristics of this sector.
- To understand the Real Estate industry, investment, market crash, in addition to Indian Real Estate market in brief.
- To focus on the way Artificial Intelligence is changing the Real Estate sector and developing it further.
- To consider some significant case studies and future scope of Real Estate industry incorporating advanced technologies like AI.

MOTIVATION

The author performed this study on Real Estate and AI, because in the recent decade a lot of projects and construction activities are taking place in his home country, India. He observed that the Central Government of India has been focusing on Residential Real Estate projects in the recent years as well as planned to do the same in the coming years, to meet the demands of the increasing population of the country. However, at the same time there has been significant construction of Commercial Real Estate in India, especially in the urban areas, to meet the growing demand of people residing in newly developed Residential Real Estate in those areas. Even though, the Special Purpose Real Estate and Industrial Real Estate have also shown noticeable growth over the years, but the above two types of Real Estate are not only showing drastic growth but also significant affecting the lives of the people in India. At the same time the author, noticed the growth of Artificial Intelligence and other advanced technologies at a fast rate in his home country, as well as some form of AI technology being used in the search portals of Real Estate sector in India. Hence, the author decided to go for a detailed study on the ways AI could affect the transformation of Real Estate industry.

Why the Growth and Development of Real Estate Industry is Important?

For any country to progress, the development as well as growth of Real Estate industry is essential as quality of life, infrastructure along with economy of that country is directly affected by this industry. The property sales, construction activities involved in the land development or Residential Real Estate or Industrial Real Estate or Commercial Real Estate or Special Purpose Real Estate, as well as associated services, rentals and so on leads to massive contributions towards the GDP of a country. Further, engineers, lawyers, architects, urban planners, as well as labourers are needed during the development of Real Estate, which creates significant employment opportunities in a country. Then large-scale Real Estate projects are able to bring Foreign Direct Investment, provided they are transparent, high quality and meets global standards. This brings not only foreign capital to the said country, but the urban development as well as infrastructure are boosted considerably. Thus, the urbanization becomes more structured and future centric along with better living conditions, as well as economically active areas in the country, which is particularly needed for the growth of developing countries like India but the same can be said for underdeveloped countries in the world which needs more help from Real Estate industry to improve themselves. At the same time even developed countries need to rely on Real Estate industry to maintain their lead in the development in the world. This is why countries like India, China, South Korea, and other developing countries are having high urbanization in the recent decades, to become developed countries as early as possible, but of course the Real Estate industry is playing an important role in that plan in those countries. Then developed countries like Japan, USA and so on, are also paying attention to Real Estate industry as the said industry is a major pillar of their economy, and living conditions and day to day lives of their citizens are significant affected by Real Estate market. Plus, the quality and affordable housing are a must for all middle class and lower-class citizens in a country, as they are the dominant population, especially those countries which have a growing population. Better living standards, urban equity and decrease of the growth of slums, could be achieved with the development of Real Estate industry, which will also increase the Happiness Index of any country. Then as earlier mentioned in the chapter, that Real Estate is considered as an asset class, so for organizations and individuals, it could help them in dodging inflation. Plus, the long-term generation of income which increases the wealth of the involved people in Real Estate industry, could not be ignored. Then allied sectors like banking, facility management, logistics, and insurance are also helped with the development of Real Estate sector in a country. Then the current Real Estate industry is showing signs of technological integration and supports sustainable development, which is crucial for the development and future long-term growth of countries in the world.

Is There Any Need to Apply AI in the Development of Real Estate Sector?

The Real Estate sector plays an important role in the country's development as mentioned above in the last subsection, but normally this sector before the development of computers was completely manual, and with internet in the world, the Real Estate sector improved considerably. However, many of the activities in the Real Estate sector are still manual, which consumes a lot of time, leading to delays. Then, several activities are repetitive in nature, which not only consumes time, but also increases the chances of human error. This causes unnecessary inconvenience to the involved parties and makes the Real Estate sector significantly inefficient. Plus, the lack of transparency, scammers, illegal work, and lack of knowledge in Real Estate sector affect the customers greatly. Advanced technologies like Artificial Intelligence have been greatly improving many disciplines as well as areas over the years, and the glimpse of its positive effects could recently see in the Real Estate sector as well. Singh (2023) discussed that sustainable practices could be encouraged by AI in the construction industry, which means that construction activities becoming sustainable could lead to sustainable Real Estate projects in a country, which is what India and other countries need in today's time. Then, Singh (2025) explained how Civil Engineering could be evolved with the incorporation of AI and IoT, which is crucial for Smart Real Estate projects not only in the present scenario but also in the future. Then the Built Environment in India has been going sustainable development in addition to planning, as explained by Singh (2023), which means that India needs AI implementation in its Real Estate sector to support its Built Environment through the sustainable AI assisted Real Estate projects. Then Singh (2023) talked about ANN use in Civil Engineering, which is a subfield of AI, and could help in the predictive analytics while performing valuation of Real Estate, demand forecasting, analysis of the Real Estate market, and so on. Then, Singh (2023) shows the digitalization of the construction field, which is essential, but this will also a necessity for Real Estate projects as construction activities being digitalized will help in smooth implementation of AI technology in Real Estate projects. Then a number of infrastructure projects are being build in India in the recent years for development as explained in Singh (2023), where Real Estate sector will properly complement those infrastructure projects in India's development. As India has already became the most populus country in the world in the year 2023, and the population is still growing in the country, so Residential Real Estate, Commercial Real Estate, Special Purpose Real Estate and Industrial Real Estate projects will be essential to compliment the Infrastructure projects in the country for a high-speed development in the coming years. However, AI implementation of Real Estate sector at the same time will increase India's development at a considerable rate, as valuation of properties will be enhanced, customer services will be personalized, property management will be more efficient, recommendation and property search will be improved as well as faster, predictive maintenance, and so on will be possible. Of course, some form of AI technology has recently already entered the Indian Real Estate market, but more execution of AI technology will be essential in the Indian Real Estate sector. Similarly, other countries in the world to either develop further or to maintain their development, AI application in Real Estate sector will be important.

LITERATURE REVIEW

Gardes (2025) studied the professionals in the Real Estate sector, and she then proceeded with the apparent threat to the independence in services due to Artificial Intelligence, in order to understand whether AI could be considered as enemy or friend. Tysiac et al. (2024) studied principles of good governance in the case of real estate, in order to fuse algorithms led by AI technology with unmanned aerial vehicles measurement in their studied area, to get fruitful results. Haurum et al. (2024) Studied LangChain, and they worked on an agent on the basis of their studied area, in order to assist AI technology in the real estate sector. Bao et al. (2024) studied GA-SVM algorithm, and they then worked on supply chain in the context of real estate on the basis of their studied area, in order to aid the model on managing financial credit risk, where they thoroughly compared conventional model with AI based model, to get desired results. Xiao and Zhang (2025) studied debt risk of the government at the local level, and they worked on regulating the market of real estate. Manahov and Li (2025) studied the market of real estate, and they worked on the digitalisation of their studied area, as well as considered the major attacks from crypto hackers to accomplish their research. Guarini et al. (2025) studied the pricing of real estate, in order to understand their extrinsic in addition to intrinsic characteristics, and they further proceeded with strategies related urban planning in a sustainable manner, to get fruitful outcomes. Shah et al. (2025) studied the stock market of real estate in the case of Europe, and they then worked on environmental changes effects of fintech market in addition to renewable energy on their studied area. Ma and Sang (2024) studied the prices of real estate, and they worked on understanding whether carbon tax policy could encourage their studied area or not. Perera et al. (2024) studied the development of real estate, and they worked on solutions which are on the basis of nature to be permitted justice in their studied area. Dell'Anna (2025) studied real estate, and he worked on the effectiveness of evaluation energy performance certificate where ML framework was used, finally towards his studied area. He also chose private residential market of Turin in Italy as his case study to get desired outcome. Lenaers and De Moor (2023) studied rent prediction in real estate, and they worked on improving interoperability as well as transparency of model with the assistance of XAI techniques, to be used in their studied area. Elnagar (2024) studied websites of real estate, and she worked on improving the quality of electronic service by utilizing DL technique, which was further applied to her studied area, to get desired the desired results. Nyoni et al. (2023) studied the sector of real estate, and they considered the institutional as well as organizational perspective while considering the sustainability action in their studied area. De Paola et al. (2024) studied the investment risk investment in the case of real estate, and they worked on a model to classify their studied area which could be used for Naples administrative municipalities. Cardone et al. (2024) studied price estimation of real estate, and they worked on genetic algorithm which could lead a fuzzy partition to assist their studied area. Wolifson et al. (2024) studied the digital real estate, and they worked on discrimination in the time of their studied area by explaining Australian rental schema. Aungkulanon et al. (2024) studied the management of real estate, and they worked on the way the maintenance responsibility distribution could be optimized in the case of their studied area. Unel and Yalpir (2023) studied land management, and they worked on the appraisal of mass real estate utilization with the assistance of design of sustainable tax system, which could help in their studied area. Allam and Allam (2024) studied the development of real estate, and they worked on the metaverse as well as in the direction of their studied area to get fruitful outcomes. Ullah et al. (2021) studied smart real estate in the context of Australia, and they worked on hurdles which comes on the path of innovation as well as digitalisation of their studied area. Yukun and Yan (2022) studied commercial real estate, and they worked on short lasting period of rights on land utilization with the discussion of their studied area, on examination ideas of the income method, to accomplish their research. Ullah et al. (2021) studied the online platforms of real estate, and they worked on modelling the perception of users of their studied area in the context of environment which is considered digitally disruptive in nature. Renigier-Bilozor et al. (2022) studied the sustainable development of the market of real estate, and they worked on necessary automatic modernized tools acceptance worldwide in the case of their studied area. Lin et al. (2021) studied the industry of real estate, and they worked on disruptive technology with model based on spatial neural network, which could appraise property in their studied area. Ullah et al. (2021) studied online platforms of real estate, and they worked on investigating the perception of the users using DEMATEL method in the context of their studied area, to get desired results. Pu et al. (2022) studied the transactions in real estate, and they worked on examining the asymmetrical financial information in the country China in the context of their studied area. Potrawa and Teterewa (2022) studied the market of real estate, and they worked on model based hedonic pricing which was powered by ML technique to be used for their studied area. Jamil et al. (2025) studied Brunei Darussalam, and they worked on 3 decades of data related to residential property starting from records to Artificial Intelligence, to be used in their studied area, to obtain desired results. Al Bahsh and Hosseinian-Far (2021) studied UAE based real estate developer as part of case study which could be considered qualitative in nature, and they worked on the way big data analytics affect the competitive intelligence, to accomplish their study. Aljohani et al. (2024) studied property market while considering volatility transmission during not only the recent pandemic crisis but also the financial crisis in the world during the period of 2008-09, to get favourable outcomes. Zheng (2023) studied ML method, and he worked on house prices as well as community resilience, to obtain fruitful outcomes. Sahbi et al. (2024) studied textual advertisements including boats as well as real estate, to proceed with programmed ontology population, while comparing semantic and LLM methods, to accomplish their study. Doan and Nguyen (2025) studied the relationship between counterproductive work behavior, in addition to AI awareness, as well as turnover interaction across six Vietnamese enterprises which comprised real estate, transportation, etc., and they worked on a mediation model which was moderated to assist in their studied area. Erricsson et al. (2024) studied building materials reutilization in the construction as well as real estate industry, while considering clients perspective in Sweden, to get desired results. Gonzalez et al. (2025) studied the way environmental performance of REITs affecting the sustainability associated interest rates on loans to help in the direction of green gains, which would be beneficial in real estate sector's decarbonization. Sahbi et al. (2025) studied ontology populations from advertisements in France which comprised of boats, real estate in addition to restaurants, and they considered Claude vs KOnPoTe as the case study for their studied area, where they used method on the basis of LLM vs semantics, to obtain favourable outcomes. Yu and Zhang (2025) studied cleaner waterways, and they worked on its value on housing prices, businesses, etc., while considering the proof from the water program named Black-and-Odorous, which will cover the associated real estate market in that area in China. Tukur et al. (2024) studied the applications, technologies, in addition to techniques from past studies in the digital environs particularly the metaverse one, where they considered eleven industries which included real estate, education, etc., to observe the effects of the earlier mentioned applications, in order to get desired results. Li et al. (2023) studied the proof from cost behaviour of corporate, to proceed with unintentional results of financing on the basis of collateral, where their results displayed gain in the real estate assets of a company which eases the cost stickiness of the said company.

RESEARCH GAPS

After going through several published papers in journals and research books, only the manuscripts published in the past few years were considered further. However, more research was done from China or developed countries like USA, Australia and so on, in comparison to other countries in the world. Research from India was lacking in among the recent publications selected in this study compared to other countries. Since, India is among the top 3 countries recently in the list of total asset value in the Real Estate sector in the world, so more research must be done to incorporate AI in Real Estate development. Further, this study considered numerous AI tools, and major case studies where Artificial Intelligence is being incorporated in the Real Estate industry, which will be helpful for research in this direction, to develop the said sector further.

MAIN FOCUS OF THE CHAPTER

This book chapter focuses on explaining the concept of Real Estate and the five major types of Real Estate in existence. Then, the chapter discusses pillars of Real Estate, as well as the characteristics of Real Estate sector in detail. It also talked about the outline of Real Estate industry, and the current situation in the country India in the context of Real Estate. Further, Artificial Intelligence in Real Estate sector was introduced, and the reason behind this study by the author were talked about. Then, importance of the development as well as growth of Real Estate industry was discussed, in along with the necessity of implementing AI technology in the Real Estate sector was explained in detail. Moreover, issues and problems related to AI implementation in Real Estate were mentioned, along with applications of Artificial Intelligence in the said sector, were covered. The chapter also mentioned various major AI tools which could be beneficial to the people involved in the Real Estate sector. Additionally, incorporating AI technology in the existing Real Estate system in India as well as major case studies across the globe, were discussed in this chapter. Then, gaps in the literature and the possible situation and future of AI applied Real Estate sector were explained.

ISSUES, CONTROVERSIES, PROBLEMS

Even though Real Estate sector is important for any country's growth and investment in it looks very promising with stable regular long-term returns normally. However, the issues and problems related to Real Estate sector could not be overlooked, which will be discussed in this section. The investment return from the Real Estate sector, where Real Estate fund is being talked about, takes longer time from other funds, so it is not suitable if the investors want quick income. Even though Real Estate fund depends upon the property type, but in some cases the return takes extremely long time, which might be highly uncomfortable for certain investors who are not having multiple sources of income generation. Then Real Estate market is also unpredictable, as interest rates, economy, government policies as well as demand, affect it. This causes fluctuations in Real Estate market, and the graph also goes down at times. Then the wrong location of the property will lead to undesirable return from Real Estate fund. Additionally, transaction cost is high in the case of Real Estate funds, which might reduce the return on the invested amount or profit will be lower than expected. Further, vacancy rates and maintenance requirements are high in Real Estate projects, which are also major problems. Then low liquidity of Real Estate funds in compared to other types of funds along with cases of negative cash flow, are other issues related to Real Estate sector. Plus, problems with the implementation of Artificial Intelligence in Real Estate sector must also not be ignored. The problem of automated leaning abilities with AI often occurs wherever AI is implemented, and the same could be said for Real Estate sector, so AI specialists are must who could solve such problems. Data security and protection against cyber crimes are other issues which happens when AI is executed in Real Estate sector. Then, compliance with the local, state and national laws must be considered when implementing AI in Real Estate industry, to avoid legal problems. Moreover, transparency with high quality and detailed data in high volume are a must for AI to analyse and operate efficiently. However, building profiles, rents information, lease agreements, sales history, tenants' details, and so on in the case of Commercial Real Estate as well as Residential Real Estate might be missing or not much in detail, which might hinder the analysis and operation of AI in Real Estate sector. So, the above problems and issues must be handled properly to enjoy the numerous benefits of Real Estate sector and when the same sector implements Artificial Intelligence to develop further to offer more benefits.

ARTIFICIAL INTELLIGENCE APPLICATIONS IN REAL ESTATE

AI is developing the Real Estate sector in an innovative way where it could meet the technological advancements and grow to the next level. In this section, some major applications of executing AI technology in the Real Estate Industry will be discussed which are:

- Automated Property Management

- Intelligent Property Search
- Automated Property Inspections and Valuation Model
- Intelligent Virtual Staging
- Real Estate Portfolio Management with Risk Assessment
- Smart Energy Management
- Lease Abstraction
- AI-powered Fraud Detection
- Virtual Property Tours and 3D Modeling
- Advanced Analytics
- Legal and Regulatory Compliance
- Building Health Monitoring
- AI Predicted Urban Planning
- AI Real Estate Advisors

Automated Property Management: Due to Artificial Intelligence, reminders as well as record keeping in property management have become automatic. This increases the tenant experience, enhanced efficiency, along with modernized operations and automation of many things to business owners, which includes rent collection, maintenance request management, and so on.

Intelligent Property Search: With the implementation of AI technology, the recommendation systems as well as property search engines have become tailored in nature for users based on their browsing patterns. The system analyses budget of the users, location preferences, past searches, as well as property type to give suitable results to the users.

Automated Property Inspections and Valuation Model: The value of properties are assessed in addition to their evaluation, based on statistical analysis, mathematical models, as well as computer vision abilities, with the assistance of AI technology implementation in the Real Estate sector. Large datasets are used for the training purposes, ML models, along with advanced algorithms, makes assessment of property highly accurate as well as fast, which prevents the risk of under-pricing or overpricing and saving of time to visit the property again and again.

Intelligent Virtual Staging: The incorporation of AI helps in the process of virtual staging optimization, as computer vision as well as generative AI are able to automate the turning outdated photos or empty photos of properties into fully furnished as well as visually appealing photographs.

Real Estate Portfolio Management with Risk Assessment: With large datasets being analysed by AI, such as economic indicators, property performance, in addition to market trends, the management of property portfolios become easier. Then risk assessment like property depreciation, fluctuating market conditions, in addition to tenant defaults, could be done through predictive analytics of AI.

Smart Energy Management: When Implementing AI in the operating system of Real Estate, in the case of energy management, real-time data analysis on the consumption of energy in order to detect inefficiencies to proceed with usage optimization in the operations become possible, making the move environmentally sustainable.

Lease Abstraction: Here, AI is able to summarize the main information coming from the lease agreement, which is converted by it into standard format, making the management of lease highly efficient as well making sure that it is according to the accounting standards as well as terms.

AI Powered Fraud Detection: There are thousands of frauds happening in the Real Estate sector leading to millions of USD losses, but with assistance from AI, the anomalies could be detected, transaction patterns could be analysed, detection of forgeries or alterations in bank statements as well as pay stubs to authenticate documents. Further, identifying unauthorized alterations in property records which could signal towards title or deed fraud through property document analysis could be performed easily.

Virtual Property Tours and 3D Modeling: Many times, physical tours in the properties which is essential, is not possible by buyers or tenants, which is also time consuming in nature. However, AI helps in 3D models as well virtual property tours creation, which improves the experiences of the customers greatly, as well as accelerated sales of the properties at the end.

Advanced Analytics: Human mind has limit to the amount of data it can analyse as well as speed and quality of analysis decreases with higher amounts of data. However, AI integration in Real Estate sector, aids in advanced analytics in the case of predictive property value trends, predictive analytics for property investment, quality analysis of neighbourhood life, and behavioural analysis as well as screening of tenants.

Legal and Regulatory Compliance: AI aids Real Estate professionals in detecting regulatory or legal risks linked to the development of any property, monitoring regulatory changes in real-time, loophole in legal documents, making sure that accuracy is present in the data of the reports, and allows programmed compliance reports to be generated.

Building Health Monitoring: In Real Estate, AI is able to help in the identification of smoke and fire, detection of leakage of water, monitoring of escalator as well as elevator, scanning electrical systems, HVAC, signs of wear as well as tear, plumbing problems and so on, to ensure the overall health of the property or building.

AI Predicted Urban Planning: High volumes of datasets are analysed by AI algorithms, which comprise of traffic patterns, social trends, demographics, and environmental factors, which are important in making decisions related to city planning. This leads to liveable as well as sustainable cities and enhancing quality of life for people in urban areas while focusing on pressing challenges related to urban planning at the same time.

AI Real Estate Advisors: AI does not only analyse the market more accurately and quickly but at the same time give insight through predictive modeling, ML, as well as data analysis. It detects profitable ventures, market dynamics through historical data analysis, along with virtual assistants as well as chatbots to provide customer support 24/7, investment risk evaluation, and so on, way better than human advisors in Real Estate sector.

ARTIFICIAL INTELLIGECNE TOOLS

This section will discuss major AI based tools which could considered useful for agents involved in the Real Estate sector. Even though there might be many more Artificial Intelligence tools made for Real Estate agents but the ones which might be bugged, or underperforming will be ignored as well as considering the scope of this chapter, some major ones are:

- CINC
- Top Producer
- Lofty
- Saleswise
- REimagineHome
- Wise Agent
- Alma
- Sidekick
- Scout
- RealGrader
- Styldod
- Write.homes
- Canva
- Otter.ai
- Grammarly
- ChatGPT
- Xara
- Tidio
- Midjourney
- DALL-E 2

CINC: This is a platform which is thoroughly designed particularly for the professionals involved in the Real Estate sector. Artificial intelligence is utilized by CINC to connect and detect with the clients, in addition to enhance the process of conversation.

Top Producer: This is a Customer Relationship Management system which is improved with Artificial Intelligence. It aids agents in raising as well as detecting possible clients in particular areas in a geographic context, market insights led by AI technology and so on.

Lofty: This AI tool is for professionals in the Real Estate sector, and it concentrates on optimization of pipelines as well as improved client relationship management.

Saleswise: Another tool supported by AI technology, which aids in not only visualization of the transformations in rooms, but also in email writing, and performs comparative market analysis.

REimagineHome: An AI tool which helps in visualizing properties and aids the agents to display those properties in the desired light, to get better results in deals.

Wise Agent: This platform is powered by AI technology to automate marketing, and customer relationships management in the Real Estate sector. It assists the agents in modernizing their efforts in marketing, tasks to be automated, as well as contacts to be managed properly.

Alma: This AI tool was developed DealMachine which acts as an investment tool in the Real Estate industry. It aids investors as well as agents to analyse the opportunities in investment in addition to off-market properties with potential to be detected.

Sidekick: For agents in the Real Estate sector, this AI tool is a virtual assistant, which assist those agents in communications management, routine tasks to be automated as well as workflow of the agents to be modernized.

Scout: This tool utilizes Artificial Intelligence, to help agents in Real Estate sector to not only connect but also detect customers at around ten times speed faster than the norm.

RealGrader: This tool is supported by AI technology, to help agents in Real Estate industry in reputation as well as social media management.

Styldod: This platform is led by AI technology, helps in the creation of content, virtual staging, as well as marketing of properties.

Write.homes: This AI tool is for generation of content, and it aids agents in efficiently as well as swiftly marketing materials, posts to be blogged, and convincing property descriptions to be made.

Canva: This platform is a generic design one supported by AI technology, but could help agents in promotional images, flyers, graphics, in addition to social media posts creation.

Otter.ai: This AI tool is general purpose one for taking notes as well as transcription, but could assist agents in taking property notes, and client meetings to be transcribed as well as recoded.

Grammarly: This is another generic AI tool which helps in writing, by correcting syntax, spellings, as well as grammar. It could help agents in writing emails, blog posts, in addition to listing descriptions.

ChatGPT: This is a famous generic AI tool which is for almost all-purpose use instead of anything specific, but it could help agents in online ads, listing descriptions to be generated, email writing, and so on.

Xara: This is AI tool, which is for content marketing, but agents could also use in printing materials, tasks related to content creation, as well as social media posts to be automated.

Tidio: This tool is powered by AI for customer services which is beneficial for different kinds of businesses. The users involved in Real Estate sector, could use it to get instant responses, real-time connection with possible clients, in addition to scheduling displays.

Midjourney: This AI tool is for generation of images, and agents in Real Estate industry could use it in 3 dimensional renderings to be generated, visual staging to be created, as well as photos of properties to be edited.

DALL-E 2: This tool is supported by Artificial Intelligence for generation of images from texts. However, agents in the Real Estate sector could use it make images which are customized, like virtual staging, and visuals of properties.

AI IMPLEMENTATION IN EXISTING SYSTEM IN INDIA

This section will cover the execution of Artificial Intelligence in the already old existing platforms in the Real Estate sector. Here, some of the major platforms used in India will be discussed to understand the development of Real Estate sector in India with the incorporation of Artificial Intelligence, which are:

- MagicBricks
- Housing.com
- 99acres.com

MagicBricks

A major leading portal in Indian Real Estate sector, which came into existence in the year 2006. PropIndex was introduced by the portal in the recent years, which is an application that has the ability to utilize AI technology in order to give price indices of properties in real-time. The said application of MagicBricks separates high volume of information on the activity of the users, conditions of the Real Estate market, analyses of property rates, property listings, in addition to current trends.

Figure 14. Satellite map of MagicBricks headquarters in sector 16a, Noida, Delhi NCR, U.P., India

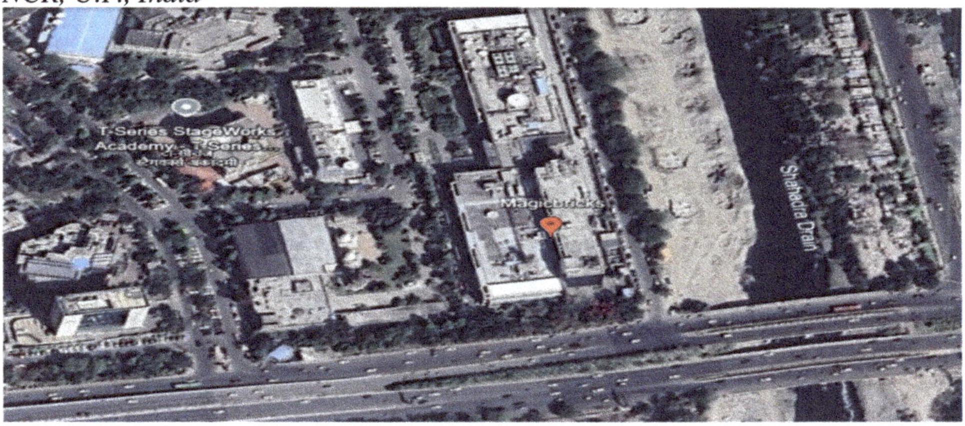

Source: Google Earth

Housing.com

Housing.com is another major search portal in the Indian Real Estate sector, which came into existence in the year 2012. Image recognition technology-based system was introduced by Housing.com, which uses Artificial Intelligence in order to improve the site for the users, so that the posted properties have higher clarity. Through this technology, images of the properties are tagged as well as categorized independently, which makes the listings of the properties more useful as well as highly accurate. The users have better search experience, as images which are of poor quality or not related, are excluded with the assistance of the Image Recognition based AI technology. That's why the property searches by the users have become not only highly accurate but also efficient.

Figure 15. Satellite map of Housing.com registered office in Sector 32, Gurugram, Delhi NCR, Haryana, India

Source: Google Earth

99acres.com

99acres.com is another major leading portal in the Real Estate market in India, which was launched in the year 2005. Chatbot using Artificial Intelligence was later added by 99acres.com, to meet with help search as well as inquiries of the clients in the context of properties, to improve their service. Natural language queries are processed by the chatbot of 99acres.com, in order to recommend properties to the users along with other useful information.

Figure 16. Satellite map of 99acres.com head office in sector 132, Noida, Delhi NCR, U.P., India

Source: Google Earth

SOME MAJOR CASE STUDIES IN THE WORLD

In this section some major case studies across the world where Artificial Intelligence implementation in the Real Estate sector could be observed. Real Estate companies mentioned here uses AI technology in investing, selling as well as buying of properties, which are as follows:

- Royal London Asset Management
- Zillow Group, Inc.
- Move, Inc.
- Redfin Corporation
- Crexi

Royal London Asset Management

An Investment company which is a leading one in UK, called as Royal London Asset Management, which came into existence in the year 1988. It is said that the company experienced considerable enhancements in energy efficiency in addition to its Heating, Ventilation and Air Conditioning in a commercial office building of area of 11,600 sq. mt. The company not only was able to decrease the emission of carbon by 500 MT/year but also fifty nine percent of energy savings, as well as seven hundred and eight percent of return on investment, with the incorporation of JLL's Hank technologies led by Artificial Intelligence.

Figure 17. Satellite map of the headquarters of Royal London Asset Management in London, UK

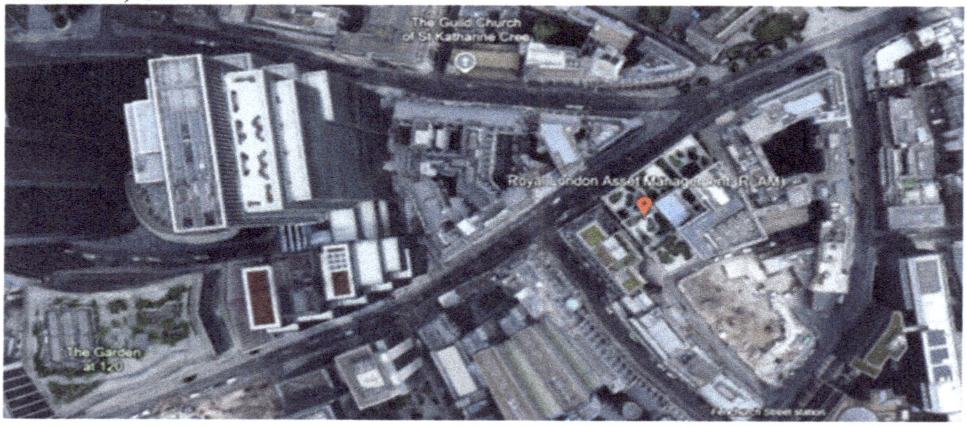

Source: Google Earth

Zillow Group, Inc.

The company Zillow, Inc. was founded in the year 2006 in USA, with the registered headquarters in Seattle, Washington. However, later on after acquiring other companies, it was changed to Zillow Group, Inc. The company uses Zestimates which is on the basis of AI technology, to estimate the value of properties based on digital photos. It particularly depends on neural networks, where it was trained by millions of photographs to have only 2.4% of national median error rate. Zillow Group in the year 2019 launched three-dimensional home tours which was supported by AI technology to be used in USA as well as Canada. The company has subsidiaries like StreetEasy, Trulia, in addition to HotPads. Trulia, Inc. was founded in the year 2005 but from 2015 onwards it became a subsidiary company of Zillow Group, Inc. Trulia uses Artificial Intelligence to make the experience of the users customized in nature, based on what the user browses as well as the search terms used related to properties, the learning of AI about the user takes place, to show the suitable recommendations to the user.

Figure 18. Satellite map of Zillow Group, Inc. Headquarters in Seattle, USA

Source: Google Earth

Figure 19. Satellite map of the headquarters of Turlia, Inc. in San Franciso, USA

Source: Google Earth

Move, Inc.

The American Real Estate company Move, Inc. was founded in the year 1993 in with the headquarters at present in California's Santa Clara. The company has the famous Real Estate website known as Realtor.com which has its own registered headquarters in Austin currently. It offers standard Real Estate processes like rentals, preapprovals, purchases, as well as sales of properties, with the assistance of AI, the listings are examined for lead generation, which could be later on sold to partner agents who are directly under the respective homeowners.

Figure 20. Satellite map of the headquarters of Move, Inc. in Santa Clara, USA

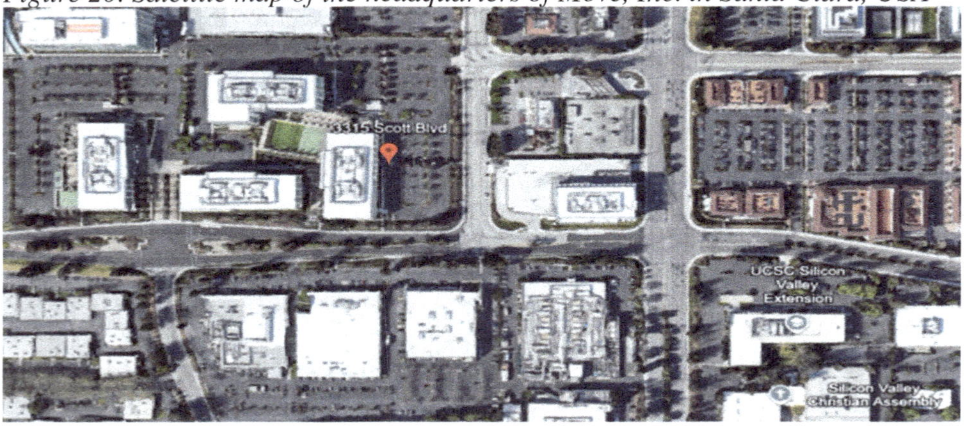

Source: Google Earth

Figure 21. Satellite map of the headquarters of Realtor.com in Austin, USA

Source: Google Earth

Redfin Corporation

Another American company in the Real Estate industry, which has the headquarters in Seattle and came into existence in the year 2004. The agents are augmented with Artificial Intelligence in the company, where part of their work-intensive is taken care by AI. The process of recommendation was automated which includes the matchmaking tool powered by AI, to generate better property results.

Figure 22. Redfin Corporation headquarters in Seattle, USA

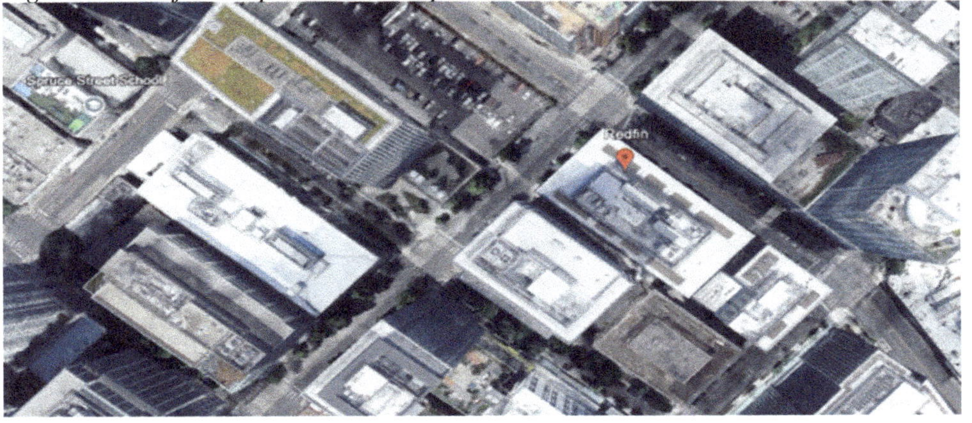

Source: Google Earth

Crexi

This company came into existence in the year 2015 and is known as Commercial Real Estate Exchange, Inc. The company is suitable for Commercial Real Estate type and offers data analytics, integrative as well as sophisticated technology for the same. The Commercial Real Estate solutions are assisted by AI technology of the company which aids agents to show the properties which are not only aesthetically pleasing but also detailed, to help millions of users on a monthly basis.

Figure 23. Satellite map of the headquarters of Crexi in Los Angeles, USA

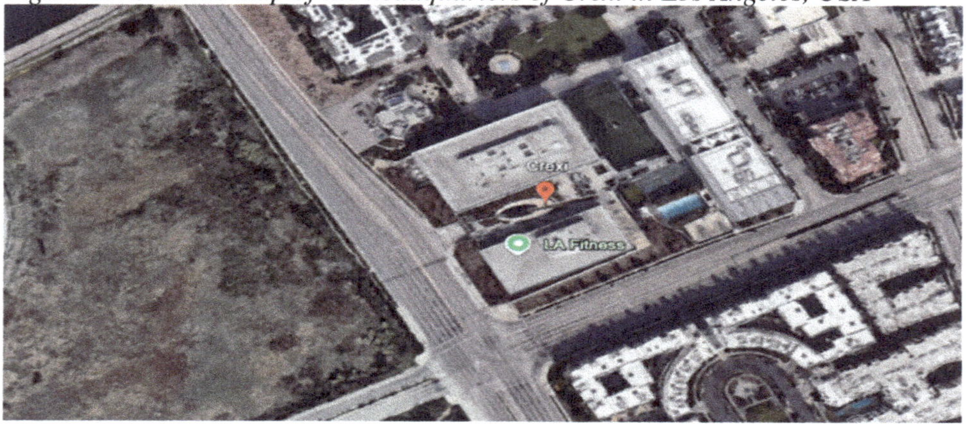

Source: Google Earth

METHODOLGY

In this section the problems and obstacles observed in the above case studies will be discussed, which are as follow:

- Royal London Asset Management
- Zillow Group, Inc.
- Move, Inc.
- Redfin Corporation
- Crexi

Royal London Asset Management: The company faced major issues like legal as well as compliance problems, pressures related to diversity as well as inclusion, so on. However, challenges like data integrity and transparency of AI driven models were some concerns which the company has to deal while implementing Artificial Intelligence.

Zillow Group, Inc.: The company faced problems such as regulatory changes, market volatility, high market competition and etc. Then the company's iBuying which was heavily dependent on AI, which failed, and the company was also criticized for accuracy along with ethical concerns during its operating life.

Move, Inc.: The company faced considerable obstacles like market sensitivity as buyer demand as well as inventory levels fluctuates on the basis of economic factors, and then intense competition with major Real Estate companies like Redfin

Corporation as well as Zillow Group, Inc. The company had noticeable concerns regarding AI implementation were data gaps, lagging behind Zillow's sophisticated AI technology, and so on.

Redfin Corporation: This company had significant obstacles to face which were debt obligations, legal settlements, financial problems due to decline in market share, and so on. Moreover, the company had to face challenges like AI model transparency, possible operational AI risk because higher dependence on it in the case of volatile market, as well as fair housing compliance.

Crexi: The company is comparatively new in comparison to large established Real Estate companies like Redfin Corporation, Move, Inc., Zillow Group, Inc. and etc. So, the company had to face challenges like adoption as well as monetization of its platform in the Real Estate industry, and market competition. Furthermore, gaining trust of users in its automation, data standardization, and so on were other major obstacles on the company's path while implementing Artificial Intelligence.

RESULTS AND DISCUSSION

In this section, data from different sources will be collected to perform graphical analysis in order to understand Real Estate sector.

Table 1. Total asset value real estate market

Market Value of Real Estate	In (USD)
USA in 2025 (Expected)	136,600,000,000,000
World in 2025 (Expected)	645,390,000,000,000

Source: Adapted from data in Statista

Table 2. Comparison of APAC and rest of the world in the share of luxury real estate projects in 2023

Share of Luxury Real Estate Projects	In (%)
APAC in 2023	19.5
Rest of the World in 2023	80.5

Source: Adapted from data in Statista

Table 3. Global market growth in terms of total asset value of real estate sector

Market Growth of Real Estate Sector in World	In (USD)
Market Value in Year 2025	654,390,000,000,000
Market Value in Year 2029	727,800,000,000,000

Source: Adapted from data in Statista

Table 4. Comparison of total asset value of real estate market globally on the basis of types

Global Market Value of Real Estate Based on Type	In USD
Residential Real Estate in 2025 (Expected)	534,370,000,000,000
Other Types (Example: Commercial, Industrial, Special Purpose) in 2025 (Calculated)	120,020,000,000,000

Source: Adapted from data in Statista

Residential Real Estate in 2025 (Expected) (Statista) = 534,700,000,000,000 USD

Global Real Estate in 2025 (Expected) (Statista) = 654,390,000,000,000 USD

Hence, other types of Real Estate in 2025 = 645.39 trillion USD – 534.7 trillion USD = 120.02 trillion USD

Table 5. Comparison of total asset value of real estate market in India on the basis of types

Market Value of Real Estate in India	In (USD)
Residential Real Estate in 2025 (Projected)	40,200,000,000,000
Other Types of Real Estate in 2025 (Calculated)	6,560,000,000,000
Overall Real Estate in 2025 (Projected)	46,760,000,000,000

Source: Adapted from data in Statista

Residential Real Estate in 2025 in India (Expected) (Statista) = 40,200,000,000,000 USD

Overall Real Estate in 2025 in India (Expected) (Statista) = 46,760,000,000,000 USD

Hence, other types of Real Estate in 2025 in India = 46.76 trillion USD – 40.2 trillion USD = 6.56 trillion USD

Table 6. Growth of total asset value of real estate market in India

Market Growth of Real Estate Sector in India	In (USD)
Market Value in Year 2025 (Projected)	46,760,000,000,000
Market Value in Year 2029 (Expected)	51,540,000,000,000

Source: Adapted from data in Statista

Table 7. Comparison of total asset value of real estate market among top 3 countries in the world

Market Value of Real Estate Based on Country	In (USD)
USA in 2025 (Projected)	136,600,000,000,000
China in 2025 (Projected)	133,470,000,000,000
India in 2025 (Projected)	46,760,000,000,000
Rest of the World in 2025 (Calculated)	337,560,000,000,000

Source: Adapted from data in Statista

Real Estate in USA in 2025 (Statista) = 136.6 trillion USD

Real Estate in China in 2025 (Statista) = 133.47 trillion USD

Real Estate in India in 2025 (Statista) = 46.76 trillion USD

Real Estate market value of top 3 countries in the world in 2025 = 136.6 trillion USD + 133.47 trillion USD + 46.76 trillion USD = 316.83 trillion USD

Global Real Estate in 2025 (Statista) = 654.39 trillion USD

Rest of the World in 2025 = 654.39 trillion USD – 316.83 trillion USD (calculated) = 337.56 trillion USD

Table 8. Growth of annual market output of real estate in India

Annual Market Output of Real Estate in India	In (USD)
Year 2020	180,000,000,000
Year 2047 (Expected)	5,800,000,000,000

Source: Adapted from data in Knight Frank

Table 9. Total units sold and launched in top 3 Indian cities in real estate sector in FY 2022-23

Top 3 Cities in India in Sales & Launches in Real Estate Sector	FY 2022-23 (Units)
Mumbai	175,603
Delhi NCR	121,693
Bengaluru	96,783

Source: Adapted from data in IBEF (Calculated and Modified)

Table 10. Share of office space based on industries in India in 2023

Share of Industries in Office Space in India in 2023	In (%)
Banking, Financial Servies, and Insurance	22
Technology	21
Others (Engineering & Manufacturing, Flexible Office Space, etc.) (calculated)	57

Source: Adapted from data in IBEF

Table 11. Growth of commercial space demand in top 8 cities in India

Commercial Space Demand in Top 8 Indian Cities	In (square feet)
Year 2017	29,000,000
Year 2023	38,000,000

Source: Adapted from data in IBEF

Table 12. Progress of Pradhan Mantri Awas Yojna (PMAY)

PMAY Progress in India	In (Units)
Completed Houses in Urban Areas (PMAY-Urban)	9,261,000
Completed Houses in Rural Areas (PMAY-Rural)	26,714,866

Source: Adapted from data in Ministry of Housing and Urban Affairs, Government of India (PMAY-Urban), and PIB Delhi, Ministry of Rural Development (PMAY-Rural)

Figure 24. Total asset value Real Estate market

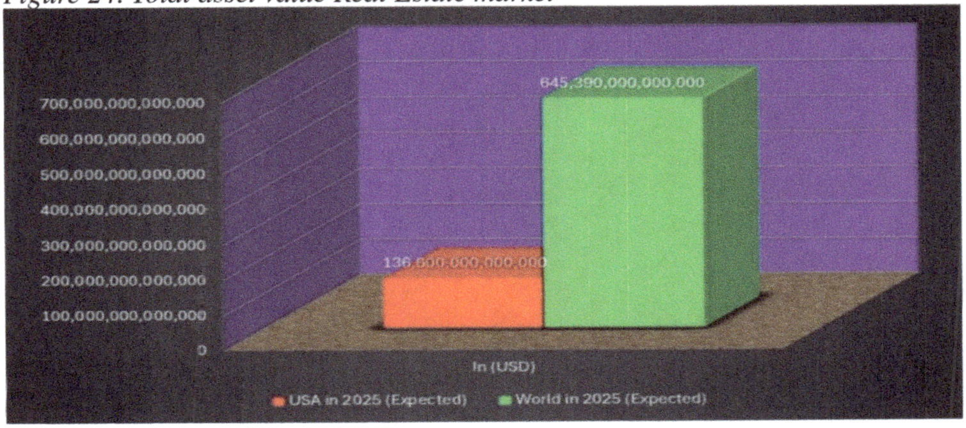

Source: Adapted from data in Statista

Figure 25. Comparison of APAC and rest of the world in the share of luxury Real Estate projects in 2023

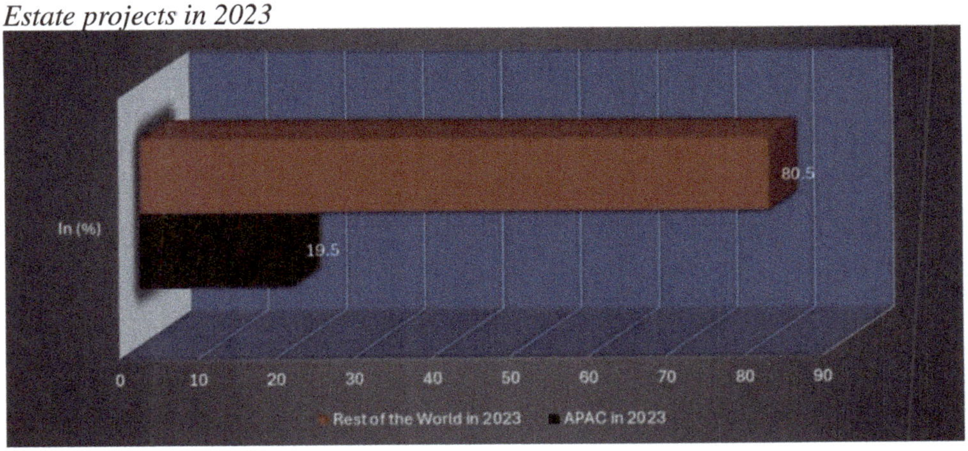

Source: Adapted from data in Statista

Figure 26. Global market growth in terms of total asset value of Real Estate sector

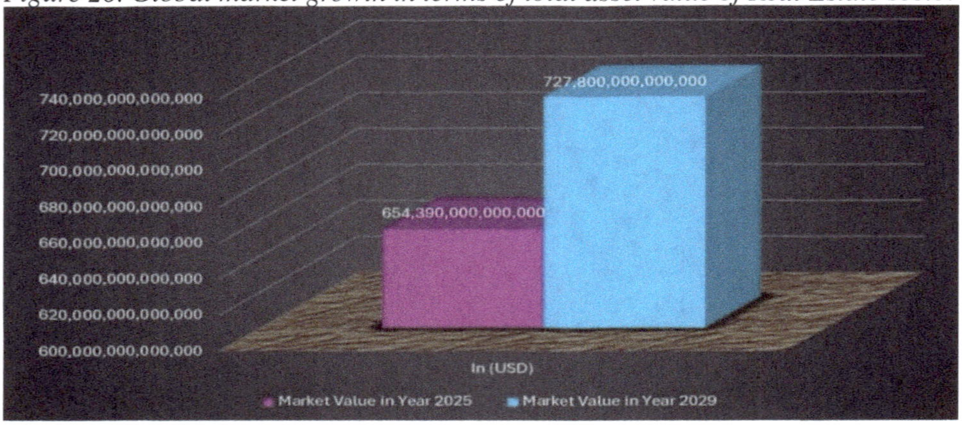

Source: Adapted from data in Statista

Figure 27. Comparison of total asset value of Real Estate market globally on the basis of types

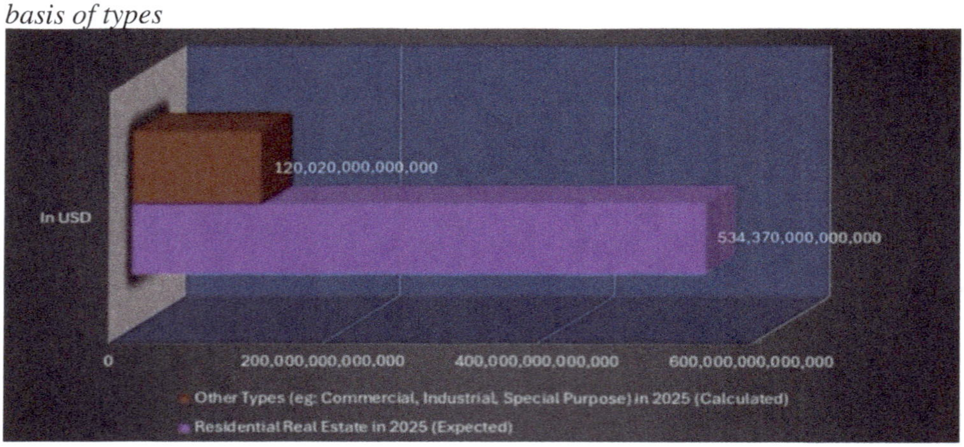

Source: Adapted from data in Statista

Figure 28. Comparison of total asset value of Real Estate market in India on the basis of types

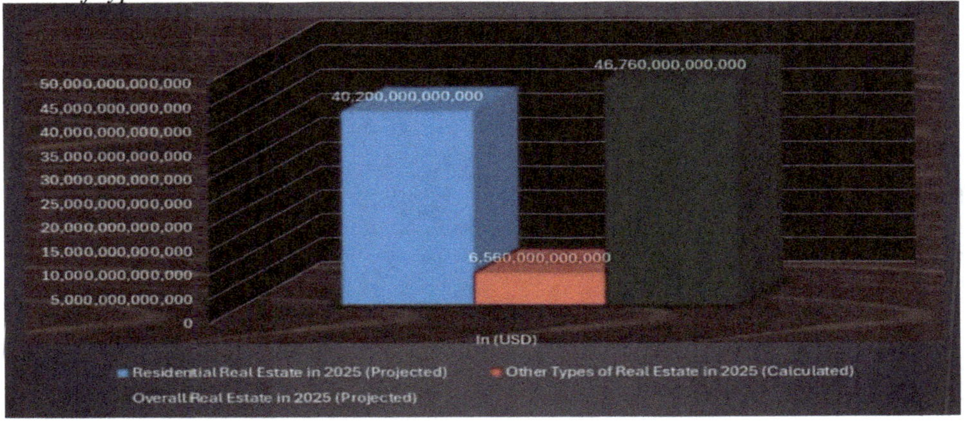

Source: Adapted from data in Statista

Figure 29. Growth of total asset value of Real Estate market in India

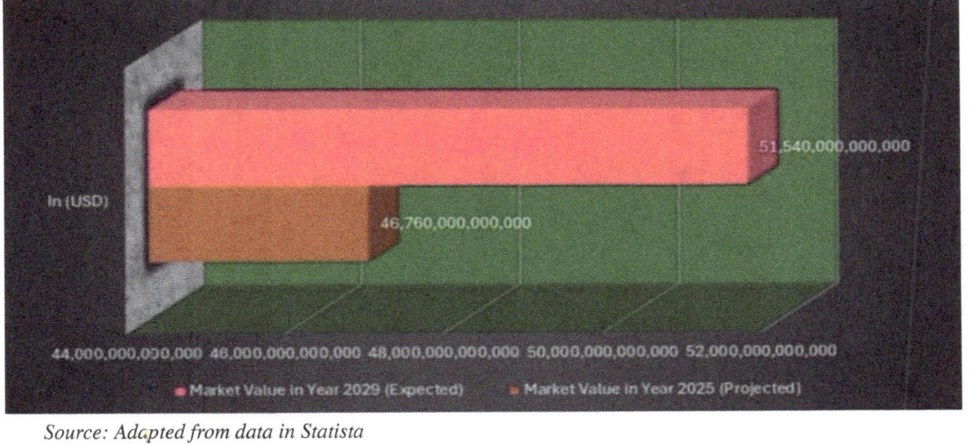

Source: Adapted from data in Statista

Figure 30. Comparison of total asset value of Real Estate market among top 3 countries and the rest of the world

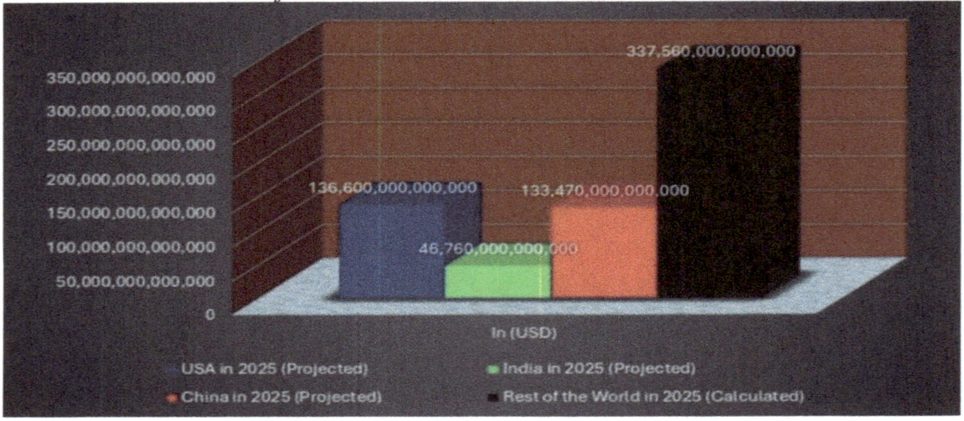

Source: Adapted from data in Statista

Figure 31. Growth of annual market output of Real Estate in India

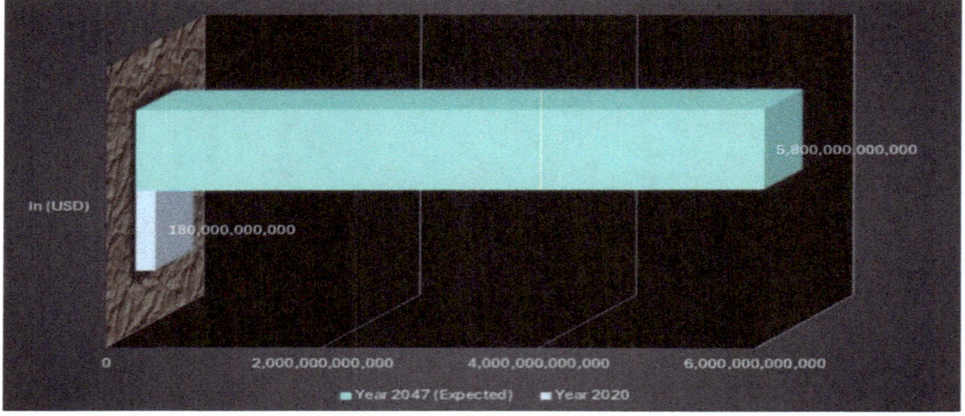

Source: Adapted from data in Knight Frank

Figure 32. Total units sold and launched in top 3 Indian cities in Real Estate sector in FY 2022-23

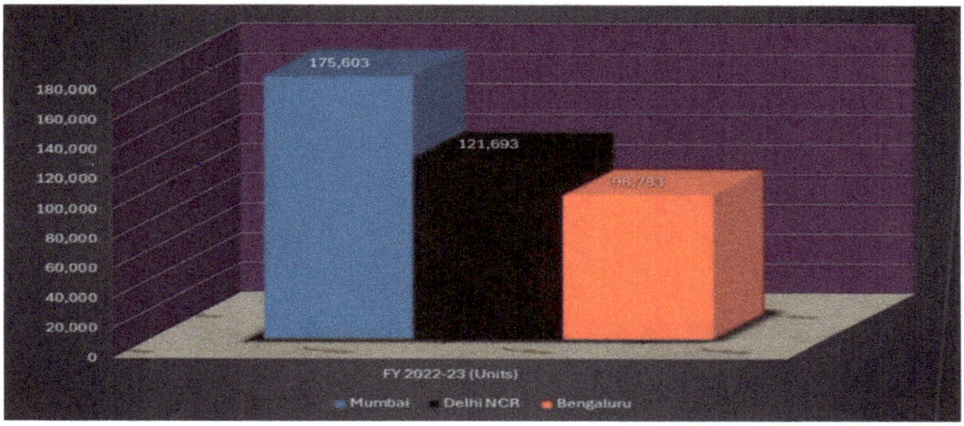

Source: Adapted from data in IBEF (Calculated and Modified)

Figure 33. Share of office space based on industries in India in 2023

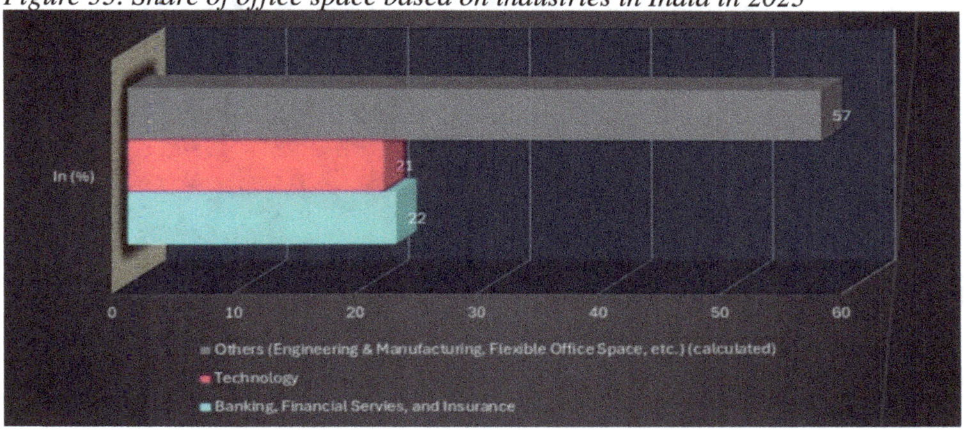

Source: Adapted from data in IBEF

Figure 34. Growth of commercial space demand in top 8 cities in India

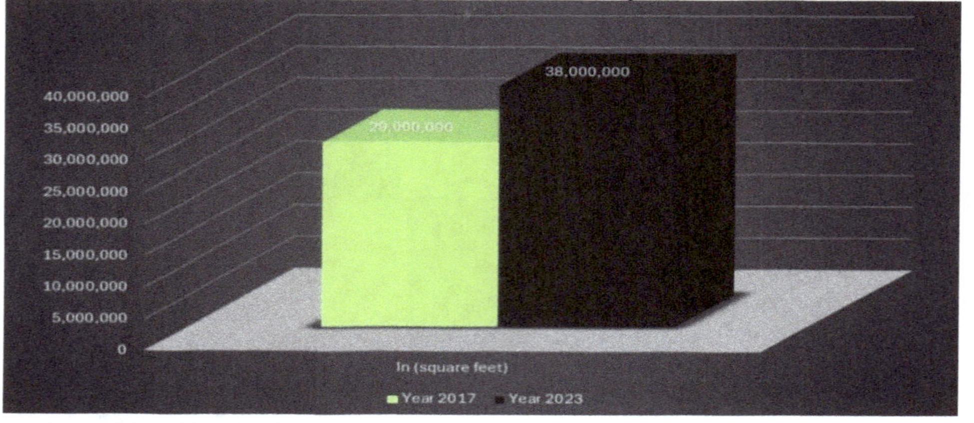

Source: Adapted from data in IBEF

Figure 35. Progress of Pradhan Mantri Awas Yojna (PMAY)

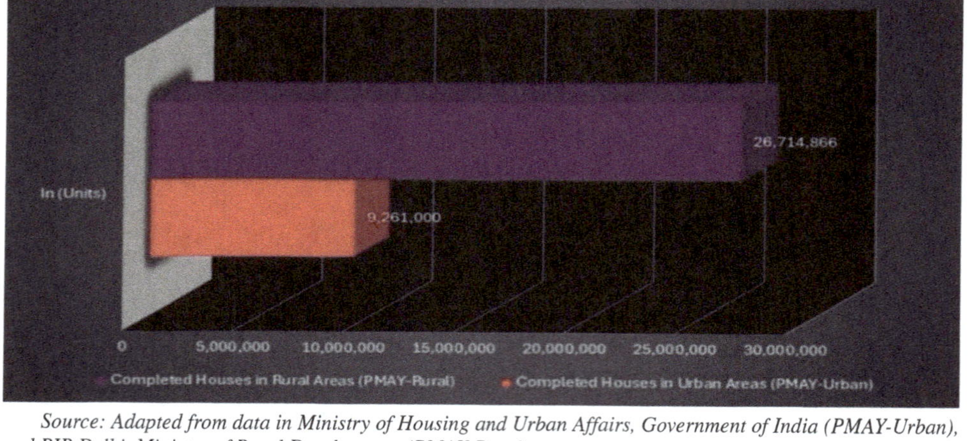

Source: Adapted from data in Ministry of Housing and Urban Affairs, Government of India (PMAY-Urban), and PIB Delhi, Ministry of Rural Development (PMAY-Rural)

Major Advantages and Challenges of Real Estate

In this subsection, based on the above graphical analysis, the advantages and disadvantages of Real Estate sector will be discussed in order to get an idea of its future, which are as follows:

- According to Figure 24, the total asset value of Real Estate market in USA is the highest among all the countries in the world in 2025, with roughly 21.17% of the world's Real Estate market. It is safe to say that the advanced country like USA which is also a superpower is maintaining its development up to certain extent by its lead in the Real Estate market in the world.
- As per Figure 25, Asia Pacific Region shares 19.5% of share in luxurious Real Estate projects in the year 2023. Even though Asia Pacific Region dominates the world with its high population, but majority of its countries lie in the middle-income group, due to which its share is not comparably high compared to other regions but still significant in the world.
- Based on Figure 26, the growth of total asset of global Real Estate market is considerable till year 2029. So, it is safe to say that the coming years Real Estate market will see noticeable growth worldwide.
- According to Figure 27, the total asset value of global Real Estate based on types, is dominated by Residential Real Estate which is over 4 times larger than other types of Real Estate. This means that high demand and bright future of Residential Real Estate in the world will be there over other types of Real Estate.
- As per Figure 28, the total asset value of Real Estate in India on the basis of types of Real Estate, is highly dominated by Residential Real Estate in the country over other types of Real Estate. It is safe to say that the current policies of Government of India will help the Residential Real Estate to grow further, while considering the high demand of India's growing population despite becoming the most populus country in the world.
- Based on Figure 29, the total asset value of Real Estate market in India is expected to almost five trillion USD in the coming four years by 2029. This means that the contribution of Real Estate in India's GDP will remain significant.
- According to Figure 30, the total asset value of Real Estate market in the world is led by USA and China with a narrow difference between the two countries. India in the recent decade has developed its Real Estate sector significantly to reach the list of top 3 countries at present. However, to catch up with USA and China, India needs more time and consistent growth in Real Estate sector in the future. However, the top 3 countries are having almost 48.42% of the total asset value of the global Real Estate market, meaning that the other countries are lagging behind with a wide margin in the Real Estate sector particularly in the case of total asset value.
- As per Figure 31. the annual market output of Real Estate in India in 2018 was $180 billion USD but by 2047, it is expected to reach $5.8 trillion USD. This means that there will be a drastic increase in the annual market output of

Indian Real Estate sector in the future, which will be beneficial for the country's development and in Real Estate's contribution in country's GDP.

- Based on Figure 32, the top 3 cities in India based on sales and launch of Real Estate, are Mumbai, Delhi NCR and Bengaluru. Here, Mumbai, Delhi, and Bengaluru have noticeable margin despite being tier 1 cities. However, it can be said that Mumbai being the financial capital of the country and Delhi being the Capital of India shows major lead among the tier 1 cities of the country. Even though Bengaluru entered the list of tier 1 city of the country in 2007, but in terms of sales and launch of Real Estate, it has surpassed most of the tier 1 cities already, showing its rapid development and growth.
- According to Figure 33, Banking, financial services and insurance along with technology are acquiring 43% of office space in the country compared to other industries. This means that Real Estate professionals should target the banking, financial services, and insurance industry first when selling or renting office space and then technology industry in India, followed by other industries, to earn profit.
- As per Figure 34, the commercial space demand in top 8 cities in India has grown drastically from 29 million square feet to 38 million square feet in the span of 6 years only by 2023. This means that the commercial space demand in the coming years in top 8 cities will remain high. So, Real Estate professionals have to take that into account, while dealing with their customers.
- Based on Figure 35, the progress of PMAY in India has been significant, and Pradhan Mantri Awas Yojna – Rural has shown drastic progress in terms of number of houses built in comparison to Pradhan Mantri Awas Yojna – Urban. This might be because the Central Government of India might be focusing on proving affordable houses more to the people living in rural areas in the country. However, it can be said that the Residential Real Estate in rural areas as well as urban areas in India in the coming years will have bright future, in order to meet the growing population and demand for housing facilities in the country. Since, the AI market is also growing, so it will be a wise idea to invest more in AI technology to apply them in the Real Estate sector to develop the sector further and enjoy the benefits of AI more than in present situation.

LIMITATIONS

In this book chapter, a number of AI tools were discussed but none were covered in detail. Then, some major case studies across the world where Artificial Intelligence was incorporated by companies in the Real Estate sector, but not in depth. Plus, there are many more case studies in the world which the chapter couldn't cover. The Indian scenario was talked about regarding the status of Real Estate market incorporating AI technology, but the book chapter didn't mention much about other countries in the world at the same time.

SOLUTIONS AND RECOMMENDATIONS

AI application in the Real Estate sector brings numerous benefits, but the challenges and problems can't be ignored. Data protection, data integrity, safety against cyber threats as well as data theft, and many more are the hurdles which must be taken seriously while incorporating AI technology in the Real Estate sector to develop and advance further.

FUTURE RESEARCH DIRECTIONS

The current progress of executing AI technology in the Real Estate sector is looking promising, but more research must be done in the future to expand the applications of AI in the Real Estate industry. Then, cyber crimes are also increasing more with the passage of time, so it is essential to perform research in the direction to ensure data integrity, safety of personal data of users, as well as protection of data theft, etc. in the context AI powered Real Estate sector.

CONCLUSION

The book chapter explained the concept of Real Estate and its types thoroughly. It further covered the major pillars of the Real Estate sector as well as the characteristics of the sector, which were majorly divided into social and economic characteristics. This chapter talked about Real Estate industry in brief to provide a rough idea and then the Indian Real Estate conditions. Then, gaps in the current literature were mentioned in addition to the necessity of AI technology in the Real Estate sector and AI tools were discussed. The chapter also explained numerous applications of Artificial Intelligence in the Real Estate sector. Then, how AI technology is being used in the existing Real Estate system in India was explained and some major case studies in the world were covered. Furthermore, some major problems and issues related to Real Estate and implementation of AI technology in the Real Estate industry were also discussed. Then, graphical analysis was performed on the basis of available data from different sources, to understand the market and future of Real Estate not only in India but in the world. It is safe to say that the advanced country like USA which is also a superpower is maintaining its development up to certain extent by its lead in the Real Estate market in the world. Even though Asia Pacific Region dominates the world with its high population, but majority of its countries lie in the middle-income group, due to which its share is not comparably high compared to other regions but still significant in the world. It is safe to say that the current policies of Government of India will help the Residential Real Estate to grow further, while considering the high demand of India's growing population despite becoming the most populus country in the world. It can be said that the Residential Real Estate in rural areas as well as urban areas in India in the coming years will have bright future, in order to meet the growing population and demand for housing facilities in India. However, the global trend also shows high dominance of Residential Real Estate over other types, meaning that in the coming years Residential Real Estate might contribute significantly to the GDP and growth of the countries in the world. Since, the AI market is also growing, so it will be a wise idea to invest more in AI technology to apply them in the Real Estate sector to develop the sector further and enjoy the benefits of AI more than in present situation.

ACKNOWLEDGEMENT

This research received no specific grant from any funding agency in the public, commercial, or not-for-profit sectors.

REFERENCES

AlBahsh, E. R., & Hosseinian-Far, A. (2021). *16 - The implication of big data analytics on competitive intelligence: A qualitative case study of a real estate developer in the UAE.* Strategy, Leadership, and AI in the Cyber Ecosystem. The Role of Digital Societies in Information Governance and Decision Making.

Aljohani, B.M., Fadul, A., Asiri, M.S., Alkhathami, A.D., & Hasan, F. (2024). Volatility transmission in the property market during two inflationary periods: The 2008–2009 global financial crisis and the COVID-19 crisis. Research in International Business and Finance, Volume 70, Part B, 102413.

Allam, Z. (2024). *The metaverse and real estate development.* The Metaverse and Smart Cities Urban Environments in the Age of Digital Connectivity Smart Cities. DOI: 10.1016/B978-0-443-22351-8.00010-6

Aungkulanon, P., Hirunwat, A., Atthirawong, W., Phimsing, K., Chanhom, S., & Luangpaiboon, P. (2024). Optimizing maintenance responsibility distribution in real estate management: A complexity-driven approach for sustainable efficiency. *Journal of Open Innovation, 10*(1), 100239. DOI: 10.1016/j.joitmc.2024.100239

Bao, W., Xu, K., & Leng, Q. (2024). Research on the Financial Credit Risk Management Model of Real Estate Supply Chain Based on GA-SVM Algorithm: A Comprehensive Evaluation of AI Model and Traditional Model. *Procedia Computer Science, 243*, 900–909. DOI: 10.1016/j.procs.2024.09.108

Cardone, B., Di Martino, F., & Senatore, S. (2024). Real estate price estimation through a fuzzy partition-driven genetic algorithm. *Information Sciences, 667*, 120442. DOI: 10.1016/j.ins.2024.120442

Doan, X. H., & Nguyen, T. P. L. (2025). Moderated mediation model of relationship between artificial intelligence awareness and counterproductive work behavior, turnover intention. *Journal of Organizational Change Management, 38*(2), 501–519. DOI: 10.1108/JOCM-07-2024-0398

Gardes, N. (2025). Ally or adversary? AI and the perceived threat to autonomy in services – The case of real estate professions. *Journal of Strategy & Innovation, 36*(1), 200533. DOI: 10.1016/j.jsinno.2025.200533

Gonzalez, T. A., Romero, L. C., Karmaziene, E., & Yuan, X. (2025). Green gains: The impact of REITs' environmental performance on sustainability-linked loan interest rates. *Finance Research Letters*, *71*, 106415. DOI: 10.1016/j.frl.2024.106415

Haurum, K. R., Mia, R., & Long, W. (2024). Real Estate with AI: An agent based on LangChain. *Procedia Computer Science*, *242*, 1082–1088. DOI: 10.1016/j.procs.2024.08.199

Jamil, H., Noorosmawie, A. B., Rabu, H. W., & Razak, L. A. (2025). From archives to AI: Residential property data across three decades in Brunei Darussalam. *Data in Brief*, *60*, 111505. DOI: 10.1016/j.dib.2025.111505 PMID: 40235696

Li, T., Lu, C., & Chen, Z. (2023). The unintended consequence of collateral-based financing: Evidence from corporate cost behavior. *Journal of Contemporary Accounting & Economics*, *19*(1), 100338. DOI: 10.1016/j.jcae.2022.100338

Lin, R. F.-Y., Ou, C., Tseng, K.-K., Bowen, D., Yung, K. L., & Ip, W. H. (2021). The Spatial neural network model with disruptive technology for property appraisal in real estate industry. *Technological Forecasting and Social Change*, *173*, 121067. DOI: 10.1016/j.techfore.2021.121067

Ma, L., & Sang, D. (2024). Would carbon tax policy promote real estate prices? International Review of Financial Analysis, Volume 96, Part A, 103668.

Manahov, V., & Li, M. (2025). The digitalisation of the real estate market: New evidence from the most prominent crypto hacker attacks. International Review of Financial Analysis, Volume 103, 104166. Guarini, M.R., Roma, A., Sabatelli, E., & Segura-de-la-Cai, A. (2025). Intrinsic and extrinsic attributes in real estate pricing: Insights for sustainable urban planning strategies. *Land Use Policy*, *153*, 107543.

Nyoni, V., Piller, W. B., & Vigren, O. (2023). Sustainability action in the real estate sector — An organizational and institutional perspective. *Cleaner Production Letters*, *5*, 100049. DOI: 10.1016/j.clpl.2023.100049

Perera, C., Toxopeus, H., Klein, S., & Merfeld, K. (2024). Enabling justice for nature-based solutions in real estate development. Nature-Based Solutions, Volume 6, 100148. Dell'Anna, F. (2025). Machine learning framework for evaluating energy performance certificate (EPC) effectiveness in real estate: A case study of Turin's private residential market. Energy Policy, Volume 198, 114407. Lenaers, I., & De Moor, L. (2023). Exploring XAI techniques for enhancing model transparency and interpretability in real estate rent prediction: A comparative study. Finance Research Letters, Volume 58, Part A, 104306. Elnagar, S. (2024). Using deep learning to enhance electronic service quality: Application to real estate websites. *Intelligent Systems with Applications*, *21*, 200330.

Potrawa, T., & Tetereva, A. (2022). How much is the view from the window worth? Machine learning-driven hedonic pricing model of the real estate market. *Journal of Business Research*, *144*, 50–65. DOI: 10.1016/j.jbusres.2022.01.027

Pu, G., Zhang, Y., & Chou, L.-C. (2022). Estimating financial information asymmetry in real estate transactions in China - An application of two-tier frontier model. *Information Processing & Management*, *59*(2), 102860. DOI: 10.1016/j.ipm.2021.102860

Renigier-Bilozar, M., Zrobek, S., Walacik, M., Borst, R., Grover, R., & d'Amato, M. (2022). International acceptance of automated modern tools use must-have for sustainable real estate market development. *Land Use Policy*, *113*, 105876. DOI: 10.1016/j.landusepol.2021.105876

Sahbi, A., Alec, C., & Beust, P. (2024). Automatic Ontology Population from Textual Advertisements: LLM vs. Semantic Approach. *Procedia Computer Science*, *246*, 3083–3092. DOI: 10.1016/j.procs.2024.09.364

Sahbi, A., Alec, C., & Beust, P. (2025). Semantic vs. LLM-based approach: A case study of KOnPoTe vs. Claude for ontology population from French advertisements. *Data & Knowledge Engineering*, *156*, 102392. DOI: 10.1016/j.datak.2024.102392

Shah, W. U., Younis, I., Missaoui, I., & Liu, X. (2025). Environmental transitions effect of renewable energy and fintech markets on Europe's real estate stock market. *Renewable Energy*, *243*, 122603. DOI: 10.1016/j.renene.2025.122603

Singh, A. (2023). The Present and Future Role of Artificial Intelligence in the Promotion of Sustainable Practices in the Construction Sector. In Kahraman, C., Sari, I. U., Oztaysi, B., Cebi, S., Cevik Onar, S., & Tolga, A. Ç. (Eds.), *Intelligent and Fuzzy Systems. INFUS 2023. Lecture Notes in Networks and Systems* (Vol. 759). Springer., DOI: 10.1007/978-3-031-39777-6_2

Singh, A. (2023). The Recent Trend of Artificial Neural Network in the Field of Civil Engineering. In Vasant, P. (Eds.), *Intelligent Computing and Optimization. ICO 2023. Lecture Notes in Networks and Systems* (Vol. 855). Springer., DOI: 10.1007/978-3-031-50158-6_32

Singh, A. (2023). The Current Trends and Need of Infrastructure Projects in the Country India. In Popescu, C., Yu, P., & Wei, Y. (Eds.), *Achieving the Sustainable Development Goals Through Infrastructure Development* (pp. 235–261). IGI Global Scientific Publishing., DOI: 10.4018/979-8-3693-0794-6.ch010

Singh, A. (2023). The Current Scenario of Sustainable Planning and Development of the Indian Built Environment. In Cobîrzan, N., Muntean, R., & Felseghi, R. (Eds.), *Circular Economy Implementation for Sustainability in the Built Environment* (pp. 106–138). IGI Global Scientific Publishing., DOI: 10.4018/978-1-6684-8238-4.ch005

Singh, A. (2023). The Significance of Digitalization of the Construction Sector. In Vasant, P., Weber, G. W., Marmolejo-Saucedo, J. A., Munapo, E., & Thomas, J. J. (Eds.), *Intelligent Computing & Optimization. ICO 2022. Lecture Notes in Networks and Systems* (Vol. 569). Springer., DOI: 10.1007/978-3-031-19958-5_100

Singh, A. (2025). The Evolution of Civil Engineering Field With the Emergence and Incorporation of Artificial Intelligence and Internet of Things. In Tariq, M. (Ed.), *Merging Artificial Intelligence With the Internet of Things* (pp. 431–486). IGI Global Scientific Publishing., DOI: 10.4018/979-8-3693-8547-0.ch014

Tukur, M., Schneider, J., Househ, M., Dokoro, A. H., Ismail, U. I., Dawaki, M., & Agus, M. (2024). The Metaverse digital environments: A scoping review of the techniques, technologies, and applications. *Journal of King Saud University. Computer and Information Sciences*, *36*(2), 101967. DOI: 10.1016/j.jksuci.2024.101967

Tysiac, P., Janowski, A., & Walacik, M. (2024). UAV measurements and AI-driven algorithms fusion for real estate good governance principles support. *International Journal of Applied Earth Observation and Geoinformation*, *134*, 104229. DOI: 10.1016/j.jag.2024.104229

Ullah, F., Sepasgozar, S. M. E., Shirowzhan, S., & Davis, S. (2021). Modelling users' perception of the online real estate platforms in a digitally disruptive environment: An integrated KANO-SISQual approach. *Telematics and Informatics*, *63*, 101660. DOI: 10.1016/j.tele.2021.101660

Ullah, F., Sepasgozar, S. M. E., Thaheem, M. J., & Al-Turjman, F. (2021). Barriers to the digitalisation and innovation of Australian Smart Real Estate: A managerial perspective on the technology non-adoption. *Environmental Technology & Innovation*, *22*, 101527. DOI: 10.1016/j.eti.2021.101527

Ullah, F., Sepasgozar, S. M. E., Thaheem, M. J., Wnag, C. C., & Imran, M. (2021). It's all about perceptions: A DEMATEL approach to exploring user perceptions of real estate online platforms. *Ain Shams Engineering Journal*, *12*(4), 4297–4317. DOI: 10.1016/j.asej.2021.04.023

Unel, F. B., & Yalpir, S. (2023). Sustainable tax system design for use of mass real estate appraisal in land management. *Land Use Policy*, *131*, 106734. DOI: 10.1016/j.landusepol.2023.106734

Wolifson, P., Maalsen, S., & Rogers, D. (2024). Discrimination in the time of digital real estate: Illustrating a rental schema in the Australian setting. *Digital Geography and Society*, *6*, 100088. DOI: 10.1016/j.diggeo.2024.100088

Xiao, Z., & Zhang, H. (2025). Real estate market regulation and local government debt risk. *Finance Research Letters*, *80*, 107362. DOI: 10.1016/j.frl.2025.107362

Yu, Y., & Zhang, Q. (2025). The Value of Cleaner Waterways: Evidence from the Black-and-Odorous Water Program. *Journal of Environmental Economics and Management*, *132*, 103159. DOI: 10.1016/j.jeem.2025.103159

Yukun, G., & Yan, Z. (2022). Discussion of Commercial Real Estate with A Short Remaining period of Land Use Rights on Evaluation Thoughts of Income Approach. *Procedia Computer Science*, *199*, 103–109. DOI: 10.1016/j.procs.2022.01.013

Zheng, Y. (2023). Community resilience and house prices: A machine learning approach. Finance Research Letters, Volume 58, Part B, 104400.

Zillow, https://www.zillowgroup.com/about-us/business/

Chapter 11
Financial Literacy and Organizational Sustainability for Green and Renewable Energy in Mexico

Aishwarya Arya
http://orcid.org/0009-0000-4236-3983
CSJM University, Kanpur, India

Shivansu Sachan
http://orcid.org/0000-0002-9893-8018
CSJM University, Kanpur, India

ABSTRACT

This study examines the crucial role of financial literacy in enhancing the organizational sustainability of green and renewable energy enterprises in Mexico. As the country accelerates its transition towards cleaner energy sources, financial management skills and awareness among stakeholders become vital for securing investments, managing risks, and ensuring long-term viability. The research investigates the link between financial literacy levels and sustainable business practices in small and medium-sized renewable energy firms. A mixed-methods approach is employed, combining survey data from industry professionals with case studies of successful Mexican renewable energy projects. Findings reveal that higher financial literacy contributes significantly to better strategic planning, efficient resource allocation, and stronger stakeholder engagement, which collectively bolster

DOI: 10.4018/979-8-2600-3038-7.ch011

ght © 2027, IGI Global Scientific Publishing. Copying or distributing in print or electronic forms without written permission of IGI Global Scientific Publishing is prohibited. Use of this chapter to generative artificial intelligence (AI) technologies is expressly prohibited. The publisher reserves all rights to license its use for generative AI training and machine learning model development.

organizational sustainability. The study highlights the need for targeted financial education programs and supportive policy frameworks to empower entrepreneurs and managers in this sector.

INTRODUCTION

In recent years, the world has been moving more quickly toward renewable energy sources. This is because people are worried about climate change, technology, and making money. Mexico, the second-largest economy in Latin America, has both problems and opportunities for change. Mexico has a lot of solar, wind, and geothermal resources, which puts it at the front of the line for renewable energy potential. However, different sectors are ready to take advantage of these opportunities in different ways.

Financial literacy, especially when it comes to investing in green and renewable energy, is a big factor in how well organizations do and how long they last. Organizations that know a lot about green financing vehicles, the economics of renewable energy, and sustainability indicators are better prepared to handle the complicated web of environmental rules, market fluctuations, and stakeholder pressure. People who don't have this kind of knowledge might not be able to see chances, figure out hazards, or put into action effective sustainability programs.

Organizational sustainability resilience is the ability of an organization to plan for, respond to, and recover from shocks to the environment, society, and economy while still meeting its operational and strategic goals. As Mexico's energy landscape changes, companies need to be able to comprehend and use green energy financial tools, evaluate renewable energy investments, and make financial decisions that take sustainability into account in order to be more resilient.

There are a number of reasons why the link between green energy financial literacy and organizational resilience is especially important in Mexico. The energy industry in the country has gone through a lot of changes in the rules, such the 2013 energy reform that opened up power markets to private companies. Second, Mexico has set big goals for the climate, such as cutting greenhouse gas emissions by 50% by 2050 compared to levels in 2000. Third, global markets and trade nations are putting more and more pressure on the country to be environmentally responsible and exhibit good sustainability performance.

The purpose of this study paper is to give a thorough look at how green and renewable energy-related financial literacy affects the long-term viability of organizations in Mexico. The research for the Mexican market talks about real-world data, theoretical frameworks, and what they mean for organizations. By looking at these relationships, stakeholders may come up with better ways to promote sustainable development and make organizations more resilient in the face of environmental and economic difficulties.

The study finds that Mexican organizations lack some important knowledge about investing in renewable energy, which might make it harder for them to take advantage of future possibilities and protect themselves from climate change-related hazards. The suggested measures include creating targeted financial education programs, policy support infrastructure, and incentive systems that will help people learn more about green energy and make organizations more resilient. This study adds to the growing corpus of research on sustainable finance and organizational resilience. It also gives Mexican policymakers, business leaders, and banks helpful practical advice.

This study's significance goes beyond only academics. For Mexican businesses, having more knowledge about green energy may provide them a competitive edge when it comes to getting green financing, lowering operational expenses, building better connections with stakeholders, and managing risks better. Policy makers may use their knowledge of these links to create educational programs, rules, and incentive systems that help both organizations become more resilient and people learn more about money.

FRAMEWORK OF IDEAS

What is Green and Renewable Energy Financial Literacy? Green and renewable energy financial literacy is the set of skills, information, and abilities needed to comprehend, analyze, and handle financial challenges related to sustainable energy businesses and investments. This level of technical financial literacy goes beyond basic financial literacy to include knowing about carbon markets, environmental laws, sustainability measures, and green finance options.

There are several mainwyns that make up the idea. Technical financial expertise involves knowing how to pay for green energy projects with things like power purchase agreements (PPAs), green bonds, carbon credits, and government incentives. To be good at risk analysis, you need to be able to look at how changes in technology, regulations, the market, and the environment might affect green energy investments. Being good at strategic planning means being able to include sustainability challenges in long-term financial planning and decision-making.

Another crucial part is market intelligence, which includes knowing about the energy markets, how prices work, how demand changes, and how competitive the renewable energy sector is. Being aware of regime consciousness means knowing about environmental regulations, reporting requirements, compliance costs, and policy efforts that might have an impact on investments in green energy. Being conscious of how environmental performance affects relationships with investors, customers, employees, and the community is what stakeholder management is all about.

The Organizational Sustainability Resilience Framework. Organizational resilience for sustainability is a complicated idea that includes a number of related skills. Adaptive capacity is the ability of a business to alter how it works, its structures, and its plans in response to changes in the economy and the environment. This involves being able to change how resources are used, make plans for the future, and run operations. Absorptive capacity is the ability of a company to handle shocks and disruptions while keeping its essential activities and performance levels up.

Transformative capacity is the organization's ability to change its business model, value proposition, or way of doing things in a big way because of big changes in the market or environment. This is especially true for the energy transition, when companies may need to completely rethink how they do business in order to be competitive and sustainable.

Recovery capability is how well an organization can bounce back from problems, learn from them, and come out stronger than before. It involves being able to handle crises, talk to stakeholders, and learn. Innovation capacity is the ability of a company to create new goods, services, or processes that improve its sustainability performance and provide it a competitive edge.

Making the link between financial literacy and organizational resilience stronger

There are several ways that organizational resilience and green energy financial literacy are connected. Organizations that are better at handling money are better at processing information, which helps them understand market signals, changes in regulations, and the needs of stakeholders. More information processing makes people better at making decisions and preparing strategically, which makes them more resilient in the end.

Another important factor is making the best use of resources. Companies who are good at managing their money when it comes to green energy may make the best use of their resources for sustainability projects that will provide them the best return on investment and risk management. These include optimizing an energy portfolio, managing capital investments, and using financial tools to help reach sustainable goals.

Companies are better able to manage risk when they know a little bit about green energy financial instruments and markets. This level of understanding makes it possible to do more complex risk assessments, hedging, and planning for the worst. Businesses may get better at predicting and becoming ready for a range of events, including as changes in the law, changes in the market, and changes in technology.

Companies' strategic posture is better when they can see and take advantage of new chances in the green energy industry. These include finding new markets, making new goods and services, and getting an edge over the competition by being more environmentally friendly. Companies may better analyze these possibilities and take steps to improve long-term resilience when their employees are financially literate.

A REVIEW OF THEORETICAL LITERATURE

Theoretical Foundations of Financial Literacy Financial literacy research is based on ideas from many different disciplines of study, including as economics, psychology, and organizational behavior. Traditional economic theory says that people and businesses make smart choices when they have all the information and know all their possibilities. But research in behavioral economics has shown that cognitive limits, prejudice, and not having all the facts can have a big effect on how people make financial decisions.

Lusardi and Mitchell's (2014) first study on financial literacy created a theoretical framework that said there were three main ideas of financial knowledge: being able to grasp interest rates, inflation, and how to spread out risk. After that, other academics have used this framework to include more complex topics including risk assessment, portfolio management, and investment analysis. When it comes to green energy, these basic ideas need to be backed up with knowledge of genuine renewable energy technology, environmental rules, and policies that promote sustainability.

Ajzen's (1991) planned behavior theory is another key idea that helps us understand how financial literacy affects how organizations act. The idea says that attitudes toward the activity, subjective standards, and perceived behavioral control all have an effect on behavioral intentions. Companies that know more about money in the green energy investment industry will also be more likely to support sustainable investments, understand what stakeholders want, and be more sure that they can correctly carry out green energy projects.

Bandura (1986) came up with the social cognitive theory, which says that observation, self-efficacy, and contextual influences all have a role in how people respond. The notion says that organizations learn about money through several ways, such as formal schooling, watching their peers, and having experiences. Self-efficacy, or the belief that you can do certain things, is very important in deciding whether or not businesses would try to use green energy policies.

The theory of organizational resilience. The idea of organizational resilience has come from several fields, such as systems theory, ecology, and psychology. In his early work on ecological resilience, Holling (1973) came up with the idea of adaptive cycles, which show how systems recover from disturbance in cycles of growth, conservation, release, and rearrangement. This has been used in business settings to talk about how companies deal with problems and come out stronger.

Teece et al.'s (1997) dynamic capabilities theory is significant for showing how organizations become more resilient by building up certain skills. Dynamic capabilities are the skills that allow you to combine, build, and change your own and others' skills to adapt to changing situations. In terms of sustainability, these skills might include being able to look at environmental hazards, find green investment possibilities, and change how things are done to meet new rules.

Barney's (1991) study led to the development of resource-based theory, which focuses on how unique resources and skills may provide a company an edge over its competitors. Companies that improve their green energy finance literacy may build valuable, uncommon, and hard-to-copy resources that provide them a long-lasting competitive edge and make them more resilient. This thesis says that financial literacy for green energy may be a strategic resource that helps organizations do better and be more flexible.

Another important way to look at organizational resilience is through complexity theory. According to complexity theory, organizations are complex adaptive systems that need to change all the time to stay alive in a changing world. Using this method, green energy financial literacy is a way for organizations to learn how to adapt to changes in the environment and rules while still being able to run their business efficiently.

The Theory of Sustainable Finance. Sustainable finance theory is a new area of research that combines traditional finance theory with social and environmental issues. The efficient market hypothesis is one of the main ideas behind classical finance theory. It says that asset prices reflect all the information that is out there in the financial markets. However, real-world research in the field of sustainable finance has demonstrated that market pricing don't always adequately represent environmental and social issues. This means that companies may have better green energy financial knowledge.

Freeman's (1984) stakeholder theory says that organizations should think about all of the people who are affected by their decisions, not just shareholders. This frame of view is highly important for investments in green energy since these kinds of decisions often entail difficult trade-offs between social, environmental, and financial factors. Companies who are good at managing their money when it comes to green energy are in the greatest position to make these trade-offs and choose options that benefit all stakeholders.

Markowitz (1952) came up with modern portfolio theory, which explains diversification as a way to lower the risk of investments while keeping the expected returns. The hypothesis says that companies may make themselves stronger by using more renewable energy sources in their energy mix. This will lower the risk of fossil fuel price changes and regulatory issues.

Environmental disclosure theory says that corporations should provide environmental information on their own to show that they are committed to sustainability and to level the playing field with stakeholders. Green energy firms that know more about money are better at seeing the benefits of being open about their environmental impact and using communication tactics that develop strong relationships with stakeholders and make the company more resilient.

A REVIEW OF EMPIRICAL LITERATURE

Evidence from throughout the world on sustainability and financial literacy. In the last few years, there has been a huge increase in empirical research on the link between sustainability and financial literacy. This study has given us important information on how financial literacy affects decisions regarding the environment. A major study by Klapper et al. (2019) looked at financial literacy in 140 countries and found that people and institutions who are better at managing their money are more likely to make long-term investments that take the environment into account. This study finds that being financially literate is a basic need for planning and carrying out sustainability in a way that works.

Weber's (2017) study of European companies showed that management teams in companies that were better at handling money would use clean technology and start comprehensive sustainability programs. The study found that managers who knew a lot about money could better assess the long-term financial value of investments in sustainability. This led to increased use of green practices and technologies. Renneboog et al. (2008) did a study on socially responsible investing and found that institutional investors were more likely to take ESG issues into account when making investment decisions. This study suggests that an investor who is better off financially may better assess risks and opportunities associated to sustainability, which can lead to improved investment success.

Eccles et al. (2014) did a long-term study to look at how sustainability reporting affects the financial performance of publicly traded corporations. The study indicated that organizations that did more advanced-level sustainability reporting, which requires a high level of financial literacy, did better financially in the long run. This backs up the idea that knowing how to use green energy may help a business stay strong and do well in the long run.

In Latin America. Research on Latin American financial literacy and sustainability has found that the area is different from the rest of the globe in certain important ways. Garcia-Meca et al. (2015) looked at how common sustainability practices are among businesses in Latin America. They discovered that financial literacy was typically lower than in more developed markets, which might make it harder for businesses to use green energy technology and practices.

The Inter-American Development Bank (2018) did a thorough study of how Latin America was using renewable energy. They discovered that financial literacy was a major predictor of how many organizations will join green energy markets. Organizations that were better at managing their money were more likely to know about and use funding options like green bonds and development bank loans to pay for renewable energy projects. Villanueva et al. (2019) studied how Latin American companies report on their environmental impact. They discovered that companies with management teams who were better at handling money wrote better environmental reports and did better on environmental performance. This shows that knowing about money may help with better environmental management and telling stakeholders about it.

Lopez et al. (2020) did a research comparing the use of renewable energy in six Latin American nations. They found that Mexico was in the center when it came to organizations' awareness of green energy finances. The study found that carbon market knowledge, green financial instruments, and sustainability measures are all areas where there is a lot of room for development.

Proof from the Mexican market. Research on the Mexican market gives us useful information on how green energy financial literacy and organizational resilience are related. Hernandez and Rodriguez (2021) looked at how Mexican industrial companies were adopting renewable energy. They discovered that financial literacy was the best predictor of membership in Mexico's Clean Energy Certificate program. Companies that knew a lot about money were more likely to see the program's economic benefits and make arrangements for as many people as possible to take part.

The study by Gonzalez et al. (2020) on how Mexico's energy reform affected how businesses responded showed that enterprises who were better at managing their money were better able to deal with the shift to competitive power markets. These kinds of companies were better at managing their energy costs, organizing their energy portfolios, and responding to rules.

The Mexican Business Council for Sustainable Development (2019) did a thorough survey and found that just 35% of Mexican enterprises have the right level of financial knowledge to make green energy investments. The study found that Mexican companies had big gaps in their understanding on how to finance renewable energy, carbon markets, and sustainability indicators.

Morales et al. (2021) found that companies in Mexico that were better at managing their money were more likely to get green bonds and get good loan terms. This shows that companies with better financial literacy may improve their capital structure and lower the cost of borrowing for activities that are good for the environment.

Evidence for a Specific Sector

Studies of some parts of the Mexican economy have shown that the link between organizational resilience and financial knowledge about green energy is quite different in different areas. Ramirez et al. (2020) looked at the industrial sector and found that organizations that were better at managing their money were more likely to use renewable energy technologies and adopt energy-saving techniques. These businesses were better able to handle costs and follow the rules.

Torres and Silva (2021) looked at the service industry and found that organizations who wanted to use sustainability performance to set themselves apart needed to be financially literate. Companies that were more financially literate were more likely to come up with integrated sustainability plans and tell stakeholders about their environmental performance.

A research by Mendoza et al. (2020) on the agricultural business indicated that those who were financially literate about green energy were more likely to use sustainable farming practices and renewable energy technologies. These companies were better able to handle resources and adjust to the effects of climate change.

Castro and Lopez (2021) looked at the financial services industry and found that institutions that were better at understanding green energy were more likely to provide green financing goods and services. They were better at dealing with environmental concerns and taking advantage of chances in new markets.

When comparing research results, it's important to keep in mind that the empirical studies on organizational resilience and green energy finance literacy also have significant methodological issues. Some studies look at basic financial knowledge, while others look at more complicated topics like investment analysis and risk management. This is the biggest difference in how they assess financial literacy. Because of these disparities, it is hard to compare study results and come to solid conclusions regarding how financial literacy affects organizational outcomes.

Another key methodological concern is causality. Many studies have found links between financial literacy and how well an organization does, but finding causal links needs more advanced analytical methods.

Longitudinal research and natural experiments give better evidence for causal linkages, although there aren't many of these kinds of studies in the literature. Selection bias might affect the outcomes of the study since the organizations that take part in green energy financial literacy research may be different in a systematic way from those that don't. It might make people think that financial literacy and organizational resilience are more closely related than they really are. Cultural and organizational factors might change the relationship between financial literacy and organizational outcomes, which could make it hard to apply the results to other settings. It might not be easy to apply advanced market research to emerging markets like Mexico since the institutions and market conditions are different.

How to Do It

The design and approach of the research. The current study uses a mixed-methods approach, which means it uses both quantitative analysis of available data and qualitative input via case studies and literature reviews. The main goal of the research's design is to describe and analyze the level of green energy financial literacy in Mexico and how it affects the resilience of organizations that want to be sustainable. The research uses an in-depth literature review method that integrates real-world and theoretical studies from peer-reviewed journals, industry papers, and government publications. The databases checked contain a number of them and have literature from 2010 to 2024, which covers the complete spectrum of the most recent information on the subject.

Sources of data and how it was gathered. Government papers, databases from international organizations, corporate analysis, and scholarly articles are all examples of primary sources. Scopus, Web of Science, JSTOR, and Google Scholar are the main databases used. They use combinations of the keywords "Mexico," "sustainability," "organizational resilience," "financial literacy," and "green energy." Secondary data sources include reports from international organizations like the International Energy Agency, the World Bank, and the Inter-American Development Bank, as well as Mexican government agencies like the Ministry of Energy (SENER) and the National Institute of Statistics and Geography (INEGI).

Consulting companies, research institutions, and industry groups also publish industry reports that give further information about the state of the market and how organizations work in Mexico. Some of them are reports from companies like McKinsey & Company, Deloitte, PwC, and Mexican research institutions.

The analytical framework uses ideas from several fields to give a complete picture of how green energy financial literacy and organizational resilience are related. The framework uses theories from financial literacy, organizational resilience, and sustainable finance to create a complete model that shows how financial literacy affects the sustainability performance of an organization. The analysis uses a methodical way to combine evidence from different research to look for patterns, trends, and gaps in the literature. Research done in developing countries and emerging markets is given more weight since these settings are more likely to be useful for the Mexican example.

Things to think about and limitations. There are certain problems with this study method that need to be thought about. When using secondary data sources, analysis is limited by the quality and availability of the research that is already out there. There aren't many real-world studies that focus on Mexico, thus researchers may have to use studies from other places and not get a full picture of the Mexican market. Because both green energy markets and research on financial literacy change so fast, some of the findings will quickly become out of date. To get over this problem, the study includes fresh studies and points out new patterns that will affect trends in the future.

Language barriers can make it hard to read material published in Spanish or other languages, and they may also miss out on useful information from academics and organizations in other regions. The study tried to get around this problem by using information from Mexican government organizations and regional research institutions.

ANALYSIS OF OUTCOMES

The current level of Mexico's financial literacy on green energy. The research demonstrates that Mexico's financial literacy about green energy is still in its early stages, and there are big differences between industries, company sizes, and locations. Multinational enterprises in Mexico seem to be better at understanding green energy finances since they are exposed to worldwide norms and practices. These companies have technological know-how, dedicated departments for sustainability, and other resources that let them do more in-depth research on green energy investments.

Medium-sized enterprises have different levels of financial expertise. Some are very good in certain areas but don't know much about the broader green energy financial landscape. Most of these companies have taken basic steps to save energy and may have found ways to use renewable energy, but they tend to make decisions based on simple rather than detailed risk and opportunity assessments.

Developing green energy financial literacy is hardest for small and medium-sized businesses (SMEs). These kinds of organizations often can't come up with the skills and knowledge they need to make good green energy investment decisions because they don't have enough resources, don't have enough specialized skills, or have too many conflicting demands. This is especially hard because small and medium-sized businesses (SMEs) make up a large part of the Mexican economy and have a lot of potential to help with the energy transition.

It's clear that there are differences in financial understanding based on where you live. firms in big cities like Mexico City, Guadalajara, and Monterrey are more likely to be financially literate about green energy than firms in smaller towns and rural regions. This difference is because people have different levels of access to training resources, professional networks, and the market.

Patterns that are exclusive to a sector. The industrial sector also has a lot of financial knowledge about green energy, which comes from the fact that they have to deal with worldwide supply chain challenges and operations that use a lot of energy. Most manufacturing organizations have also bought advanced energy management systems and have a history of procuring renewable energy. The degree of financial literacy is very different across the industry. Export businesses tend to have better financial literacy than those that focus mostly on the domestic market.

The services industry shows different tendencies. Financial services, telecoms, and retail companies tend to be more cognizant of green energy finances than other service industries. The variance is caused by the fact that various service sectors have distinct stakeholders, regulatory bodies, and competitive pressures. Even though there is a lot of promise for using renewable energy, the agriculture industry has a hard time getting access to green energy financial knowledge. It has trouble because it can't get enough money, the market is broken up, and it focuses on short-term cash flow management, among other things.

The building industry shows a stronger interest in green energy finance knowledge since there is a rising need for green building practices and certification for sustainable buildings. Most building businesses, on the other hand, don't have the financial knowledge to look at complicated green energy systems and finance options.

Effects of the size and structure of an organization. There are several reasons why large organizations are more likely to be financially aware of green energy. These include access to professional services, training and development facilities, worldwide marketplaces and practices, and stakeholders including investors, consumers, and regulators. These kinds of businesses are likely to hire experts in sustainability and have formal ways to look at green energy investments.

Some medium-sized businesses are really good at certain things but don't have a full knowledge of green energy funding. A lot of these kinds of businesses have taken small steps to be more environmentally friendly, but they do not have the advanced analytical skills needed for more complex green energy projects. Small businesses also have a hard time becoming financially literate about green energy because there aren't enough people with the right skills, management priorities are different, there aren't enough resources for training and development, and they don't know enough about green energy financing tools.

Even with these problems, some small businesses have been able to use renewable energy solutions successfully with the right help and advice. There are a lot of family enterprises in Mexico's economy, and they have their own unique ways of understanding green energy finances. These firms have long-term attitudes that align with sustainable investment but lack technical skills and financial acumen to appraise innovative green energy initiatives.

Effect on the organization's ability to bounce back. The study shows that green energy finance literacy may help organizations in Mexico become more resilient in a number of ways. When energy prices are volatile, there are regulatory obligations, and supply chains are disrupted, organizations that are better at managing their money are better at managing their risks. The groups can better identify and reduce the dangers that come with climate change and the shift to renewable energy.

Organizations with better green energy finance skills are much better at strategic planning. They are better able to come up with comprehensive sustainability plans that make environmental issues a part of their main business planning efforts. They are better at seeing and taking advantage of future possibilities in the green energy industries.

Companies that know more about green energy finances can make their operations more efficient. These groups are more likely to utilize energy-saving methods, install renewable energy technology, and make the most of their energy use. Part of being resilient and competitive as a business is conserving money.

Capitalized green energy financial literacy organizations have better access to capital markets. These groups have a fair chance of getting green finance tools including green bonds, loans tied to sustainability, and money from development banks. Access to funding facilitates implementation of broader-scale sustainability initiatives and gives financial flexibility during rough times.

Problems and obstacles. There are certain big problems that make it hard for people in Mexico to learn about green energy finance. There aren't many training and instructional materials available that focus primarily on green energy financing, especially in Mexico and Spanish-speaking countries. Most of the materials being made for foreign markets haven't really looked at the specific rules and market situations in Mexico.

Organizations that want to teach people about green energy finance have problems since policies change and regulations are unstable. Changes in rules and regulations at energy and regulatory institutions happen all the time, which makes it hard for businesses to make long-term strategies and may make them less likely to invest in building specialized knowledge and skills. Problems with market structure and competition may make it harder for people to learn about green energy finances. In certain areas, there may not be enough competition or there may be rules that make it harder for companies to build up their understanding of complex green energy investments. Also, when a few industries have a lot of market power, it might be hard to get information and learn about best practices.

Institutional and cultural factors may affect how people learn about green energy and money. Sustainability may not be fully taken into account in traditional business and financial models, thus companies need to come up with innovative ways to make decisions. People who don't want to take risks or make changes may not use new green energy alternatives.

Opportunities and Possible Answers. Even with these obstacles, there are many chances to teach people in Mexico about green energy finance. Policy and government initiatives may be quite important in helping people learn about money. This includes funding for education programs, technical help, and financial incentives for businesses who invest in green energy training. Schools and professional groups may help people learn about green energy finances by offering training, certification, and professional development. Partnerships between schools and businesses can assist make sure that programs deal with actual problems and chances.

Financial institutions have a lot of room to improve people's understanding of green energy through product creation, customer education, and market-making operations. Banks and other financial institutions can learn a lot about green energy funding and help their clients with it. Some of the things that make it hard to learn about green energy finance can be fixed with technology and digitalization. Digital training platforms, decision support tools, and data analysis may help enterprises of all sizes get expert knowledge at a lower cost and with less effort.

TALK

Theoretical Effects. The study of green energy financial literacy and organizational sustainability resilience in Mexico provides a number of key theoretical additions to the body of research. The study shows that standard ways of measuring financial literacy need to be changed to better reflect the complexity of investments in green energy and sustainable financing. To make good decisions on green energy, you need to know more than just fundamental financial ideas. You also need to know about environmental laws, technical hazards, and how stakeholders interact with each other.

The results support and add to the notion of organizational resilience by demonstrating that financial literacy is one of the most important dynamic qualities that help businesses deal with economic and environmental uncertainty. The study implies that green energy financial literacy is a “meta-capability” that makes other organizational competencies better, such as strategic planning, risk management, and stakeholder engagement. The study adds to the notion of sustainable finance by showing how important it is to be financially literate in order to make good decisions for the environment.

The paper shows that green energy markets may not be as efficient because of differences in information and financial literacy. However, companies that have better knowledge and skills may have more opportunities. The study also helps us understand how cultural and institutional factors affect the growth and use of financial literacy. The Mexican environment has unique difficulties and opportunities that theories based on other institutional setups don't handle.

What this means for businesses in real life. The results of the study have important real-world effects for businesses in Mexico. If businesses want to be more resilient in the face of environmental changes, they need to make sure that their management teams and key decision makers are well-versed in green energy finance. This involves spending money on training programs, hiring people with unique skills, and improving analytical skills. Organizations must have a deep understanding of renewable energy markets, funding, and regulatory frameworks in order to carefully analyze the potential and dangers of green energy in their strategic planning processes.

Companies also need to think about how climate change and the shift to cleaner energy will affect their business models and competitive edge in the long run. Environmental and sustainability risks must be added to risk management frameworks, and companies must be able to find and reduce climate-related hazards. This means knowing about the physical risks of climate change, the transition risks of changes in policy and the market, and the reputational risks of what stakeholders anticipate. Companies should also think about how investing in green energy might help them compete by lowering costs, lowering risks, and creating value for stakeholders. This means that you need to carefully look at different investment possibilities and financing arrangements, as well as keep up with changes in the market and the rules that govern it.

What this means for policy. The results of the study have substantial policy implications for Mexican policymakers who want to encourage sustainable development and make organizations more resilient. To fill up the gaps in people's understanding of green energy finances, the government has to make policies that include focused educational programs, technical help, and initiatives to enhance people's skills. Regulatory frameworks should be set up to make green energy markets more open and liquid. This would reduce information asymmetries and help firms make better decisions. This involves making sustainability reports more consistent, creating green taxonomy frameworks, and adding market-making tools for green financial instruments.

Green bonds, sustainability-linked loans, and special financing vehicles for renewable energy projects are all examples of green finance products and markets that the financial industry could support. This includes rules that help markets grow and financial institutions learn how to manage risk. Business school classes, professional training programs, and criteria for continuous professional development should all include sustainability and green energy financing. This means that government organizations, colleges, and business groups need to work together to produce training materials that are useful and relevant.

Limitations and Future Research. There are a few things with this study that should be noted. The utilization of secondary data sources and literature review research methods indicates that the discussion may not represent the actual condition of green energy financial literacy in Mexico. To get more useful and up-to-date information, future research should involve collecting primary data through surveys, interviews, and case studies. The green energy markets and financial tools are evolving so swiftly that studies will shortly become out of date. Longitudinal studies that look at how financial literacy and organizational resilience change over time must be part of future study. These studies should also look at new trends and technologies that are coming up.

Focusing on the organizational level of analysis could miss important information about individual decision-makers and how they help people learn about green energy. Future studies need to look at the knowledge, attitudes, and actions of the most influential people in businesses, such as senior executives, finance managers, and sustainability experts. There should be further research on the differences in green energy financial literacy across regions and sectors to find out what causes these differences. Future research should look at how Mexico compares to other growing economies in terms of regions and sectors, as well as how Mexico compares to other countries.

Suggestions for those involved. The study results lead to a number of suggestions for each group of stakeholders. The first thing companies should do is create comprehensive green energy financial literacy programs that include both the technical and strategic sides of sustainable financing. This includes spending money on training and development, hiring people with specific skills, and putting together teams from different departments to work on sustainability issues. Suggestions for policymakers include creating national plans to improve people's understanding of green energy finances, setting up education and technical help programs, and putting in place rules that let the market flourish and be open.

To make policy intervention more effective and consistent, it has to be coordinated between different levels of government and agencies. For banks, this means creating special green energy financing products and services, educating and developing people, and setting up special green finance teams. Banks also need to take the lead in developing the market and spreading best practices across the sector. Some suggestions for colleges are to incorporate green energy finance in business school classes, provide special training courses and certificates, and start research programs in sustainable finance and organizational resilience.

In conclusion. This in-depth look of green energy financial literacy and organizational sustainability resilience in Mexico creates a complicated and ever-changing situation with high implications for both sustainable development and economic competitiveness. The study shows that understanding green energy finances has important effects on an organization's capacity to deal with environmental and economic risks and uncertainties while still meeting its operational and strategic goals.

Mexico's understanding of green energy money is quite different in different industries, sizes of organizations, and areas of the country. Some companies, especially international organizations, have a good grasp of green energy finance. However, most Mexican enterprises do not have the skills and expertise needed to properly assess and carry out green energy investments. The unevenness makes it hard to grow Mexico's economy in a way that is good for the environment, but it also creates chances for it to happen.

There are a few fundamental ways that green energy finance literacy affects the resilience of organizations. Organizations that are better at managing money have better risk management, strategic planning, operational efficiency, and access to capital markets. These skills make companies more resilient when faced with environmental and economic challenges, allowing them to adapt, recover, and grow even when things change. Limited educational resources, unclear regulations, problems with the market structure, and cultural considerations are some of the main things that are holding back the growth of green energy financial intelligence in Mexico. However, there are big chances to lower these obstacles if government, schools, banks, and business groups work together.

The work's theoretical contributions build on earlier frameworks for financial literacy and organizational resilience by adding environmental aspects to the mix. The results show that we need more complex models that show how complicated green energy investment is and how it might make organizations more resilient. The need to make green energy financial literacy a top goal for development, include sustainability factors in strategic planning, and make risk management frameworks bigger so they address environmental challenges are all examples of practical organizational applications. When looking at investments in green energy, you should also think about how they can help you get more business by saving money, managing risk, and creating value for stakeholders.

Policy implications include the need for broad strategies to improve people's understanding of green energy finances through educational programs, new rules, and initiatives to grow the market. When making conditions right for long-term investment and innovation, policymakers should focus on closing the gaps in knowledge and skills that have been found. Later studies need to get beyond these study limits by obtaining primary data, looking at things across time, and looking more closely at differences between sectors and regions. Studies might also look at how the link between organizational resilience and financial literacy changes as individual decision-makers, new technology, and changing market circumstances come into play.

The results of this study show how important it is for organizations in Mexico to be financially literate about green energy in order to be able to survive and thrive. As the country progresses toward a more sustainable energy system, it will be important to help organizations increase their financial literacy so they can participate in this change in a meaningful way and benefit from it. Mexico's efforts to promote sustainable development will only be successful if its institutions can properly evaluate, assess, and pursue green energy investments that help both the environment and the economy.

The findings of this study have ramifications beyond Mexico for other developing countries who are also having trouble building sustainable energy systems and making their organizations more resilient. The ideas, findings, and suggestions from this study can help countries throughout the world make their organizations better at promoting sustainable development and being good stewards of the land.

The link between financial literacy and organizational resilience will become more and more significant as the world economy moves toward being more environmentally friendly and sustainable. Companies that teach their employees about green energy finance will be better prepared to deal with the problems and chances that come with the sustainable economy. Companies that don't do this, on the other hand, might lag behind.

This report's work is a big step toward figuring out how financial literacy, sustainability, and organizational resilience all affect one other in complicated ways. But there is still a lot of work to be done to properly understand these interactions and come up with effective plans for long-term growth and organizational resilience in Mexico and other emerging countries. For goals of sustainable development and environmental preservation to be met together with economic growth and competitiveness, research, policy making, and field implementation will need to continue.

The COVID-19 epidemic has made it even more clear how important it is for businesses to be able to bounce back from shocks and disruptions and have business models that are sustainable. Companies who were very financially savvy in green energy were better able to deal with the problems caused by the epidemic, such as supply chain problems, fluctuating energy prices, and shifting stakeholder expectations. This lecture stresses how investing in financial literacy skills may help an organization do well in the long run and be more resilient.

There will be a number of new developments that will affect how green energy financial literacy and organizational resilience are connected in Mexico in the years to come. Increasing sophistication of green financial products like sustainability-linked bonds, green loans, and carbon credit markets will need enterprises to strengthen their analytical skills. The increased incorporation of ESG issues in investment choices will bring with it new possibilities and difficulties for firms who desire to show sustainable performance. New innovations in renewable energy, energy storage, and smart grids will open up new investment opportunities and make it necessary for businesses to learn about these technologies and the money that supports them. Digitalization of the market and the rise of new business models like energy-as-a-service and peer-to-peer energy trading will mean that companies need to adapt their financial literacy abilities to new market structures and chances.

Climate change impacts and growing frequency of extreme weather events will bring new risks and possibilities for organizations, necessitating higher-level climate risk analysis and adaptation planning capabilities. Companies will need to set up new ways to analyze data and make climate a part of their financial planning routines as climate scenarios and risk models become more complex and changeable.

The European Union's list of sustainable operations and the Task Force on Climate-related Financial Disclosures standards are two examples of global developments that will affect how Mexican businesses are increasing green energy financing and sustainability reporting. Groups that teach people how to manage their reducible energy finances will be able to address these changing requirements and get into global markets and money. The study's real-world findings reveals that Mexico has a lot of potential to make its organizations more resilient and sustainable by putting money into teaching people about green energy. To make this happen, numerous groups, such as government ministries, training institutions, financial institutions, and industry associations, will need to work together.

To reach sustainable development goals while keeping the economy competitive, it will be important to construct detailed national plans that meet both short-term demands and long-term ambitions. In conclusion, understanding how to use green energy is an important skill for organizations in Mexico to be able to stay strong. Organizations need to learn how to interpret and analyze things in a complex way since green energy markets, financial structures, and regulatory frameworks are all connected and always changing. People who can acquire these abilities will be in a good position to succeed in the sustainable economy, while those who don't invest in learning how to manage their money will lag behind.

The way forward requires all parties to be committed to closing gaps in financial literacy and building institutional frameworks that promote sustainable growth. The success of such projects would not only ensure the resilience of specific institutions but also the sustainability and competitiveness of the Mexican economy during the global transition to a low-carbon future

REFERENCES

Ajzen, I. (1991). The theory of planned behavior. *Organizational Behavior and Human Decision Processes, 50*(2), 179–211. DOI: 10.1016/0749-5978(91)90020-T

Bandura, A. (1986). *Social foundations of thought and action: A social cognitive theory*. Prentice-Hall.

Barney, J. (1991). Firm resources and sustained competitive advantage. *Journal of Management, 17*(1), 99–120. DOI: 10.1177/014920639101700108

Castro, M. J., & Lopez, R. A. (2021). Green finance growth in Mexican banks: Problems and chances. *Latin American Business Review, 22*(3), 245–267. DOI: 10.1080/10978526.2021.1895432

Eccles, R. G., Ioannou, I., & Serafeim, G. (2014). The impact of corporate sustainability on organizational processes and performance. *Management Science, 60*(11), 2835–2857. DOI: 10.1287/mnsc.2014.1984

Freeman, R. E. (1984). *Strategic management: A stakeholder approach*. Pitman.

Garcia-Meca, E., Pucheta-Martinez, M. C., & Cuadrado-Ballesteros, B. (2015). The influence of ownership structure on the costs of corporate debt in the context of intellectual capital. *Journal of Business Research, 68*(4), 860–869. DOI: 10.1016/j.jbusres.2014.11.040

Gonzalez, A., Martinez, L., & Sanchez, P. (2020). The transformation of the energy market and corporate adaptation in Mexico: Evidence from the industrial sector. *Energy Policy, 145*, 111750. DOI: 10.1016/j.enpol.2020.111750

Hernandez, J. C., & Rodriguez, M. A. (2021). Financial literacy, renewable energy certification, and business sustainability in Mexico. *Renewable Energy, 165*, 789–801. DOI: 10.1016/j.renene.2020.11.089

Holling, C. S. (1973). Resilience and stability of ecological systems. *Annual Review of Ecology and Systematics, 4*(1), 1–23. DOI: 10.1146/annurev.es.04.110173.000245

Inter-American Development Bank. (2018). *Financing renewable energy in Latin America: Challenges and opportunities*. IDB Publications.

Lopez, C., Rivera, J., & Torres, S. (2020). Renewable energy adoption in Latin America: The role of financial literacy and institutional factors. *Energy Economics, 89*, 104785. DOI: 10.1016/j.eneco.2020.104785

Lusardi, A., & Mitchell, O. S. (2014). The economic importance of financial literacy: Theory and evidence. *Journal of Economic Literature*, *52*(1), 5–44. DOI: 10.1257/jel.52.1.5 PMID: 28579637

Markowitz, H. (1952). Portfolio selection. *The Journal of Finance*, *7*(1), 77–91. DOI: 10.1111/j.1540-6261.1952.tb01525.x

Mendoza, R., Silva, A., & Vargas, E. (2020). Financial literacy and the adoption of sustainable agriculture and renewable energy practices in Mexico. *Agricultural Systems*, *185*, 102943. DOI: 10.1016/j.agsy.2020.102943

Mexican Business Council for Sustainable Development. (2019). *Green finance survey: Current status and opportunities in Mexico*. CESPEDES.

Morales, P., Gutierrez, D., & Ramos, F. (2021). Green bond markets in Mexico: Determinants of corporate participation and financing costs. *Journal of Sustainable Finance & Investment*, *11*(2), 156–178. DOI: 10.1080/20430795.2020.1762911

Ramirez, S., Fernandez, M., & Castillo, J. (2020). Financial capabilities and renewable energy adoption in Mexican manufacturing: Implications for energy efficiency. *Journal of Cleaner Production*, *267*, 122089. DOI: 10.1016/j.jclepro.2020.122089

Renneboog, L., Ter Horst, J., & Zhang, C. (2008). Socially responsible investments: Institutional aspects, performance, and investor behavior. *Journal of Banking & Finance*, *32*(9), 1723–1742. DOI: 10.1016/j.jbankfin.2007.12.039

Teece, D. J., Pisano, G., & Shuen, A. (1997). Dynamic capabilities and strategic management. *Strategic Management Journal*, *18*(7), 509–533. DOI: 10.1002/(SICI)1097-0266(199708)18:7<509::AID-SMJ882>3.0.CO;2-Z

Torres, L., & Silva, R. (2021). Financial literacy and stakeholder management as differentiators between sustainable and non-sustainable service firms in Mexico. *Business Strategy and the Environment*, *30*(4), 1745–1759. DOI: 10.1002/bse.2712

Villanueva, C., Moreno, J., & Perez, A. (2019). Environmental disclosure and firm performance in Latin America: The moderating role of financial literacy. *Corporate Social Responsibility and Environmental Management*, *26*(5), 1167–1179. DOI: 10.1002/csr.1746

Weber, O. (2017). Corporate sustainability and financial performance of Chinese banks. *Sustainability Accounting. Management and Policy Journal*, *8*(3), 358–385. DOI: 10.1108/SAMPJ-09-2016-0066

World Bank. (2020). *Mexico country climate and development report*. World Bank.

Chapter 12
Metaheuristic Optimization Approaches for Decision-Making in Sustainable Dairy Farming:
A Multi-Objective Framework for Feed Efficiency and Profitability

Demet Çanga Boğa
Niğde Ömer Halisdemir University, Turkey

Mustafa Boğa
Niğde Ömer Halisdemir University, Turkey

ABSTRACT

This chapter explores the implementation of metaheuristic optimization techniques to support decision-making in sustainable dairy farming systems. By applying methods such as genetic algorithms, particle swarm optimization, and other bio-inspired approaches, the authors aim to address challenges related to feed efficiency, profitability, and environmental sustainability. The chapter outlines a multi-objective framework integrating economic and ecological indicators and provides insights into real-world applications in livestock production. This chapter contributes to the growing field of computational agriculture by highlighting how intelligent optimization methods can improve resource allocation and support long-term sustainability in dairy operations.

DOI: 10.4018/979-8-2600-3038-7.ch012

ght © 2027, IGI Global Scientific Publishing. Copying or distributing in print or electronic forms without written permission of IGI Global Scientific Publishing is prohibited. Use of this chapter to generative artificial intelligence (AI) technologies is expressly prohibited. The publisher reserves all rights to license its use for generative AI training and machine learning model development.

INTRODUCTION

Dairy cattle farming plays a pivotal role in ensuring global food security, supporting rural economies, and promoting sustainable agricultural systems. In recent years, the sector has faced mounting challenges including climate change-related production risks, escalating feed costs, growing consumer demand for environmentally responsible practices, and increasing pressure to reduce greenhouse gas emissions. These complex issues have spurred significant interest in Precision Livestock Farming (PLF) - an innovative approach leveraging sensor technologies, big data analytics, and artificial intelligence to enable data-driven decision making in livestock management (Kılıç et al., 2023)

Within this context, metaheuristic optimization techniques have emerged as powerful computational tools for addressing the multi-objective, high-dimensional nature of modern livestock systems. For instance, feed ration optimization requires balancing cost efficiency with nutritional adequacy, while genetic selection programs must simultaneously enhance productivity and animal welfare. Conventional deterministic models often prove inadequate for solving such nonlinear, constrained optimization problems, creating a pressing need for more flexible and adaptive computational strategies.

The expanding integration of Internet of Things (IoT) technologies and big data analytics in modern dairy operations has significantly broadened the applicability of metaheuristic algorithms. Continuous monitoring of animal health, feed intake, environmental conditions, and production performance through sensor networks generates vast datasets. Metaheuristic models applied to this data can optimize feeding regimens, enable early disease detection, and improve genetic selection strategies. This convergence of digital agriculture with intelligent optimization presents unprecedented opportunities to develop more sustainable, adaptable, and economically viable livestock production systems.

Globally, dairy farming remains a fundamental agricultural activity that substantially contributes to sustainable food production and economic development. However, it involves numerous complex decision-making processes spanning animal nutrition, genetic improvement, herd logistics, and environmental management. In such multifaceted operational environments, traditional approaches frequently lack the necessary flexibility and precision to achieve optimal outcomes, highlighting the growing importance of advanced computational techniques, particularly metaheuristic optimization algorithms.

Nutritional management and optimization strategies are particularly crucial in dairy production for enhancing both economic efficiency and long-term sustainability. Recent research has demonstrated the practical benefits of metaheuristic applications in this domain. For example, Atıcı and Elen (2024) successfully employed genetic algorithms to reduce feed costs while developing an innovative optimization framework for cost-effective feeding. Reynolds et al. (2015) further emphasized the strategic importance of livestock systems in achieving sustainability goals and strengthening food security frameworks.

Research on feed quality and forage management remains central to nutritional optimization. Budak and Budak (2014) conducted comprehensive analyses of quality parameters in forage crops, while Karabulut et al. (2000) and the Turkish Feed Industrialists Association (2019) examined structural developments in Turkey's compound feed industry. Parlar and Koç (2020) evaluated the impact of feeding practices like Total Mixed Rations (TMR) on farm-level performance, demonstrating how feeding strategies influence overall productivity. Boğa and Çevik (2012) contributed significantly through their development of an information-based compound feed formulation program for ruminants, underscoring the transformative potential of technological integration in modern feeding systems.

These collective studies provide a robust theoretical and empirical foundation for exploring optimization in dairy nutrition. Metaheuristic optimization techniques, inspired by natural phenomena and employing heuristic search strategies, prove particularly effective for solving complex, nonlinear, high-dimensional optimization problems that challenge conventional mathematical methods.

This review examines applications of metaheuristic optimization techniques in dairy farming, with particular emphasis on animal nutrition and genetic improvement, while evaluating both their potential benefits and implementation challenges. The study investigates widely-used algorithms including Genetic Algorithms (GA), Particle Swarm Optimization (PSO), and Ant Colony Optimization (ACO), along with other prominent techniques such as Simulated Annealing, Tabu Search, Artificial Bee Colony, and Differential Evolution.

By offering a multidimensional perspective on how these algorithms can address real-world challenges in livestock systems, this work aims to serve as both a comprehensive reference and catalyst for future research and technological innovation in precision dairy farming. The integration of these advanced computational methods with emerging agricultural technologies promises to revolutionize decision-making processes and operational efficiency across the dairy value chain.

1. METAHEURISTIC OPTIMIZATION TECHNIQUES

Metaheuristic optimization techniques are advanced computational strategies developed to find high-quality solutions to complex optimization problems that are often nonlinear, non-convex, and multidimensional. These algorithms are commonly inspired by natural phenomena such as evolution, swarm behavior, or thermodynamics. Unlike exact methods, metaheuristics do not guarantee finding the global optimum but can provide near-optimal solutions in a reasonable computational time, especially for large-scale and NP-hard problems. Their ability to escape local optima and flexibly adapt to different objective functions and constraints makes them highly suitable for real-world agricultural applications, including dairy cattle farming.

1.1. Genetic Algorithm(GA)

The Genetic Algorithm (GA) is a metaheuristic optimization method inspired by the principles of natural selection and genetics. It is particularly effective for solving complex, non-linear, and multi-dimensional optimization problems where traditional methods may struggle. In dairy cattle farming, GA is applied to optimize feed rations, design genetic improvement programs, and enhance herd management. GA uses genetic operators such as natural selection, crossover, and mutation to evolve a population of potential solutions toward an optimal or near-optimal solution. The main steps of GA are as follows:

1. **Initialization** – Randomly generate an initial population of candidate solutions.
2. **Fitness Evaluation** – Compute a fitness score for each individual based on the objective function.
3. **Selection –** Select parents from the current population based on their fitness (e.g., roulette wheel or tournament selection).
4. **Crossover** – Combine the selected parents to form offspring by exchanging parts of their genetic material.
5. **Mutation** – Introduce random changes to the offspring to maintain genetic diversity.
6. **Replacement** – Form the new population by selecting the best individuals from parents and offspring.
7. **Termination** – Repeat the cycle until a stopping criterion is met (e.g., maximum generations or satisfactory fitness).

In dairy cattle farming, GA has been successfully applied in feed formulation, herd structure optimization, and multi-trait genetic selection problems. For instance, when designing a feed ration, the GA evaluates combinations of feed ingredients under nutritional and cost constraints to select the most cost-effective formulation.

1.2. Particle Swarm Optimization (PSO)

Particle Swarm Optimization (PSO) is a population-based optimization technique inspired by the social behavior of bird flocks and fish schools. In PSO, each particle represents a potential solution and moves through the search space by benefiting from its own experience and the experiences of other particles in the swarm. The core steps of PSO include:

1. **Particle Initialization** – Assign initial positions and velocities randomly.
2. **Fitness Evaluation** – Evaluate each particle's fitness value.
3. **Update Personal Best (pBest)** – If the current position yields better fitness than previous, update pBest.
4. **Update Global Best (gBest)** – Identify the best-performing particle across the swarm.
5. **Velocity and Position Update** – Update each particle's velocity and position using the equation (1) and (2).

$$v_i(t+1) = \omega \,.\, v_i(t) + c_{1.} \,.\, r_i \,.\, \left(p.Best\, t_i - x_i(t)\right) + c_2 \,.\, r_2 . \left(\text{g. Best} - x_i(t)\right) \tag{1}$$

$$x_i(t+1) = x_i(t) + v_i(t+1) \tag{2}$$

Where ω is the inertia weight, c_1 ve c_2 are learning coefficients, and r_1 ve r_2 are random numbers in [0,1].

PSO has been shown to converge faster than GA in some cases and is particularly effective in continuous optimization problems such as feed cost minimization and selection index weighting in genetic evaluations (Kennedy & Eberhart 1995; Özsağlam & Çunkaş, 2008; Marini & Walczak, 2015).

1.3. Ant Colony Optimization (ACO)

ACO is based on the foraging behavior of ants, which deposit pheromones along their paths to communicate favorable routes. In optimization, artificial ants build solutions by moving probabilistically through a graph, influenced by pheromone concentration and heuristic information. The basic cycle of ACO is:

1. **Initialize pheromone trails and parameters.**
2. **Construct solutions** – Each ant probabilistically chooses components based on pheromone levels and desirability.
3. **Evaluate solutions** – Compute objective value for each solution.
4. **Update pheromones** – Increase pheromones on better solutions and evaporate globally to prevent stagnation.
5. **Check termination** – If not met, repeat.

ACO is especially effective in routing and combinatorial problems. In dairy operations, it is used for optimizing milk collection routes and scheduling problems where traditional linear methods are impractical (Spangler et al. 2009).

1.4. Overview and Selection Criteria

Choosing the right metaheuristic technique depends on the problem's nature:

- For continuous, high-dimensional problems with cost minimization, PSO is advantageous.
- For discrete or combinatorial problems like route optimization, ACO performs well.
- GA provides robust performance for multi-objective problems and offers flexibility through custom encoding schemes.

A comparison of metaheuristic optimization techniques is presented in Table 1. This table summarizes the advantages, disadvantages, application areas in dairy cattle farming, convergence speeds, and the number of parameters of various methods of techniques. See Table 1 for a detailed comparison of metaheuristic techniques in dairy cattle farming.

Additionally, Table 2 includes an evaluation based on performance criteria such as solution quality, convergence speed, ease of implementation, parameter sensitivity, and computational cost. Refer to Table 2 for a performance evaluation of these techniques.

The application areas of metaheuristic optimization techniques in dairy cattle farming are visually summarized in Figure 1. See Figure 1 for a flowchart illustrating the application domains of metaheuristic techniques in dairy cattle farming, including animal nutrition, genetic improvement, herd management, and logistics management.

Table 1. Comparison of metaheuristic optimization techniques in dairy cattle farming, summarizing their advantages, disadvantages, application areas, convergence speeds, and number of parameters.

Technique	Advantages	Disadvantages	Application Areas in Dairy Cattle	Convergence Speed	Parameter Sensitivity
Genetic Algorithm (GA)	Effective in multi-dimensional problems, avoids local optima, allows parallel processing, effective in large search space	Needs parameter tuning, high computational cost, slow convergence	Feed ration optimization, genetic improvement programs, multi-trait selection, culling decisions	Medium	High
Particle Swarm Optimization (PSO)	Fast convergence, simple implementation, fewer parameters, effective in continuous optimization	Risk of getting stuck in local optima, struggles in high-dimensional problems, premature convergence	Genetic parameter estimation, selection index optimization, feed formulation, heritability estimation	Fast	Low
Ant Colony Optimization (ACO)	Effective in dynamic problems, positive feedback, distributed solution, successful in combinatorial problems	Slow convergence, parameter sensitivity, weak in continuous problems	Milk collection/ distribution routing, herd management, resource allocation, genotype selection	Slow	Medium
Simulated Annealing (SA)	Avoids local optima, simple implementation, single-solution based	Slow convergence, sensitive to parameter tuning, needs cooling schedule, limited to complex constrained problems	Optimization of production structure, supply chain management	Slow	Medium
Tabu Search (TS)	Avoids local optima, uses memory, flexible structure	Needs parameter tuning, requires memory, complex tabu list management	Resource allocation, scheduling problems, labor planning, equipment utilization	Medium	Medium
Artificial Bee Colony (ABC)	Effective in high-dimensional search space, balanced exploration/ exploitation, simple structure	Parameter tuning required, variable convergence speed, sensitive to problem size	Feed formulation, production planning, resource allocation, optimization problems	Medium	Medium

Additionally, Table 2 includes an evaluation based on performance criteria such as solution quality, convergence speed, ease of implementation, parameter sensitivity, and computational cost.

Table 2. Performance evaluation of metaheuristic optimization techniques based on solution quality, convergence speed, ease of implementation, parameter sensitivity, and computational cost.

Criterion	GA	PSO	ACO	SA	TS	ABC
Solution Quality	★★★★☆	★★★☆☆	★★★★☆	★★★☆☆	★★★☆☆	★★★☆☆
Convergence Speed	★★☆☆☆	★★★★★	★★☆☆☆	★★☆☆☆	★★★☆☆	★★★☆☆
Ease of Implementation	★★☆☆☆	★★★★☆	★★★☆☆	★★★★☆	★★☆☆☆	★★★☆☆
Parameter Sensitivity	★★☆☆☆	★★★★☆	★★☆☆☆	★★☆☆☆	★★☆☆☆	★★★☆☆
Computational Cost	★★☆☆☆	★★★★☆	★★★☆☆	★★★★☆	★★★☆☆	★★★☆☆

2. APPLICATIONS OF METAHEURISTIC OPTIMIZATION IN ANIMAL FEEDING

Animal feeding is one of the most significant cost items in dairy farming, accounting for up to 60–70% of total production expenses. Ensuring balanced, cost-effective, and sustainable feeding strategies is vital to maximizing milk yield, maintaining herd health, and achieving long-term economic viability (El Bilali et al. 2021). However, feed formulation involves solving complex, multi-constraint optimization problems, where the nutrient requirements of animals must be met using diverse feed ingredients under cost and availability constraints. In recent years, metaheuristic optimization techniques have emerged as valuable tools to tackle this challenge more flexibly and efficiently than conventional linear programming models.

2.1. Feed Ration Optimization Using Genetic Algorithm

Genetic Algorithm (GA)-based feed ration optimization aims to minimize the total feed cost while fulfilling the nutritional requirements (e.g., energy, protein, fiber, minerals, and vitamins) of dairy cattle at different physiological stages. A standard objective function in such problems is as shown equation (3) and (4):

$$min \sum_{i=1}^{n} C_i \cdot x_i \tag{3}$$

Subject to:

$$\sum_{i=1}^{n} N_{ij} \cdot x_i \geq R_j, \text{ for all } j = 1,2,\ldots, m \tag{4}$$

where:

- x_i is the amount of feed ingredient i,
- Ci is the cost per unit of ingredient i,
- Nij is the amount of nutrient j in ingredient i,
- Rj is the required amount of nutrient j.

In a study by Atıcı and Elen (2024), cost-effective feed rations were prepared for dairy cows using GA. The study considered dairy cows in the 10th lactation week, weighing between 450 and 600 kg, with a target milk yield of 20 and 25 kg of 3.5% fat milk. The results of GA were compared with the June 2023 raw milk production costs published by the National Milk Council, showing approximately 20% savings in feed cost. The proposed model was compared with results obtained from Particle Swarm Optimization and Ant Colony Optimization algorithms.

2.2. Feed Ration Optimization Using Particle Swarm Optimization

PSO provides an alternative to GA by operating over continuous search spaces and often converging more quickly. In PSO-based ration optimization, each particle represents a candidate feed mix. Particles move within the solution space, guided by both their individual historical best performance and the swarm's best-known configuration.

A key advantage of PSO is its adaptability in problems with numerous nutritional constraints and variable feed availability due to seasonal or market fluctuations. PSO has been applied in studies (e.g., Saxena & Khanna, 2015; Auliya et al., 2021) where it successfully reduced feed costs while preserving milk output. Moreover, the flexibility of PSO allows integration with multi-objective formulations, such as optimizing both feed cost and methane emission reduction.

2.3. Environmental Considerations and Sustainability Metrics

Beyond cost efficiency, modern dairy farms are increasingly required to reduce their environmental footprint. Metaheuristic methods enable the integration of sustainability metrics—such as carbon footprint, water usage, and nitrogen excretion—into feed optimization models.

For instance, a multi-objective GA may simultaneously minimize feed cost and estimated enteric methane emission (CH_4), which is a potent greenhouse gas. Nutrient-based emission models (e.g., IPCC Tier II) can be embedded into the optimization process, enabling farms to generate climate-smart feed strategies without sacrificing performance. Environmental priorities can be integrated into the model through the β weight coefficient in Equation (5).

$$\min[\ \alpha\ ,\ \text{Feed Cost} + \beta.\ \text{Methane Emission}] \tag{5}$$

where α and β are weights reflecting the farm's economic and environmental priorities. Such integrated models can also help farms comply with evolving environmental regulations and consumer expectations regarding sustainable animal production systems.

2.4. Metaheuristics in Decision Support Systems (DSS)

Many of these algorithms have been incorporated into practical decision support tools for ration formulation. Web-based or mobile applications powered by GA or PSO allow nutritionists or farm managers to input constraints and instantly receive optimized rations. These systems also allow for real-time updates based on feed prices or herd performance data, making them responsive and adaptive.

In conclusion, metaheuristic optimization techniques have significantly advanced the capabilities of dairy feed management. By enabling flexible, scalable, and adaptive solutions, these methods not only reduce feeding costs but also promote resource-efficient and sustainable animal nutrition strategies.

3. APPLICATIONS OF METAHEURISTIC OPTIMIZATION IN GENETIC IMPROVEMENT

Genetic improvement in dairy cattle aims to enhance economically significant traits such as milk yield, fat and protein content, reproductive performance, and disease resistance. Metaheuristic optimization techniques provide effective tools for optimizing complex breeding strategies, particularly under multi-trait selection scenarios. These techniques help balance genetic gain, inbreeding control, and economic objectives, often beyond the capacity of traditional deterministic methods.

3.1 Optimization of Dairy Cattle Breeding Programs Using Genetic Algorithms

Dairy cattle breeding programs involve multiple decision points: which individuals to select, how much weight to assign to each trait, and how to structure mating plans. Genetic Algorithms (GA) can be used to optimize these decisions by simulating the evolutionary process of selection.

The selection index can be optimized using a linear combination of EBVs and economic weights, as shown in Equation (6):

$$I = \sum_{i=1}^{n} \omega_i . EBV_i \tag{6}$$

Where:

I: selection index,

ω: economic weight of trait i,

EBV_i: estimated breeding value for trait i.

GA can optimize the wiw_iwi weights to maximize the expected genetic gain across generations. Additionally, GA can be used to simulate and compare different selection strategies, including mass selection, genomic selection, and progeny testing.

In dairy herds, genetic algorithms can model constraints such as minimum milk yield thresholds, maximum inbreeding coefficients, or targeted allele frequencies. For example, when selecting sires and dams for mating, GA can help ensure optimal pairing that maximizes offspring merit while controlling inbreeding.

Several studies have demonstrated the effectiveness of GA in designing optimal breeding strategies. The flexibility of GA allows it to handle nonlinear objective functions, multiple constraints, and real-world genetic data, making it suitable for commercial herd improvement programs (Bellingeri et al. 2020).

3.2. Estimation of Genetic Parameters Using Particle Swarm Optimization(PSO)

PSO has also been effectively applied to genetic improvement problems, particularly in optimizing selection index weights and estimating variance components in statistical models.

One area where PSO is beneficial is in the optimization of multi-trait selection indices. Traits in dairy cattle, such as milk yield, fat percentage, and somatic cell count, often exhibit genetic correlations. PSO allows for simultaneous optimization of the index weights while considering these correlations and constraints on trait emphasis.

In this context, the fitness function may be based on as shown in Equation (7):

$$I = \frac{a'G_g}{\sqrt{a'P_a}} \tag{7}$$

Where:

- G: expected genetic gain,
- a: vector of index weights,
- Gv: genetic variance-covariance matrix,
- P: phenotypic variance-covariance matrix.

PSO searches the weight vector aaa that maximizes the genetic gain while balancing risk and genetic diversity. Koç et al. (2018) has shown that PSO converges faster and yields more consistent solutions than classical regression or iterative matrix approaches.

PSO has also been used to estimate the genetic parameters of statistical models, such as heritability and repeatability, using biological data. The flexibility of the algorithm allows integration with maximum likelihood or Bayesian frameworks to improve parameter estimates in animal models.

In dairy cattle breeding programs, determining the weights of selection indices is a critical step to maximize the rate of genetic progress. PSO is used to evaluate various selection index weights and identify the optimal combination that maximizes genetic gain. In this process, the PSO algorithm begins with a randomly generated set of solutions and iteratively updates the velocity and position of particles to search for the best solution. This approach enhances the effectiveness of breeding program (Gümüştaş & Pehlivanlı et al., 2021; Önder et al., 2023). For instance, PSO can be used to optimize selection index weights for traits such as milk yield and milk fat percentage. This ensures that breeding programs can achieve both economic and genetic objectives. The fast convergence characteristic of PSO provides an advantage over traditional methods in such optimization problems.

The PSO algorithm is effectively employed in the optimization of selection index weights within genetic improvement programs. By evaluating different combinations of index weights, PSO identifies the optimal set that maximizes the rate of genetic gain. This strategy enhances the efficiency of dairy cattle breeding programs by increasing genetic progress.

3.3. Herd Management with Ant Colony Optimization

Ant Colony Optimization (ACO) is a heuristic optimization algorithm inspired by the cooperation and pathfinding behaviors of natural ant colonies (Dorigo & Gambardella, 1997). It evaluates a large number of potential solutions and converges toward optimal ones, making it highly effective for complex decision-making processes in dairy farming. ACO has proven particularly useful for herd renewal strategies, including determining which animals to retain or cull, setting replacement rates, and grouping animals in robotic milking systems (Spangler et al., 2009).

In robotic milking farms, optimizing the number of animals to be maintained within system capacity is a multifaceted problem influenced by yield, animal behavior, and machine utilization. ACO provides a flexible framework for evaluating these combinations under multiple constraints.

Key applications of ACO in herd management include:

- Maximizing the genetic merit of retained animals
- Minimizing production cost per unit of milk
- Balancing age distribution within the herd

Keskintürk and Söyler (2006) demonstrated that ACO-based strategies outperform rule-based culling methods in profitability and herd performance. In scenarios involving robotic milking, where capacity is limited, ACO helps determine optimal herd composition, enhancing both efficiency and yield (Auliya et al., 2021).

Beyond herd renewal, ACO has also been applied in milk logistics. Efficient milk collection and distribution are crucial for reducing transportation costs and maintaining product quality. Gusti & Aryanny (2023) applied ACO to optimize milk delivery routes in the dairy industry, achieving notable improvements in route efficiency and cost reduction. These solutions are particularly valuable in rural areas with dispersed collection points.

Originally developed for single-objective problems, ACO has been successfully adapted to multi-objective optimization (MOO) tasks. Livestock applications often require balancing competing objectives such as cost reduction and productivity enhancement. Alaya et al. (2007) pioneered the adaptation of ACO to MOO problems, enabling the identification of Pareto-optimal solutions. Muddada et al. (2023) further highlighted ACO's versatility in a distinct field—multi-document text summarization—using a Multi-Objective ACO (MOACO) framework.

In dairy farming, this multi-objective capability is beneficial for evaluating trade-offs between variables such as milk yield, feed intake, productive lifespan, and animal health. ACO-based decision support tools can help farmers achieve sustainable and profitable operations. For instance, Özkan & Hill (2015) emphasized the potential of ACO in aligning economic profitability with goals like animal welfare and environmental stewardship.

By incorporating diverse and sometimes conflicting objectives into its search process, ACO enables more holistic and informed operational decisions in dairy herd management.

4. APPLICATIONS OF OTHER METAHEURISTIC OPTIMIZATION TECHNIQUES IN DAIRY CATTLE FARMING

Dairy cattle farming presents numerous complex optimization challenges characterized by nonlinear relationships, discontinuous solution spaces, and multiple competing objectives. These challenges often arise in areas such as feed formulation, herd composition, breeding strategies, disease management, waste control, and facility layout. Traditional optimization methods frequently fall short when dealing with the high dimensionality and stochastic nature of these problems.

While Genetic Algorithms (GA), Particle Swarm Optimization (PSO), and Ant Colony Optimization (ACO) are among the most commonly applied metaheuristic techniques in livestock systems, a variety of alternative methods have also shown significant promise. Techniques such as Differential Evolution (DE), Simulated Annealing (SA), Tabu Search (TS), Harmony Search (HS), Cuckoo Search (CS), and Firefly Algorithm (FA) have been successfully employed in diverse decision-making problems within dairy farming.

4.1. Simulated Annealing (SA): Precision in Thermal Optimization

Simulated Annealing is a single-solution-based metaheuristic inspired by the annealing process in metallurgy. This method explores the solution space by probabilistically accepting worse solutions in early iterations to escape local optima, gradually reducing the likelihood of such acceptances over time. In dairy cattle systems, SA has been applied in the following areas:

- Production Structure Optimization: SA has been used to optimize the allocation of cow groups (e.g., based on lactation stages) to barns. For instance, assigning cows in different lactation periods to barns based on capacity and feeding requirements improves animal welfare and reduces operational costs.
- Feed Distribution Planning: SA optimizes feed distribution routes, minimizing energy consumption and time losses for tractors or automated feed deliv-

ery systems. For example, one study used SA to reduce feed delivery times by 15% under variable feed demands and barn layout constraints.

- Supply Chain Design: SA has been employed to design optimal routes for milk collection and delivery under time-window constraints and variable transportation costs. For instance, a study used SA to optimize milk collection vehicle routes, reducing fuel costs by 12% and improving delivery times (El Bilali et al., 2021).

Challenges and Advantages: The success of SA depends on properly tuning the cooling schedule, which determines convergence speed and solution quality. Rapid cooling may lead to premature convergence, while slow cooling can increase computational time. However, SA's flexibility in handling nonlinear and discontinuous problems makes it a powerful tool for complex agricultural systems like dairy farming.

4.2. Tabu Search (TS): Strategic Memory-Based Optimization

Tabu Search (TS) is a memory-based metaheuristic that enhances traditional local search by using adaptive memory structures. Central to its functionality is the "tabu list," which records recently explored solutions to prevent the algorithm from revisiting them, thus helping it escape local optima. Aspiration criteria provide flexibility by allowing these restrictions to be overridden under beneficial circumstances.

In dairy cattle farming, TS has been successfully applied to several key operational areas:

Workforce Scheduling:

TS has optimized milking shift schedules by balancing labor availability with operational needs. For example, a study implementing TS in workforce planning achieved a 20% reduction in idle time by optimizing milking parlor usage while minimizing labor costs.

Resource Allocation:

TS has proven effective in the strategic distribution of limited farm resources such as feed, water, and bedding materials. Its application in scenarios with supply constraints has led to improved efficiency in resource use across dairy units.

Facility Layout Optimization:

TS has contributed to redesigning physical layouts of barns, milking parlors, and storage units to streamline workflows. In one documented case, implementing TS reduced the travel distance between feeding and milking stations by 10%, thereby improving operational flow and reducing animal stress.

Challenges and Advantages:

TS's performance heavily relies on two factors: the length of the tabu list and the effectiveness of neighborhood solution generation. If not carefully tuned, it may either converge prematurely or waste computational resources. However, its ability to handle complex, discreet, and combinatorial problems makes it exceptionally well-suited for logistics, scheduling, and layout challenges in dairy operations.

Compared to other metaheuristics, TS often yields higher-quality solutions, particularly in multi-constraint planning environments where standard local search would fail to provide globally competitive outcomes.

4.3. Artificial Bee Colony (ABC): Swarm Intelligence in Action

Artificial Bee Colony is a metaheuristic inspired by the foraging behavior of honeybee colonies. It involves three types of bees—employed bees, onlookers, and scouts—that collaboratively explore the solution space. ABC has shown promise in dairy farming in the following areas:ABC has shown promise in:

- **Feed Formulation:** ABC has been used to create cost-effective and nutritionally balanced rations by considering the costs and nutritional values of feed components. For example, one study used ABC to reduce feed costs by 8% while meeting nutritional requirements.
- **Production Planning:** ABC has supported planning of feeding and milking processes by considering milk production targets and inventory levels, particularly useful in managing seasonal production fluctuations.
- **Nutrient Balancing:** ABC optimizes feed mixtures to meet cows' nutritional needs (e.g., protein, energy, vitamins). Its balance of exploration and exploitation makes it effective in multi-objective optimization scenarios, especially under fluctuating feed prices and supply chain disruptions.

Challenges and Advantages: ABC's simple structure and minimal control parameters make it easy to implement in agricultural optimization tools. Its balanced search capability provides significant advantages in multi-objective optimization problems, such as feed optimization.

4.4. Differential Evolution (DE): Robust Continuous Optimization

Differential Evolution (DE) is a population-based evolutionary algorithm particularly effective for solving continuous, nonlinear, and non-differentiable optimization problems. DE generates new candidate solutions by adding the weighted difference between two population vectors to a third vector, followed by crossover and selection mechanisms.

In the context of dairy cattle farming, DE has demonstrated high utility in various domains:

- Genetic Parameter Estimation: DE has been applied to optimize genetic parameters related to milk yield, reproductive performance, and animal health. In one case study, DE-enhanced selection strategies resulted in a 5% increase in milk yield.
- Economic Decision Modeling: DE has supported the development of integrated economic models that balance costs, milk revenue, and environmental penalties. These models assist farmers in making holistic, sustainability-driven decisions.
- Sensor-Based System Optimization: DE has been used to analyze real-time data from sensors (e.g., milking frequency, activity levels, temperature). For example, DE was employed to fine-tune robotic milking system parameters, leading to a 10% reduction in energy consumption.

Challenges and Advantages:

DE is particularly advantageous in noisy or uncertain environments, which are common in agricultural operations. It offers:

- Fast convergence
- Strong global search ability
- Simplicity in implementation

Compared to traditional mathematical programming or other evolutionary algorithms, DE frequently demonstrates superior performance in terms of robustness and computational efficiency.

4.5. Integrated Optimization Framework

The most effective modern dairy farming systems combine these techniques through:

Hybrid Optimization Architectures

- SA-TS hybrids for facility design
- ABC-DE hybrids for feed formulation
- GA-PSO hybrids for breeding programs

Decision Support System Integration

- Real-time data pipelines
- Visualization dashboards
- Scenario comparison tools
- Explanation interfaces

Continuous Learning Systems

- Performance feedback loops
- Parameter auto-tuning
- Learning Transfer between farms
- Anomaly detection capabilities

These advanced implementations typically deliver:

- 15-30% operational cost reductions
- 10-20% productivity improvements
- 20-40% environmental impact reductions
- Enhanced animal welfare metrics

The choice of technique depends on specific problem characteristics:

- Discrete/combinatorial problems favor TS
- Continuous parameter spaces benefit from DE
- Multi-objective challenges suit ABC
- Dynamic environments work well with SA

Ongoing research focuses on quantum-inspired variants and neuromorphic computing implementations to further enhance optimization capabilities in precision dairy farming.

4.6. Summary and Visualization of Application Areas

Each metaheuristic techniques offer unique advantages in dairy cattle farming:

- SA and TS are powerful for constrained scheduling and facility layout problems, delivering high-quality solutions for logistics and operational planning.
- ABC excels in feed mixing and resource allocation, with its simple structure facilitating implementation in multi-objective optimization scenarios.
- DE is ideal for continuous and noisy problem spaces, particularly in genetic parameter estimation and economic modeling, offering fast and robust solutions.

These techniques expand the toolbox of precision livestock farming and enable more customized, efficient, and sustainable strategies for dairy cattle operations (El Bilali et al., 2021). The flowchart in Figure 1 visually represents the application domains of metaheuristic optimization techniques in dairy cattle farming—such as animal nutrition, genetic improvement, herd management, and logistics management—along with the specific techniques used in each domain. Refer to Figure 1 for a comprehensive overview of how metaheuristic techniques are applied across various dairy farming domains.

It also highlights the benefits achieved, such as cost reduction, efficiency improvements, and environmental sustainability. For example, ABC and DE reduce costs while maintaining nutritional balance in feed optimization; SA and TS minimize delivery times and fuel consumption in logistics; and DE improves milk yield and animal health in genetic improvement.

Figure 1. Flowchart illustrating the application domains of metaheuristic optimization techniques in dairy cattle farming, including animal nutrition, genetic improvement, herd management, and logistics management, along with associated benefits

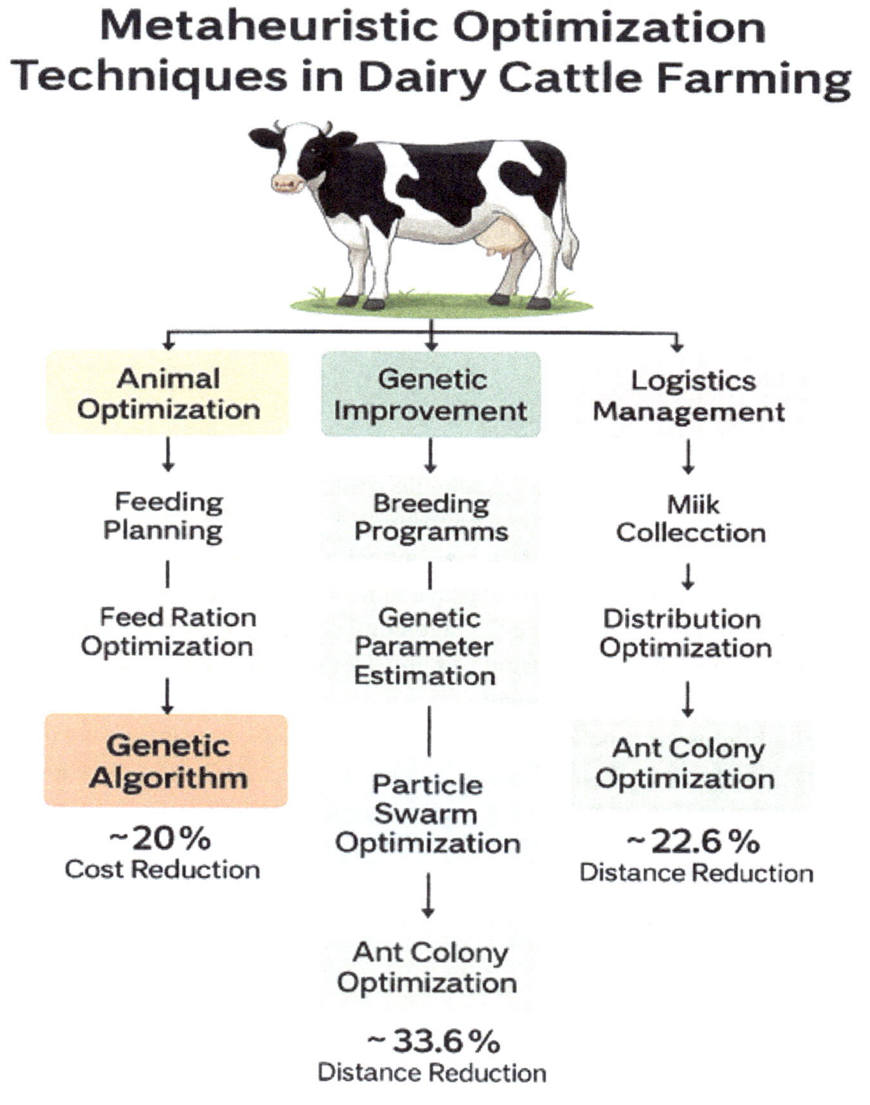

5. DISCUSSION AND CONCLUSION

A major advantage of metaheuristics is their adaptability to nonlinear and stochastic systems, which are typical of biological and agricultural settings. Unlike deterministic approaches, metaheuristics can flexibly adapt to uncertainties in animal response, weather conditions, and feed quality — providing robust solutions under real-world variability.

The increasing integration of metaheuristic optimization techniques with digital agriculture technologies, such as the Internet of Things (IoT), sensor networks, and artificial intelligence (AI), is significantly transforming the landscape of modern dairy cattle farming. These advanced computational tools allow for real-time, data-driven decision-making that supports precision agriculture principles, enhances resource efficiency, and contributes to both environmental sustainability and economic viability in the livestock sector.

One promising application is the development of real-time decision support systems (DSS) that utilize data streams from on-farm sensors, including ambient temperature, relative humidity, rumination activity, feed and water intake, and cow locomotion patterns. These platforms, when powered by metaheuristic algorithms such as PSO, Differential Evolution (DE), or Harmony Search (HS), can deliver near-optimal, adaptive recommendations for feed allocation, reproduction management, and milking schedules. For instance, Baldin et al. (2021) demonstrated that integrating such models with DSSs can significantly enhance decision quality, reduce operational costs, and minimize resource wastage in dairy systems. This aligns well with the principles of precision livestock farming (PLF), where individual-animal-level data are increasingly used for management interventions (Canga et al., 2022).

In parallel, metaheuristic approaches are playing a crucial role in addressing environmental concerns. The incorporation of ecological constraints — such as methane emissions, nitrogen excretion, and land use — into multi-objective optimization models allows for the simultaneous optimization of economic and environmental outcomes. For example, feed ration optimization models that consider carbon footprint or emission intensity contribute directly to climate-smart agriculture goals. Ma and Chen (2022) highlighted that PSO-based models can reduce greenhouse gas emissions by optimizing feed compositions, while Ben Hassen et al. (2020) emphasized that integrating low-emission dietary formulations into farm management is essential for meeting the EU Green Deal targets and the United Nations Sustainable Development Goals (SDGs). These models not only help in regulatory compliance but also support consumer demands for environmentally friendly dairy products.

Another area of advancement lies in the use of hybrid optimization techniques that combine classical linear programming with metaheuristics to create robust models capable of handling nonlinearity, uncertainty, and large solution spaces. Traditional methods often fall short when dealing with the dynamic and multi-dimensional nature of biological systems. Boğa and Çevik (2012) developed an early example of such a system in the form of a compound feed formulation program for ruminants. Contemporary advancements build upon this work by integrating machine learning algorithms (e.g., regression trees, neural networks) with metaheuristic solvers to dynamically adjust nutrient formulations according to individual animal requirements, health status, and production stage. This not only ensures optimal performance but also supports animal welfare by preventing overfeeding or undernutrition.

Moreover, intelligent logistics and animal grouping are receiving increasing attention with the deployment of multi-agent systems guided by swarm intelligence algorithms such as ACO. These systems enhance operational efficiency in complex farm environments, particularly in facilities utilizing robotic milking systems, where throughput constraints must be managed alongside animal preferences and lactation cycles (Hardhienata et al., 2024). For example, optimal cow routing in automated milking parlors can reduce waiting times and improve milking frequency, directly contributing to improved milk yield and animal comfort.

A defining advantage of metaheuristic algorithms is their adaptability to nonlinear, stochastic, and dynamic conditions — hallmarks of biological and agricultural systems. Unlike deterministic models, metaheuristics offer high flexibility in adapting to fluctuating conditions, such as variations in weather, changes in forage quality, or unpredictable animal health events. This robustness makes them particularly suitable for real-world applications, where uncertainty and variability are inevitable. In this context, the use of stochastic metaheuristics, such as Simulated Annealing or Genetic Algorithms with adaptive mutation rates, provides more resilient solutions than static optimization frameworks.

In conclusion, metaheuristic optimization techniques, especially when integrated with real-time sensing, AI models, and big data analytics, have the potential to redefine the decision-making paradigm in dairy farming. Their application across multiple domains — from nutrition and reproduction to environmental management and logistics — marks a significant step towards sustainable intensification in livestock systems. Future research should continue exploring hybrid frameworks that merge optimization, predictive analytics, and real-time monitoring to support fully autonomous, intelligent, and climate-resilient dairy operations. Stakeholder engagement, especially among farmers, technology providers, and policymakers, will be critical to ensure the scalability and practicality of these advanced decision support solutions.

5.1 Future Directions and Sustainability Perspectives

Metaheuristic optimization techniques have already shown considerable promise in improving productivity and efficiency in dairy farming. However, future research and development can further amplify their impact by integrating them with emerging digital technologies and sustainable agriculture paradigms.

5.1.1 Integration with Artificial Intelligence and Machine Learning

One of the most transformative avenues involves the hybridization of metaheuristic algorithms with artificial intelligence (AI), particularly machine learning (ML) and deep learning techniques. For example, reinforcement learning–guided Particle Swarm Optimization (PSO) or Genetic Algorithm (GA)-enhanced Deep Neural Networks (DNNs) can lead to autonomous decision-making systems capable of adapting to farm-level complexities (Shi et al., 2021). Such integrations would allow not only the optimization of feed or herd structures but also real-time response systems in robotic milking, disease detection, or behavior monitoring. These hybrid approaches may reduce human error, increase resilience to environmental fluctuations, and support smarter livestock management.

5.1.2. Operationalizing Optimization through Edge and Cloud Computing

The rise of Internet of Things (IoT) in agriculture provides a rich data stream from sensors, cameras, and environmental devices. However, to utilize this data efficiently, edge computing and cloud-based platforms are needed. Future metaheuristic models should be optimized to run in real-time on edge devices deployed directly on farms, reducing latency and dependency on high-bandwidth internet. This allows for immediate decision-making in scenarios such as detecting feeding inefficiencies or responding to animal stress events (Pal & Nag, 2022).

5.1.3. Big Data and Digital Twins for Livestock

As data volume increases from genomic sequencing, sensor data, climate records, and farm management systems, metaheuristic methods will need to scale up. This creates opportunities for applying Digital Twin models—real-time virtual replicas of animals or herds—optimized using algorithms like Differential Evolution or Tabu Search. These systems can simulate and predict the impact of feed changes, health treatments, or genetic interventions before implementation, reducing costs and risks.

5.1.4. Sustainability-Oriented Optimization

With growing emphasis on reducing environmental footprints, metaheuristic algorithms should be increasingly aligned with sustainability metrics, such as methane emission minimization, water use efficiency, and animal welfare indicators. Future models could include multi-objective optimization strategies to balance economic profitability with eco-friendly goals, guiding decision-makers toward circular and regenerative agricultural systems.

5.1.5. Interdisciplinary Collaboration and Farmer-Centric Tools

Developing user-friendly interfaces and decision-support systems remains a key challenge. Collaboration between animal scientists, computer engineers, data scientists, and agricultural economists is essential to translate algorithmic power into practical tools usable by farmers. Future studies should prioritize co-designed platforms, allowing farmers to define goals, adjust constraints, and interpret optimization results intuitively.

REFERENCES

Alaya, I., Solnon, C., & Ghedira, K. (2007, October). Ant colony optimization for multi-objective optimization problems. In *19th IEEE international conference on tools with artificial intelligence (ICTAI 2007)* (Vol. 1, pp. 450-457). IEEE. DOI: 10.1109/ICTAI.2007.108

Atıcı, E., & Elen, A. (2024). Optimization of Feed Ration Cost in Dairy Cattle by Genetic Algorithm. *Mühendislik Bilimleri ve Araştırmaları Dergisi*, *6*(1), 65–76. DOI: 10.46387/bjesr.1435749

Auliya, Y. A., Fadah, I., Zarkasi, M., Milniadi, A. D., & Baihaqi, Y. (2021). Implementation of a decision support system for optimizing feed rations in the cattle fattening industry. *International Journal of Science and Development Research*, *6*(1), 56–63.

Baldin, M., Breunig, T., Cue, R., De Vries, A., Doornink, M., Drevenak, J., Fourdraine, R., George, R., Goodling, R., Greenfield, R., Jorgensen, M. W., Lenkaitis, A., Reinemann, D., Saha, A., Sankaraiah, C., Shahinfar, S., Siberski, C., Wade, K. M., Zhang, F., & Cabrera, V. E. (2021). Integrated Decision Support Systems (IDSS) for Dairy Farming: A Discussion on How to Improve Their Sustained Adoption. *Animals (Basel)*, *11*(7), 2025. DOI: 10.3390/ani11072025 PMID: 34359153

Bellingeri, A., Gallo, A., Liang, D., Masoero, F., & Cabrera, V. E. (2020). Development of a linear programming model for the optimal allocation of nutritional resources in a dairy herd. *Journal of Dairy Science*, *103*(11), 10898–10916. DOI: 10.3168/jds.2020-18157 PMID: 32952013

Boğa, M., & Çevik, K.K. (2012). Ruminant hayvanlar için karma yem hazırlama programı. XIV. Akademik Bilişim Konferansı (Akademik Bilişim'12), Uşak Üniversitesi, 249-256.

Budak, F., & Budak, F. (2014). Yem bitkilerinde kalite ve yem bitkileri kalitesini etkileyen faktörler. *Türk Bilimsel Derlemeler Dergisi*, *7*(1), 1–6.

Canga, D., Boğa, M., Kılıç, H. N., & Bulut, M. (2022). *Information and communication technologies (ICT) and precision livestock (PLF) applications in farm conditions*. Livre de Lyon.

Dorigo, M., & Gambardella, L. M. (1997). Ant colonies for the travelling salesman problem. *Bio Systems*, *43*(2), 73–81. DOI: 10.1016/S0303-2647(97)01708-5 PMID: 9231906

El Bilali, H., Strassner, C., & Ben Hassen, T. (2021). Sustainable Agri-Food Systems: Environment, Economy, Society, and Policy. *Sustainability (Basel), 13*(11), 6260. DOI: 10.3390/su13116260

Gümüştaş, E., & Pehlivanlı, A. Ç. (2021). In-Silico Mutajenisite Tahmininde İstatistiksel Öğrenme Modeli. *Süleyman Demirel Üniversitesi Fen Bilimleri Enstitüsü Dergisi, 25*(2), 365–370. DOI: 10.19113/sdufenbed.867067

Gusti, M. L. H., & Aryanny, E. (2023). Determining The Optimal Milk Distribution Route Using The Ant Colony Optimization Method on Milk Industry. *Spektrum Industri, 21*(2), 141–148. DOI: 10.12928/si.v21i2.92

Hardhienata, M. K. D., Priandana, K., Putra, D. R., Sriatun, M., Wulandari, A. B., & Mohamed, R. (2024). Modification of the Ant Colony Optimization Algorithm for Solving Multi-Agent Task Allocation Problem in Agricultural Application. *Optimization (ACO). Journal of Advanced Research in Applied Sciences and Engineering Technology, 34*(1), 90–105. DOI: 10.37934/araset.34.1.90105

Karabulut, A., Ergül, M., Ak, İ., Kutlu, H.R., & Alçiçek, A. (2000). Karma yem endüstrisi. V. Türkiye Ziraat Mühendisliği Teknik Kongresi, TMMOB Ziraat Mühendisleri Odası, 985-1008.

Kennedy, J., & Eberhart, R. (1995). Particle swarm optimization. In Proceedings of ICNN'95 - International Conference on Neural Networks (Vol. 4, pp. 1942–1948). IEEE. DOI: 10.1109/ICNN.1995.488968

Keskintürk, T., & Söyler, H. (2006). Global Karınca Kolonisi Optimizasyonu. *Gazi Üniversitesi Mimarlık ve Mühendislik Dergisi, 21*(4), 689–698.

Kılıç, H. N., Boğa, M., & Çanga, D. (2023). Sensitive Livestock and Information Communication Technology Applications to Prevent the Spread of Covid-19. *Black Sea Journal of Agriculture, 6*(1), 104–107. DOI: 10.47115/bsagriculture.1194817

Koç, H., Dünder, E., Gümüştekin, S., Koç, T., & Cengiz, M. A. (2018). Particle Swarm Optimization-based Variable Selection in Poisson Regression Analysis via Information Complexity-type Criteria. *Communications in Statistics. Theory and Methods, 47*(21), 5298–5306. DOI: 10.1080/03610926.2017.1390129

Ma, S., & Chen, Z. (2022). Low-Carbon Development of Livestock in China: Evolution, Predicaments and Paths Ahead. *Low Carbon Economy, 13*(2), 53–69. DOI: 10.4236/lce.2022.132004

Marini, F., & Walczak, B. (2015). Particle swarm optimization (PSO): A tutorial. *Chemometrics and Intelligent Laboratory Systems, 149*, 153–165. DOI: 10.1016/j.chemolab.2015.08.020

Muddada, M. K., Vankara, J., Nandini, S. S., Karetla, G. R., & Naidu, K. S. (2023). Multi-objective Ant Colony Optimization (MOACO) approach for multi-document text summarization. *Engineering Proceedings, 59*(1), 218. DOI: 10.3390/engproc2023059218

Önder, H., Sitskowska, B., Kurnaz, B., Piwczyński, D., Kolenda, M., Şen, U., Tırınk, C., & Çanga Boğa, D. (2023). Multi-Trait Single-Step Genomic Prediction for Milk Yield and Milk Components for Polish Holstein Population. *Animals (Basel), 13*(19), 3070. DOI: 10.3390/ani13193070 PMID: 37835676

Özkan, Ş., & Hill, J. (2015). Implementing innovative farm management practices on dairy farms: A review of feeding systems. *Turkish Journal of Veterinary and Animal Sciences, 39*(1), 1–9. DOI: 10.3906/vet-1311-95

Özsağlam, M. Y., & Çunkaş, M. (2008). Optimizasyon Problemlerinin Çözümü için Parçaçık Sürü Optimizasyonu Algoritması. *Politeknik Dergisi, 11*(4), 299–305.

Pal, S., & Nag, A. (2022). Energy Harvesting in IoT-Enabled Flexible Sensors: Smart Sensing and Secure Access Control. In Nag, A., & Mukhopadhyay, S. C. (Eds.), *Flexible Sensors for Energy-Harvesting Applications. Smart Sensors, Measurement and Instrumentation* (Vol. 42). Springer., DOI: 10.1007/978-3-030-99600-0_9

Parlar, T., & Koc, F. (2020). Toplam rasyon karışımı kullanılan bir süt sığırı işletmesinin besleme açısından değerlendirilmesi. *Erciyes Tarım ve Hayvan Bilimleri Dergisi, 3*(1), 24–32.

Reynolds, L. P., Wulster-Radcliffe, M. C., Aaron, D. K., & Davis, T. A. (2015). Importance of animals in agricultural sustainability and food security. *The Journal of Nutrition, 145*(7), 1377–1379. DOI: 10.3945/jn.115.212217 PMID: 25972529

Saxena, P., & Khanna, N. (2015). Optimization of dairy cattle feed by nonlinear programming. In *Proceedings of the 2015 2nd International Conference on Computing for Sustainable Global Development (INDIACom)* (pp. 139–144). IEEE.

Shi, J., Sun, J., & Zhang, Q. (2021). Multi-objectivization inspired metaheuristics for the sum-of-the-parts combinatorial optimization problems. *Applied Soft Computing, 103*, 107157. DOI: 10.1016/j.asoc.2021.107157

Spangler ML, Robbins KR, Bertrand JK, Macneil M, Rekaya R. (2009). Ant colony optimization as a method for strategic genotype sampling. Anim Genet. Jun;40(3):308-14.. Epub 2009 Feb 10. PMID: 19220227.DOI: 10.1111/j.1365-2052.2008.01835.x

Turkish Feed Industrialists Association. (2019). *Karma Yem Sanayi Raporu 2019.* Poyraz Ofset Matbaacılık.

KEY TERMS AND DEFINITIONS

Ant Colony Optimization (ACO): A metaheuristic optimization technique inspired by the foraging behavior of ants. It is used to solve combinatorial and routing problems by simulating pheromone-based decision-making processes.

Artificial Bee Colony (ABC): A swarm-based metaheuristic mimicking bee foraging. Used in dairy farming for cost-effective feed formulation, production planning, and nutrient balancing in rations.

Dairy Cattle: Cattle breeds that are raised specifically for milk production. Effective management of dairy cattle involves nutrition, genetics, and logistics to optimize productivity and profitability.

Differential Evolution (DE): A population-based algorithm for continuous optimization. In dairy farming, it optimizes genetic selection, economic models, and sensor-based systems.

Feed Formulation: The process of combining various feed ingredients to meet the nutritional requirements of livestock at minimal cost. It plays a critical role in animal health and farm economics.

Genetic Algorithm (GA): An evolutionary optimization technique that mimics natural selection. It uses mechanisms like selection, crossover, and mutation to evolve optimal solutions in complex problem spaces.

Genetic Improvement: A systematic approach to enhance economically important traits in livestock, such as milk yield and disease resistance, through selection and breeding strategies.

Herd Management: The application of planning and decision-making methods to optimize herd structure, reproductive efficiency, and overall farm performance.

Metaheuristic Optimization: A class of high-level algorithms designed to find near-optimal solutions for complex and nonlinear optimization problems that traditional methods cannot efficiently solve.

Particle Swarm Optimization (PSO):A population-based optimization algorithm inspired by the social behavior of birds and fish. It updates candidate solutions based on individual and group experience to reach optimal results.:

Simulated Annealing (SA): A metaheuristic inspired by metallurgical annealing, accepting worse solutions early to escape local optima. Used in dairy farming for optimizing barn assignments, feed delivery routes, and milk collection logistics.

Tabu Search (TS): A memory-based metaheuristic using a tabu list to avoid revisiting poor solutions. Applied in dairy farming for milking shift scheduling, resource allocation, and facility layout optimization.

Chapter 13
Emerging Civil Engineering Strategies for Sustainable Urban Resilience:
Smart, Circular, and Adaptive Infrastructure Systems

R. Ramya
http://orcid.org/0009-0000-8016-212X
Academy of Maritime Education and Training, Chennai, India

Deepa Rajesh
http://orcid.org/0009-0008-9743-4791
Academy of Maritime Education and Training, Chennai, India

R. Vettriselvan
http://orcid.org/0009-0005-1097-0806
Academy of Maritime Education and Training, Chennai, India

J. Rengamani
Academy of Maritime Education and Training, Chennai, India

R. Swadhi
http://orcid.org/0009-0005-0247-0713
Saraswathi Institute of Medical Sciences, Hapur, India

DOI: 10.4018/979-8-2600-3038-7.ch013

ght © 2027, IGI Global Scientific Publishing. Copying or distributing in print or electronic forms without written permission of IGI Global Scientific Publishing is prohibited. Use of this chapter to generative artificial intelligence (AI) technologies is expressly prohibited. The publisher reserves all rights to license its use for generative AI training and machine learning model development.

ABSTRACT

The climatic changes, increased urbanisation, and limited available resources, urban infrastructural systems are increasingly faced by mounting pressure to employ combined approaches to sustainability and resilience. The present paper interrogates the intersection of smart technologies, the principles of a circular economy, and the adaptive design in the context of civil engineering and presents them as emerging and powerful tools in enhancing the resilience of urban environments. I expand and build on the current interdisciplinary research by presenting a systems-based analytical prism explaining how digital intelligence, life-cycle optimization, and climate-sensitive infrastructure can be logically assembled into higher-level structures. The work further refines the contexts of governance and implementation contingencies with specific reference to cities that are typified by the rapid urbanisation and increased exposure to climatic hazards and finally proposes that integrated infrastructural paradigms are feasible paths to sustainable, resilient, and habitable urban futures.

1. INTRODUCTION

The urban infrastructural systems are increasingly facing cumulative impact of climate change, swift urbanisation, environmental degradation and societal-economic transformation. Several changes in the climatic conditions of the area of the world today are escalating temperatures, abnormal precipitation, flooding, drought, rising of the sea level, and degradation of the ecosystems, which are causing pressure on transport networks, water and energy system, buildings, and the public space, as never before. These stresses are also enhanced by increased rate of urban growth in developing and emerging economies where the infrastructural development often surpasses strategic planning, governance ability and environmental protection. The constraints of conventional civil-engineering paradigms, which were mostly based on inflexible, grey infrastructure and optimised concerning short-term structural functionality, have been more and more evident as a result (Aktan et al., 2022; Ilugbusi et al., 2024; Das, 2025). Traditionally, strength, durability and efficiency have been favoured in terms of infrastructure planning due to the assumption of constant operating conditions. However, the vulnerability of them to cascading failures has been revealed by climate-impaired uncertainty and increased systems interdependencies (Swadhi 2025). Clogged travel networks, deterioration of materials due to heat and water impacts, disruption of water systems, and power blackout can also be observed as examples of how the failures at a single point can spread through the interrelated urban systems causing a massive socio-economic interference

(Asghari et al., 2023; Okem et al., 2024). As such, climate resilience has become a key goal of the modern civil engineering to require a radical paradigm shift in design, towards systems-oriented, adaptive, and performance-based models that can address long-term climatic uncertainty (Reda Taha et al., 2021; Janshevaski, 2025). Climate resilient infrastructure is not the same as shock absorbing, but the ability of urban systems to anticipate, absorb, and recover functionality, as well as evolve over a period, without compromising critical economic, social and environmental functions. The recent research emphasises that resilience should be understood as a dynamic characteristic of socio-technical systems, but not as the property of physical assets (Ramirez Lopez and Grijalba Castro, 2020; Juarez-Quispe et al., 2025). The redefined role of civil engineering is at the intersection of comprehensive urban resiliency approaches that will be robust, flexible, sustainable and inclusive.

In this changing context, the following two key streams of innovations have become salient: the digitalization of the infrastructure systems and the application of nature-based solutions (NbS) to the sustainable urban development. The concept of digital twins, digital innovations such as artificial intelligence (AI), machine learning, Internet of Things (IoT) sensing, and advanced data analytics are reinventing the notion of infrastructure, its conception, monitoring, and management. These technologies help to assess the condition in real time, model predictive risk, and make adaptive decisions to improve infrastructure performance during uncertain climatic conditions (Abbass et al., 2025a; Akhai & Abbass, 2025; Almulhim, 2025). Based on empirical findings it can be indicated that the digitally enabled infrastructure systems will be in a position to transform the civil-engineering practise into not merely reactive maintenance but to preventive and anticipatory management. The analytics SAI technologies and digital twins enable engineers and planners to simulate climate changes, test system reactions, and optimise interventions before failures come into reality (Ali & Anwar, 2024; Nyokum & Tamut, 2025). Moreover, intelligent infrastructure platforms permit cross-sectoral integration, which makes it easy to coordinate transport, water, energy and emergency systems, which is one of the requirements of resiliency in complex urban infrastructures (Abbass et al., 2025b). At the same time, nature-based solutions have become one of the keys to enhancing city-level climate resilience. The NbS is used to utilize the natural processes and ecosystem services, including wetlands, urban forests, green roofs, permeable pavements, and hybrid green-grey systems, to reduce climate risks, such as flooding, heat stress, and erosion. The NbS has natural adaptability, multifunctionality, and regenerative properties, which provide long-term resilience benefits and co-benefits to the biodiversity, human health, and social well-being compared to traditional grey infrastructure. Civil-engineering In civil-engineering terms, NbS represent a paradigm shift to infrastructure systems that adapt to environmental conditions as well as being resistant to them, and no longer isolated. Studies prove that NbS have the potential to increase system redundancy, curtail exposure to climatic changes, and add to the general livability of cities when integrated into systemic planning and engineering processes (Bayeroju et al., 2021; Ilugbusi et al., 2024). However, NbS face the difficulties of the quantification of performance, monitoring over the long term, scalability and integration into the conventional engineering processes.

Digital innovations and solutions in nature are often developed and implemented as independent spheres despite their complementary qualities. Smart technologies are frequently introduced as efficiency-oriented tools with no connexion to the ecological processes whereas NbS are regarded as additional environmental measures instead of infrastructure items. This disintegration limits their group potential in providing strong climate resiliency. There is an emerging literature that claims that meaningful and sustainable resilience is realized under a consistent system where digital intelligence and nature-based design are incorporated into a single civil-engineering system (Abbass et al., 2025a; Ajirotutu et al., 2024; Velasquez-Mendez et al., 2025). Intensive digitalnature based infrastructure systems make possible a new form of civil engineering practise where the ecological assets are engaged as performance-based components anchored by the digital monitoring, modelling, and control systems. Digital tools contribute to the measuring, predicting and adapting of NbS and ecological infrastructure increases the flexibility, sustainability and buffering capacity of urban systems that are digitally capable. Infrastructure life cycles Artificial intelligence also supports this convergence by enabling endless optimisation of resource usage, carbon trapping, risk reduction and system performance (Abbass et al., 2025b; Shanthi et al., 2025). Such integrated strategies are however subject to success based on sound governance structures, institutional capacity and policy alignment. The tools of climate-sensitive planning, adaptive regulatory mechanisms, interdisciplinary coordination, and stakeholder involvement cannot be done without to convert technological and ecological innovation into the results of resilient infrastructure (Sutriadi, 2023; Das, 2025; Wimalasena et al., 2025). Without these enabling conditions, digital and nature-based innovations may end up being disjointed, unproductively harnessed or not in line with the long-term sustainability goals. It is within this context that this chapter expresses a combined civil engineering understanding of climate resilient urban infrastructure. First, it constructs existing information on digital innovations and nature-based solutions as two opposing avenues to resilience. Second, it creates an analytical framework that connects smart technologies and sustainable design in the infrastructure planning, implementation and governance. Third, it applies this framework to application-based insights relevant to civil-engineering practice, policy-making and the management of the city. The chapter also helps in the future of the civil-engineering theory and practise because it places urban infrastructure as a socio-technical-ecological system that is resilient, sustainable, and able to withstand the future.

URBAN CLIMATE RISKS AND INFRASTRUCTURE VULNERABILITY

The modern urban infrastructural landscape is increasingly facing a complex and growing number of climate-related threats that put their functional stability, viability, and socio-economic worth at risk. Climate change has increased the occurrence, severity, and uncertainty of extreme events like floods, heat waves, droughts, storms and coastal flooding. These risks fail to work in a vacuum, but instead, they combine with quick urban growth, land-use shift, residential concentration, and decaying infrastructures to create complex vulnerability in cities. With the growth of interconnection of urban systems, perturbations in one sector of the infrastructure caused by climate easily disseminate into other sectors, producing cascading failures with socio-economic impacts that extend widely (Aktan et al., 2022; Ilugbusi et al., 2024; Das, 2025). Having a civil engineering perspective in mind, the concept of vulnerability has moved beyond the traditional paradigm of structural failure under extreme loads. Rather, it has adopted functional degradation, service failure, impaired adaptation ability and systemic weakness caused by repeated or long-lasting climatic stress. Floods and material deformation during heat may render transport networks impassable; the water supply system has to face the threat of contamination and unstable supply; energy infrastructure becomes more prone to thermal pressure and storm damage; and the construction faces a greater rate of material degradation due to excessive changes in temperature (Asghari et al., 2023; Okem et al., 2024). One notable feature of the modern city vulnerability is that it is interdependent. The present-day cities are based on strongly interconnected infrastructure networks where failures can hardly be localised. Indicatively, transport corridors might be incapacitated through flooding, and emergency response and supply chains, and digital communication systems, water pumping stations, and healthcare facilities might be disabled by power outages. These interdependencies increase climate risk and make the recovery more difficult, particularly in municipalities that are not redundant in governance and adaptive mechanisms (Reda Taha et al., 2021; Juarez-Quispe et al. 2025). This fact requires the replacement of asset-specific risk analysis with systemic vulnerability analysis that would include network effects, feedback, and cross-sectoral exposure. Socio-economic and institutional determinants are other urban factors that shape urban vulnerability. This can often be worsened by informal settlements, unequal allocation of services and disjointed governance especially in fast urbanising areas. Infrastructure projects failing to take into consideration social equity and long-term flexibility are most likely to strengthen vulnerability instead of reducing it (Ramirez Lopez and Grijalba Castro, 2020; Das, 2025). In line with this, climate-resilient infrastructure should not only be physically robust, but also inclusive, of quality governance, and capable of learning.

The conventional grey infrastructure models, which are akin to inflexible design, centralised control, and dependency on hard-engineering solutions, have not been sufficient in these new circumstances. Such systems can usually provide high-performance at the beginning, but their performance fails to be flexible and adaptive, making them susceptible to any factors outside the original design limits (Bayeroju et al., 2023; Ibrahim, 2023). Furthermore, the grey infrastructure often externalises the environmental costs, and hence leads to the ecosystem degradation that further increases the risks associated with climate, such as the enhancement of runoff through the presence of impervious surfaces or the disappearance of natural buffers in coastal and riverine ecosystems. Contemporary civil engineering scholarship in turn is more inclined to promote resilience-based design, with its priority on adaptability, modularity, redundancy, and learning as opposed to static optimisation. The frameworks based on resilience feature the infrastructures that have the capacity to absorb shocks, restructure their functionality, and evolve in reaction to the altered climatic conditions (Aktan et al., 2022; Janshevaski, 2025). This paradigm views uncertainty as normal and not an exception and encourages lifecycle planning and agile management across the infrastructure life cycle. In the framework of this, digital technologies, as well as nature-based approaches, become the most important resilience facilitators. But to know how they are applicable, it is important to first realise the shortcomings of the traditional vulnerability assessment tools. Conventional risk models tend to be based on the presence of historic climatic data, deterministic models, and sectoral analysis, which does not support the representation of nonlinear system behaviour, system behaviour in real-time, and socio-ecological feedback processes (Asghari et al., 2023; Rezvani et al., 2024).

The latest developments in the field of urban resilience recommend scenario-based, systems-oriented, and adaptive evaluation models. Scenario approaches enable engineers and planners to consider the performance of infrastructure in a variety of different plausible future conditions, as opposed to a single future projection. Systems-based approaches combine physical, digital, ecological, and social aspects to be able to represent real-world complexity. Adaptive frameworks focus on constant cheque up, education and adaptation as circumstances change (Reda Taha et al., 2021; Juarez-Quispe et al. 2025). In spite of these, the conceptual improvements, there still exist profound gaps in theory and practise. The cities are still stuck in the mesmerised planning processes, old design codes, and deficient cross-sectoral planning. In addition, the lack of coherent technologies that could help transform climate risk information to practical engineering actions limits the efficiency of resilience measures (Almulhim, 2025; Nyokum and Tamut, 2025). Such change has a premise in the increasingly recognised growth of urban infrastructure as a socio-technical-ecological system. It is based on the recognition that the performance of infrastructure is created through interactions between physical resources, digital systems, natural processes, governance patterns, and human behaviours. The issue of climate resilience cannot, however, be obtained based on unilateral interventions but through concerted actions targeting vulnerability at multiple levels at the same time (Wimalasena et al., 2025; Velasquez-Mendez et al., 2025). In this respect, the awareness of the risks of urban climate and the vulnerability of infrastructure is not an academic per se but a condition towards designing integrated solutions. Civil engineering can advance to anticipatory and not reactive resilience strategies by diagnosing the causes and routes of vulnerability including structural as well as functional, institutional and ecological. Such a diagnostic ground establishes the preconditions of further deliberations, where digital innovations and solutions related to nature are systemically combined to provide beneficial systems of vulnerability reduction and improvement of the adaptive capacity of urban infrastructure systems.

DIGITAL INNOVATIONS FOR CLIMATE-RESILIENT URBAN INFRASTRUCTURE

The digital innovation has now become central to the revolution of the civil engineering practise with respect to the development of the climate-resilient urban infrastructure. As the risks of climate get stronger, and the future of the infrastructure planning process becomes more and more uncertain, digital technologies provide the means that make the process of anticipation, real-time responsiveness, and adaptable governance. The traditional paradigms of engineering, based on the assumptions of the static design and regular inspections, are replaced with digitally enabled infrastructure systems that act as dynamic and data-based systems capable of learning and developing. This shift represents a paradigm shift in the development of infrastructure, namely, the replacement of reactive maintenance with predictive, performance-based, and resilience-oriented management (Reda Taha et al., 2021; Almulhim, 2025). Among the most powerful digital technologies in this field, the introduction of artificial intelligence (AI) and machine learning to improve infrastructure planning, operation, and risk management deserves a mentioning. Models based on AI facilitate the analysis of large and heterogeneous data, such as climate projections, sensor data, land-use data, and socio-economic indicators, and therefore would reveal the concealed patterns and nonlinear correlations that occur in more exact forecasting of climate impacts and infrastructure performance when faced with stress conditions. In the context of smart cities, AI has been used to show smarter maintenance, energy and water systems optimization, as well as the support of evidence-based urban decisions (Abbass, Abbas, Jafri, Arif, & Akhai,

2025; Abbass, Akhai, Abbas, Jafri, and Arif, 2025). The closely related concept to AI is the growing use of digital twins, which simulates physical infrastructure assets and infrastructure systems. Digital twins combine real-time data flow with the simulation models to replicate the dynamics of infrastructure provided different environmental and operational conditions. Digital twins, when used in climate resilience planning, can simulate extreme climate scenarios, e.g., a flood, heatwave, or earthquake, and can therefore be used by engineers to test system vulnerabilities and experiment with adaptive strategies before they are implemented (Ali and Anwaar, 2024; Almulhim, 2025).

The other most important element of digital innovation is the implementation of intelligent sensing networks and Internet of Things (IoT). Given the role of the climate resilience, these monitoring systems are invaluable in early warning, real-time detection of the risks, and categorical response. As an example, stress development, material wear, or uncharacteristic hydrology can be detected using sensors on bridges, drainage systems, and coastal defences, and thus, timely and effective action can be taken to reduce the chance of an uncontrolled collapse (Abbass, Akhai, Chouksey, Pathak, Abbas, and Abass, 2025; Okem et al., 2024). Outside of personal resources, digital systems make integration and coordination at systems levels - a requirement to manage interdependent urban architecture networks. Sophisticated data analytics and decision-support engines provide cross-sectoral visibility, allowing the planners and the operators to determine the potential spread of disruptions in one system to other systems. This functionality is particularly essential in cases of emergencies caused by climate, where energy, transport, water, and communication systems coordinated actions are needed to reduce cascading failures (Juarez-, Quispe et al., 2025; Nyokum and Tamut, 2025). Digital innovations can enhance resilience strategies through increasing awareness of the system on a holistic level that goes beyond asset robustness toward network stability and recovery capacity. Urban planning and other urban governance is also being transformed by digital transformation. Geospatial applications, coupled with integrated modeling systems, and participatory digital platforms enable the stakeholders to visualize the risks of climate, performance of infrastructure, and available adaptation options. The tools increase transparency and facilitate the process of decision-making between engineers, policymakers, and communities. The digital governance has been applied in smart city planning to promote the integration of the infrastructure investment with the long-run sustainability and resilience goals so that the climate adaptation actions are encompassed in larger urban development plans (Abbass, Abbas, Jafri, Arif, and Akhai, 2025; Das, 2025).

Notably, digital innovations have a role to play in climate resilience not only in the possibility of technology but in organisational learning and flexibility. Ongoing data gathering and feedback will help infrastructure personnel to analyse the efficiency of interventions, optimise operational plans and adapt to changed climatic conditions. This adaptability is a stark contrast to the conventional model of infrastructure, which can be characterised by the frequent use of assessment and strict operational procedures. Overall, resilient infrastructure systems can be characterised by the capability to learn based on real-time performance data as climate uncertainty increases (Aktan et al., 2022; Reda Taha et al., 2021). In spite of these benefits, the process of digitization creates new issues that should be solved in the sphere of civil engineering practise. Both data interoperability, cybersecurity, institutional capacity, and unequal access to digital infrastructure might not support resilience goals when ignored. Moreover, solutions that are being introduced based on technology and do not consider both environmental and social backgrounds are likely to strengthen existing weak points instead of reducing them (Sutriadi, 2023; Ibrahim, 2023). These shortcomings highlight the importance of controlling digital innovation through powerful governance structures and incorporation with the support of complementing resilience frameworks. In this respect, digital innovations can be most efficient when seen not as solutions per se but as facilitators in a resilience system. Their real worth is that they are able to create situational awareness, assist adaptive decision making and coordinate complex infra structure system in the face of climate stress. Digital technologies can enhance the robustness of the urban infrastructure when they are coupled with the principles of sustainable materials, inclusive planning, and ecological design. This acknowledgement forms the foundation to the further discussion that involves getting to know nature-based solutions as additional tools in the adaptation to climate. Digital innovations improve prediction, monitoring and control whereas nature-based solutions can provide regenerative, adaptable, and ecosystem-based protection. It is necessary to comprehend their role and what synergies exist between digital and nature-based approaches to climate-resilient civil engineering in the face of an increasing pace of environmental change.

SMART INFRASTRUCTURE AND DIGITAL TRANSFORMATION IN URBAN SYSTEMS

Urban infrastructure digitalization has become a central strategy in making cities more sustainable, operationally efficient and resilient. Smart infrastructure is a combination of digital technologies, high-tech analytics, and smart systems into the process of planning, operation, and governance of civil infrastructure assets. This transformation is a radical change in the practise of civil engineering, which has been reactive, asset-based management to information-oriented, system-focused, and future-oriented decision-making (Sertyesilisik, 2019; Almulhim, 2025). These developments are able to put the cities in a position to proactively predict risks, efficiently distribute resources, and ensure continuity of services during climatic and socio-economic unpredictability presented by infrastructure systems. Smart city planning offers institutional and spatial conditions, in which smart infrastructure systems are implemented. Recent synthesis research underlines the fact that smart-city efforts are increasingly focusing on urban resilience through the integration of digital technologies with the infrastructure governance and spatial planning practices (Almulhim, 2025). This approach does not consider technological intervention as an isolated intervention but instead, rather than relying on single systems, it aims at the coordinated intelligence of the transportation, water, energy, and built-environment systems. Detailed is the same argument by Nyokum and Tamut (2025), who argue that smart technologies cannot be avoided in realising environmentally sustainable and resilient urban infrastructure, in areas that are increasingly becoming urbanised with climate stress and service demand escalating in equal measure. Civil engineering-wise, smart infrastructure improves the ability of constant monitoring, diagnostics and predictive control. Technologies related to sensing, real-time data collection, and analysis modelling allow engineers to monitor infrastructure status and the performance on a continuous basis. The early degradation realisation by utilising emerging technologies, including structural health monitoring system, digital platform, and risk-informed analytics, stands out significantly in enhancing proactive maintenance planning and responding promptly to an extreme event (Reda Taha et al., 2021). Such capabilities minimise the chance of disastrous failures, increase the life cycle of assets, and harmonise the management of the infrastructure with long-term sustainability and resiliency goals.

Infrastructure intelligence is also not limited to single assets, but to combined digital platforms that assist in planning and governance. Connected and smart urban systems are based on interoperability and the exchange of data between infrastructure sectors and institutional demarcations. Such connectivity, as noted by Sertyesilisik (2019) supports the process of making holistic decisions taking into consideration the interdependencies of systems in urban environments and supporting orchestration of stakeholders. In this regard, smart infrastructure is not only a technical performance improvement but also an infrastructure that enables redesign of the system city in which the work of infrastructure is evaluated based on the overarching social, environmental, and economic objectives. Artificial intelligence also increases the promise of smart infrastructure by making it possible to detect patterns more effectively, predictively, and in response to changes. It has been proven that AI-driven solutions can be used in environmental surveillance, health systems controlled by infrastructures, and smart city services (Shanthi et al., 2025). Even though these applications are not limited to classical areas of civil engineering, it is important to note that through increased integration with infrastructure systems the nature of smart urban development is becoming more and more interdisciplinary.

Infrastructure intelligence plays a role in the physical resilience, as well as social resilience and urban well-being, through such integration. Nevertheless, the successfulness of smart infrastructure depends greatly on the conducive institutional and governance environment. Technological capability cannot achieve the results of resilience without additional so-called soft infrastructure, such as regulations, institutions, and human resources. Sutriadi (2023) highlights the lack of such soft infrastructure would continue to be one of the biggest impediments to the success of smart systems in many urban contexts especially in the developing regions.

Fairness and diversity are other important aspects of smart infrastructure development. Although digital technologies have significant advantages, the presence of inequality in access to data, technology and institutional capacity may increase the current socio-spatial inequalities. Almulhim (2025) emphasises that smart-city planning to be resilience-based should clearly focus on the aspect of inclusiveness as investments in digital infrastructure should not support inequality among various urban communities. This issue is particularly relevant in urban areas that are typified by divided governance systems and lopses in infrastructure delivery. The literature also suggests that smart infrastructure shall be represented as an upgrading ability instead of a rigid solution. Urban systems are dynamic in nature and the digital technologies need to be flexible and scalable to meet the changing environmental conditions, technology and risk profile. Reda Taha et al. (2021) emphasise that digital architecture should be flexible to integrate potential innovations and address the arising climate issues. This flexibility is exactly what resilience principles entail in the sense that the infrastructure systems can learn and become better over time due to the disturbances. Besides this, smart infrastructure also plays a major facilitator role in the principles of circular-economy and adaptive design in civil engineering. Digital tools can be used to optimise the performance of the resource, track the material and life-cycle assessment, which helps to use it more efficiently and achieve circular infrastructure practises. By so doing, smart infrastructure does not appear as a separate area of technology but as part of resilient and sustainable cities. Intelligent infrastructure and digital transformation is a trend in the modern civil engineering field that provides effective tools to sustain and make cities more resilient. Smart technologies allow infrastructure systems to become more responsive and adaptive through data-driven planning, early risk detection and integration of systems across the broader system. The success of their lasting performance, however, is pegged on concomitant investments on governance capacity, institutional structures and inclusive planning. These findings are critical in the analysis of how the supplementary strategies, especially the circular economy strategies and material innovation can contribute towards more robust climate-resilient urban infrastructure in the following section.

CIRCULAR ECONOMY AND SUSTAINABLE INFRASTRUCTURE MATERIALS

The integration of the principles of the circular economy into civil infrastructure is a paradigm shift that fundamentally reconsiders the way we design, use and manage materials, resources and the socio-technical structures that support the urban community. Traditionally, infrastructure development has been on a linear course whereby raw materials are mined, consumed rudely and at the end of the lifespan components are thrown out. These practises have had a very heavy burden on the environment and drained finite resources as well as increased the amount of waste in the cities. By contrast, closed-loop systems in the framework of the circular economy should promote resource efficiency, material recovery, and all-life-cycle optimisation, thus supporting the development of infrastructure along with long-term sustainability and resiliency goals (Bayeroju et al., 2021). Circularity in the urban context is not associated with material reuse, but rather integrated planning, design, and policy. According to Bayeroju et al. (2021), circular economy principles should be integrated into the policy frameworks of urban infrastructure in order to provide consistency in their design, construction, operation, and decommissioning. With this form of integration, municipalities will be able to minimise environmental footprints and also enhance the flexibility and resilience. The working heart of the circular economy implementation in civil engineering is made out of material innovation. Emerging literature indicates the creation of climate-adapted building materials with high resilience that is designed to counter increasing environmental pressures. Bayeroju et al. (2023) suggest a constructive material framework of resilience, which preempts the sustainability, environmental sustainability, and the ability to respond to climatic fluctuations. Such materials will be able to lengthen the infrastructure service life, minimize its maintenance requirements and lower its life-cycle costs, which will strengthen the sustainability and resiliency of urban systems. One of such salient trends is the growing utilisation of waste-based materials and alternative materials in sustainable material development. Anderson and Adesina (2026) report the progress in smart and sustainable construction materials of disaster resistant infrastructure, which includes recycled aggregates, industrial by-products, and novel cementitious composites. These materials indicate that it is possible to operationalise the principles of the circular economy without degrading structural performance, and they improve behaviour at extreme loading conditions and environments directly due to disaster resilience.

Critical construction materials situated on Earth have also become new feasible elements of sustainable and resilient city infrastructure. The article by Bühler et al. (2024) provides an ideological approach to the introduction of earth-based materials into urban change with low embodied energy, locality, and adaptation to climatic conditions. These materials can be strong enough to perform with high power and with significantly lower carbon emission levels during construction when designed in the right way. They are dependent on local resources and low-energy production processes, which are in line with the goals of a circular economy and reduce their sensitivity to global changes in the supply of materials. The effectiveness of the concept of the circular infrastructure is being increasingly provided by its connexion to digital and smart technologies. Although circularity is frequently talked about in material terms, the real-life application of this concept depends on the data-driven tools of the life-cycle assessment, material tracking, and performance monitoring. According to Sertyesilisik (2019), the concept of smart and connected cities can promote the objectives of the circular economy by increasing transparency, coordination, and decision-making across the infrastructure life cycle. Digital systems allow making informed decisions on material reuse, maintenance schedule, and end-of-life recovery and enhance the results of sustainability and resilience. Governance systems and policies that govern the industry are crucial in facilitating the development of infrastructure that is circular. Although the interest in circular practises is growing, the implementation of circular practises is still limited due to a lack of regulatory support and incentives. According to Bayeroju et al. (2021), facilitating policy tools, including the principles of circular procurement, revised material standards, and performance-based rules, are necessary. In the absence of such structures, circular material innovations tend to be limited to pilot projects instead of bringing systemic effects.

Circular thinking of infrastructure is specifically relevant to the fast in-city areas that face a shortage of materials, issues in waste-management, and increased vulnerability to climate. Circular solutions solve several urban issues at once by decreasing construction waste, alleviating environmental effects, and improving the security of resources in the long-term. However, the literature also has shown asymmetrical work in terms of regions and infrastructure sectors because of the limitations on the technical knowledge, disjointed supply chains, and insufficientness of aligning the engineering practise and urban policy goals. These issues require building capacity in the profession of civil engineering, more significant interdisciplinary cooperation, and more tangible application of the principles of the circular economy to infrastructure planning and engineering education (Aktan et al., 2022; Ali and Anwaar, 2024). All of these ways, circular economy strategies and sustainable infrastructure materials, are a crucial pillar of resilient urban development. Circular approaches decrease the environmental load by reorganising the material flows and optimising the life-cycle performance as well as increasing the adaptability to future uncertainties. This kind of material and policy innovation will be complimentary to smart infrastructure systems and will offer a requisite base on which adaptive and climate-responsive infrastructure approaches will be advanced in the next section.

ADAPTIVE AND CLIMATE-RESPONSIVE INFRASTRUCTURE SYSTEMS

Adaptive and climate-responsive infrastructure has become the future direction in the modern civil engineering world as urban centres face growing uncertainty due to climate change, environmental degradation, and the high pace of socio-economic change. Adaptive infrastructure, as opposed to more traditional ways of thinking about infrastructure that are based largely on a state of stability and long-term predictability under assumed baseline conditions, emphasises flexibility, learning ability and a continued capability to respond to changing risks. The given paradigm shift highlights a growingly recognised fact: an uncertain climate variability, extreme events, and slow environmental change cannot and should not be fully predicted during the design phase, which makes the infrastructure systems susceptible to progressive adaptation throughout the lifespan (Ilugbusi et al., 2024; Das, 2025). The increase in the occurrence and intensity of hazards, including floods, heat stress, storms and earthquakes has burdened the urban infrastructure networks like never before. In their review of the urban adaptation strategies, Ilugbusi et al. (2024) single out infrastructure innovation as one of the building blocks of climate resilience. Some of their suggestions include climate-sensitive architecture, redundancy of essential infrastructure, and adaptive operational regulations that enable the infrastructures to maintain necessary functions during and after disruptive incidents. Okem et al. (2024) also emphasise that one of the primary requirements to build safe and sustainable urban communities, especially in the area prone to disasters, is disaster-resilient infrastructure. One of the key traits of adaptive infrastructure systems is their ability to adjust the performance to a sudden shock as well as gradual environmental changes. An example of this principle is the set of assessment approaches based on scenarios, which explicitly introduces uncertainty into the planning and design. As a case in point, Rezvani et al. (2024) create an urban resilience index on the critical road infrastructure based on a scenario-based disaster risk assessment. Such methodologies enable the planners and engineers to test the performance of the infrastructure at various probable future conditions, and as a result, making sound and well-informed decisions. The scenario-based approaches can be seen as an excellent fit with the principles of adaptive planning by assuming that the conditions are not fixed at all. The probabilistic and risk-informed approaches have also risen to the top as an instrument of bolstering adaptive capacity in civil infrastructure. Reda Taha et al. (2021) emphasise the significance of new technologies and analytical models according to which risk, uncertainty, and resilience are incorporated into infrastructure planning and management. Such strategies facilitate a systematic presence of vulnerabilities, priorities of the interventions and strategic utilisation of resources in manners that promote long term resilience as opposed to short-term optimisation. Risk assessment introduced into infrastructure governance thus enhances the capacity of cities to be proactive in predicting and responding to threats due to climate changes.

Adaptive strategies are especially important in infrastructural areas that are especially sensitive to environmental variations such as water and transportation systems. Asghari et al. (2023) argue that resilience in urban water infrastructure is complex in nature and needs the adaptive strategies that consider how the interrelations of physical assets, institutional structures, and environmental conditions interplay. The water systems should be reliable and have the flexibility to meet the demand variability, extreme precipitation, and long droughts. Adaptive design and management practises thus help such systems to dynamically adapt operational parameters whilst the continuation of services despite the changing conditions. Special challenges and opportunities to the development of adaptive infrastructure are introduced by the Global South. As Das (2025) notes, rapid urbanisation in the developing areas is more likely to suggest compounded vulnerabilities as a result of financial scarcity, informal developmental patterns, and institutional constraints. In such settings, the policies of adaptive infrastructure should be focused on the affordability, scalability, and inclusiveness. The alternative options to increase resilience are incremental adaptation strategies, low-cost design mechanisms, and technologies that fit the local conditions without causing excessive financial strain. These reflections indicate the relevance of context-specific adaptive strategies when it comes to civil engineering practise. Systems of adaptive infrastructure are greatly dependant on governance and institutional capacity to enable them to be successfully implemented. This is because effective adaptation depends on the need to coordinate various actors, such as engineers, planners, policymakers, and communities. As Ilugbusi et al. (2024) note, the adaptive strategies need to enable governance frameworks that promote learning, information exchange, and involvement of the stakeholders. In the absence of such institutional support, adaptive measures will be vulnerable to being piecemeal or reactive thus reducing their capacity to respond to systemic urban risks.

Lastly, adaptive infrastructure is not to be viewed as a one-dimensional concept, but as part of a larger sustainable infrastructure paradigm. Adaptation strategies are the most effective when used with smart technologies that make it possible to monitor everything in real time, as well as with the principles of the circular economy that can help in making resources more efficiently used. Resistance of urban infrastructure is achieved through the combination of adaptive, intelligent and circular solutions to tackle both the short-term climate implications and the long-term sustainability issues. Based on this the adaptive and climate-responsive infrastructure systems reflect an essential development of civil engineering, which will provide a platform of combined systems that connect technological innovation, material sustainability, and governance, a development that will be discussed in the following section.

INTEGRATED SMART, CIRCULAR, AND ADAPTIVE INFRASTRUCTURE FRAMEWORKS

The increasing complexity of modern urban systems can be seen as indicating the failure of siloed approaches to infrastructure that only touch on technological innovation, material sustainability, or climate adaptation, individually. Recent academic work is beginning to underscore the fact that city urban infrastructure and systems can only be supported in a sustainable and resilient manner through the systematic incorporation of smart technologies, circular economy concepts, and adaptive design approaches into an integrated framework. This type of integration helps civil infrastructure systems to act as a whole responding to environmental, social, and technological risks, and creates the prospects of long-term urban sustainability and livability (Ajirotutu et al., 2024). Infrastructure built on systems thinking frameworks Integrated infrastructure frameworks owe their foundations to the concept of infrastructure as a network of mutually dependent parts embedded into the larger urban, environmental, and institutional contexts. This view rejects the prevailing tendency to consider infrastructure assets as discrete objects; it puts more emphasis on interactions, feedback loops, and cross-sectoral connections. Ajirotutu et al. (2024) state that optimising renewable energy infrastructure, new building materials, and smart infrastructure management is essential in future cities to increase their resilience and sustainability. This congruency creates synergies in the infrastructure sectors enabling the cities to optimize performance, decrease resource usage, and increase adaptive capacity at the same time. Unlike efficiency gains, which are mainly short in terms of scope, integrated frameworks focus on optimization of systems and resilience in the long run. The introduction of intelligent technologies plays an enabling role in the centre of this framework due to its ability to provide coordination among the life cycles of infrastructure. Online technologies facilitate real-time tracking, predictive analytics, and performance evaluation functions, which are considered necessary when managing the complex, interconnected infrastructure systems. These tools, combined with a set of principles of the circular economy, help to track materials, optimise the life-cycle, and make informed decisions regarding reuse, refurbishment, and recovery. Sertyesilisik (2019) argues that smart and connected city architectures offer the digital foundation that cannot be ignored when aligning sustainability and resilience goals in the urban infrastructure networks. Conceptual Framework for Integrated Smart, Circular, and Adaptive Infrastructure Systems presented in Figure 1.

Figure 1. Conceptual framework for integrated smart, circular, and adaptive infrastructure systems

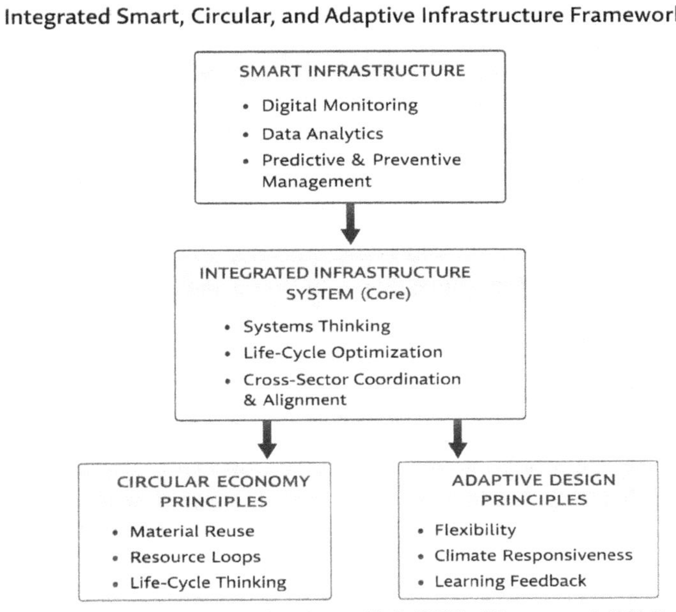

Source: Developed by the author based on Sertyesilisik (2019), Ajirotutu et al. (2024), Rezvani et al. (2024), and Juarez-Quispe et al. (2025).

The principles of adaptive design also promote integrated infrastructure structures by imparting the concepts of flexibility and learning aspect in the processes of running a system. As illustrated by Rezvani et al. (2024), scenario-driven resilience assessment helps in infrastructure planning and management in the face of uncertainty as it helps planners to exercise the performance of systems in different future conditions. A system of infrastructure that is provided with adaptive strategies and intelligent monitoring mechanisms, as well as circular material practise, attains the ability to transform with regard to the alteration of the environment conditions, the requirements of the users, and the risk profile. Such convergence enhances greater strength and maintains the possibility of long-term change. System dynamics modelling provides an imperative methodological base of operationalising integrated infrastructure frameworks. Juarez-Quispe et al. (2025) demonstrate that with the use of system dynamics tools, feedback loops, delays and non-linear interactions of urban infrastructure systems can be captured. Such models enable engineers and policymakers to determine the long-term effects of infrastructure investments and governance choices, to find points of leverage to act, and to predict unintended effects. System dynamics approaches are consistent with the concept of adaptive and resilient infrastructure management because they promote repetitive learning and continuous adaptation.

The empirical validation of integrated frameworks has become ever more sought out by the mechanisms of an experimental and participatory nature. Velasquez-Mendez et al. (2025) emphasise that urban living labs can approve the sustainable, smart, and circular urban architectures under the real-world conditions. Living laboratories offer settings where entire infrastructure solutions may be experimented, fine-tuned, and assessed in collaboration with the stakeholders. These platforms facilitate social learning, institutional adaptation, and user-centred innovation in addition to technical validation, which enhances the practical utility of integrated infrastructure strategies. Combining smart, circular, and adaptive solutions also has important implication on infrastructure governance. Its successful integration requires coordination of activities across the institutional boundaries, alignment of the policy goals and cooperation of various actors within the planning, delivery and operation. Sutriadi (2023) emphasizes that such co-ordination requires soft infrastructure such as governance capacity, regulatory frameworks, and human capital that help facilitate such co-ordination. Without favourable governance systems, integrated infrastructure systems are likely to be more theoretical than practical and this reduces their transformative nature.

Infrastructure systems that are integrated are particularly essential where there is high urbanisation rate and exposure to climatic events. In less developed areas, where urban paths are defined by infrastructure investment choices over decades, there is an opportunity with integrated approaches to achieve as many co-benefits as possible and trade-offs as few as possible. Modifying digital innovation, material efficacy and climate adjustment allows cities to create cost-effective and resilient infrastructural frameworks that meet present uncertainty and future dangers (Das, 2025; Janshevaski, 2025). Although they promise, the integrated infrastructure frameworks face ongoing problems such as institutional fragmentation, insufficient technical capacity, intersectoral coordination problems, and problems at scale. These issues cannot be mitigated by just improving technology but also professional practise, policy formulation and education of civil engineers. According to Aktan et al. (2022), civil engineers should have more expanded and systems-oriented functions that combine technical skills with strategic plans and stakeholder participation. A combination of smart, circular, and adaptive infrastructure platforms is an important development of sustainable and resilient civil engineering development. These frameworks present a holistic way of dealing with the multifaceted problems of the modern urban infrastructure through the integration of digital intelligence, resource efficiency, and climate responsiveness into a systems-based approach. The next section expands on the synthesis of this paper by looking at the challenges of governance and implementation, and the direction of the future of the integrated infrastructure practise.

GOVERNANCE, IMPLEMENTATION CHALLENGES, AND FUTURE DIRECTIONS

Although smart, circular, and adaptive infrastructure frameworks are potentially very promising in enhancing the urban sustainability and resilience, their implementation depends on the strong governance frameworks, institutional capacity, and policy action. The academic community is always arguing that technological and material innovations cannot be used with no improvements in governance and professional practise to trigger a systemic change (Sutriadi, 2023; Aktan et al., 2022). With the explosion of urban infrastructure systems in complexity, governance arrangements need to change to enable integration of sectors, scales and groups of stakeholders. Challenges in governance are common to institutional fragmentation, whereby infrastructure planning, delivery, and operation roles are divided into several agencies with weak coordination. According to Sutriadi (2023), the lack of unifying governance structures is detrimental to the effectiveness of smart and sustainable city projects because it creates silos, which hamper the sharing of information and the collective decision-making process. This type of fragmentation presents an extreme challenge to systems that depend on the interoperability and real-time data exchange, as it is in the case of modern smart infrastructure networks. Another key area in the implementation of the principles of the circular economy in the development of infrastructure entails policy alignment. Bayeroju et al. (2021) note that, despite the increased knowledge about the benefits of the circular economy, regulatory inconsistencies and inconsistencies in their procurement practises continue to act as barriers to the application of the concept to urban projects. The policy tools should motivate life-cycle thinking, material reuse and performance-based results and uphold the safety and quality standards. Without this kind of alignment, circular infrastructure strategies will be left to need pilot programmes. Unprofessional capacity and the development of skills are other obstacles to implementation. According to Aktan et al. (2022), civil engineers must more often take up interdisciplinary tasks that combine technical skills with a systems approach, ethical accountability and stakeholder involvement. However, the existing education and training models are still based on the traditional design and construction paradigm, thus limiting the willingness of the profession towards embracing integrated infrastructure methodology. The same situation is highlighted by Ali and Anwaar (2024), who insist on the necessity of continuous professional growth as the means of introducing new, innovative design solutions to infrastructure, materials, and management.

Rationality and inclusiveness are the focus of sustainable and resilient infrastructure governance. The approach to infrastructure investments in the Global South as indicated by Das (2025), should be able to tackle both the socio-economic inequalities and the vulnerability to climate. Without a conscious focus on inclusivity, intelligent and robust infrastructure projects would have the impact of solidifying pre-existing inequalities by disproportionately serving more wealthy urban areas or communities that have a greater access to digital infrastructure. Ilugbusi et al. (2024) also emphasise the need to have inclusive adaptation measures involving local communities and incorporating social vulnerability in the planning. The directions of future research indicated in the literature show the need in more context-specific and integrated frameworks of analysis. The authors Juarez -Quispe et al. (2025) support the usefulness of system dynamics methods in assessing uncertainty-related long-term infrastructure performance and policy effects. Simultaneously, Rezvani et al. (2024) address the opportunities of scenario-based resilience evaluation tools in decision-making regarding adaptive planning and priorities in investments. The extension of such methodologies to the infrastructure industries and urban settings is identified as one of the opportunities to develop the research in civil engineering. New technological and material advances are also likely to influence the future of the sustainable development of infrastructure. Ajirotutu et al. (2024) argues that the integration of renewable energy systems, emerging materials, and smart systems to manage infrastructure will become the main focus in the new urban resilience designs. Anderson and Adesina (2026) also prove that the use of sustainable construction materials can significantly raise the performance of infrastructure during the disaster conditions. It will be necessary to incorporate these innovations in governance structures that can facilitate experimentation and learning in order to scale their effects. The institutional learning and adaptive governance emerge as some of the most important facilitators of long-term resilience of infrastructure. Adaptive governance systems that promote surveillance, feedback, and progressive correction are compatible with the concepts of adaptive infrastructure systems mentioned above (Ilugbusi et al., 2024). These frameworks facilitate the process of constant improvement and provide infrastructure systems the opportunity to develop according to the altered environmental and social conditions. However, adaptive governance implementation requires political goodwill, institutional elasticity and long-term investment in data and knowledge infrastructure. The issues of governance and implementation pose serious, but not overwhelming, impediments to the adoption of smart, circular, and adaptive infrastructure frameworks that are integrated. These issues must be solved through a concerted policy effort, professional change, and a powerful dedication to equity and inclusivity. The next generation of research and practise should focus on the creation of integrative tools, governance frameworks, as well as capacity-building processes that would help civil engineering to play an active role in sustainable and resilient urban development. These deliberations give a backdrop to the finalised synthesis done in the last section.

CONCLUSION

The modern metropolitan environment is a daunting set of climatic dynamism, urbanisation, scarcity of resources, and established socio-economic inequalities. These aspects have changed the practise of civil engineering irreversibly to the extent that it has become the catalyst of negotiating the urban resilience. This chapter has located the smart, circular, and adaptive infrastructure systems in the centre of an increasingly integrated approach, holding that the simultaneous implementation of these strategies is essential to managing the demands of the multifarious pressures which plague our cities today, not considering sustainability and resilience as siloed strategies. The critical account anticipates the future in the form of how intelligent infrastructure imposes the need to change paradigms to a data-centric, anticipatory, and holistically-focused management. Digital instrumentation, in conjunction with smart surveillance networks and advanced analytical devices, improves the capacity of an urban system to predict vulnerabilities, operational exemplar, and service loyalty in a period of stochastic uncertainty. However, the chapter asserts that technological prowess cannot work as a magic bullet, it bases its effectiveness on other imperatives in governance capability, inter-institutional coordination and creation of soft infrastructural scaffolding that helps in inclusive, transparent decision-making processes. The principles of the circular economy widen the coverage of sustainability and resilience through reorganizing material flows and lifecycle activities of the civil infrastructure networks. Circular approach to material reuse and constructing inputs that are climate-adapted reduce the environmental externalities and enhance infrastructural longevity through increased resource efficiency, material reuse, and climate-adapted material engineering. It has been shown that circular strategies do not only reduce the waste streams and greenhouse emissions but also strengthen resilience by dampening the fragilities caused by the supply chain and improving the performance indicators under the harsh climatic conditions.

The need to respond to the volatile nature of urban systems is manifested as adaptive and climate-responsive infrastructure. The chapter hypothesises that there should be adaptive capacity that is infrastructural in nature, i.e. flexible, experiential and responsive governance to accommodate both the acute shocks, and the long-term environmental changes. The idea of scenario-based planning, risk-based evaluation, and adaptive frameworks of governance is explained as viable channels of operationalising resilience in the civil engineering praxis, especially those that are climate-prone and experience swift urban growth. The combination of the above-mentioned paradigms is the key intellectual input of the given discourse. The chapter displays the concurrence of the smart technologies, circularity of materials, and adaptive principles of design through the systemic lens which can be collectively tuned to produce compound results. These combined frameworks are able to simultaneously enhance efficiency, support sustainability and increase resilience to promote livability in the urban environment and realisation of long term development goals. The empirical methods, such as system dynamics modelling, urban living labs are presented as a way to experiment, develop, and implement these integrated strategies in the real urban environments. Governance and implementation barriers are also predetermined as a central success factor in the discussion. The continuing institutional fragmentation, inadequate professional competence, and policy dissonance are significant obstacles to the prevalent assimilation of integrated infrastructures. Reducing such difficulties requires a new strategic re-alignment of civil engineering education and professional practice in terms of interdisciplinary team work, systems thinking and ethical stewardship. At the same time, the infrastructural government should be grounded on equity and inclusivity so that the dividends of sustainability and resilience can reach equitably through urban demographics with specific reference to the Global South. Finally, intelligent, circular, and adaptive civil engineering systems outline a revolutionary nature of sustainable and resilient civil engineering. The combination of them provides an integrated framework to face the multidimensional and dynamic dilemmas of the 21 st century city infrastructure. With the enhancement of system-level thinking, the catalysing of innovations in the field of technology and materials as well as the strengthening of governance capabilities, civil engineering is capable of playing a central role in shaping resilient, inclusive, and sustainable urban futures. The knowledge presented in this chapter forms a framework-like of research and intervention to be adopted in the current effort to re-theorise infrastructure systems as dynamic facilitators of urban resilience and sustainability.

REFERENCES

Abbass, M., Abbas, U., Jafri, R., Arif, S. M., & Akhai, S. (2025). AI and Machine Learning Applications in Sustainable Smart Cities. In Sustainable Smart Cities and the Future of Urban Development (pp. 1-32). IGI Global.

Abbass, M., Akhai, S., Abbas, U., Jafri, R., & Arif, S. M. (2025). AI-enabled sustainable urban planning and management. In *Real-world applications of ai innovation* (pp. 233–260). IGI Global.

Abbass, M., Akhai, S., Chouksey, A., Pathak, S., Abbas, U., & Abass, S. (2025). Disaster Risk Reduction and Management With Emerging Technologies: Applications of IoT, AI, and Data Analytics for Resilient Urban Infrastructure. In Revolutionizing Urban Development and Governance With Emerging Technologies (pp. 71-110). IGI Global.

Ajirotutu, R. O., Adeyemi, A. B., Ifechukwu, G.-O., Iwuanyanwu, O., Ohakawa, T. C., & Garba, B. M. P. (2024). Future cities and sustainable development: Integrating renewable energy, advanced materials, and civil engineering for urban resilience. *Magna Scientia Advanced Research and Reviews*, *12*(2), 235–250. DOI: 10.30574/msarr.2024.12.2.0204

Akhai, S., & Abbass, M. (2025). Toward Resilient Futures: The Role of AI-Driven Strategies in Climate Adaptation. In Nexus of AI, Climatology, and Urbanism for Smart Cities (pp. 325-340). IGI Global.

Akhai, S., Abbass, M., Kaur, P., & Kaur, T. (2025). Digital Transformation Across Generations: Robotics and AI in Action. In Impacts of Digital Technologies Across Generations (pp. 23-40). IGI Global.

Aktan, A. E., Brownjohn, J. M. W., Moon, F. L., Sjoblom, K. J., Bartoli, I., & Karaman, S. G. (2022). Civil engineer for urban livability, sustainability and resilience. *Sustainable and Resilient Infrastructure*, *7*(5), 480–491. DOI: 10.1080/23789689.2021.1937776

Ali, H., & Anwaar, F. U. (2024). *The future of civil engineering: Innovations in infrastructure and urban development*. Research Corridor Journal of Engineering Science, 1(2). https://www.researchcorridor.org/index.php/RCJES/article/view/3

Almulhim, A. I. (2025). Building Urban Resilience Through Smart City Planning: A Systematic Literature Review. *Smart Cities*, *8*(1), 22. DOI: 10.3390/smartcities8010022

Anderson, S., & Adesina, A. (2026). Trends in smart, sustainable construction materials for disaster-resilient urban infrastructure. In *Recent trends in the sustainability improvement and use of wastes in cementitious materials* (pp. 587–614). Woodhead Publishing. DOI: 10.1016/B978-0-443-15670-0.00002-8

Asghari, F., Piadeh, F., Egyir, D., Yousefi, H., Rizzuto, J. P., Campos, L. C., & Behzadian, K. (2023). Resilience assessment in urban water infrastructure: A critical review of approaches, strategies, and applications. *Sustainability (Basel), 15*(14), 11151. DOI: 10.3390/su151411151

Bayeroju, O. F., Sanusi, A. N., & Nwokediegwu, Z. Q. S. (2021). Review of circular economy strategies for sustainable urban infrastructure development and policy planning. *Journal of Sustainable Infrastructure*, *9*(2), 101–118.

Bayeroju, O. F., Sanusi, A. N., & Nwokediegwu, Z. Q. S. (2023). Framework for resilient construction materials to support climate-adapted infrastructure development. *Shodhshauryam: International Scientific Refereed Research Journal*, *6*(5), 403–428.

Bühler, M., Hollenbach, P., Köhler, L., & Armstrong, R. (2024). Unlocking resilience and sustainability with earth-based materials: A principled framework for urban transformation. *Frontiers in Built Environment*, *10*, 1385116. Advance online publication. DOI: 10.3389/fbuil.2024.1385116

Das, D. K. (2025). Resilient and sustainable urban infrastructure in the Global South: Strategies to address climate change-linked vulnerabilities. *Urban Planning*, *10*, 10037. Advance online publication. DOI: 10.17645/up.10037

Ibrahim, F. J. M. (2023). Sustainable urban infrastructure: A comprehensive review of environmental safety practices in civil engineering. *i-Manager's Journal on Civil Engineering, 13*(4), 25–34. DOI: 10.26634/jce.13.4.20316

Ilugbusi, B. S., Adisa, O., Obi, O., Awonuga, K. F., Adelekan, O. A., Asuzu, O. F., & Ndubuisi, N. L. (2024). Urban resilience to climate change: A review of adaptation strategies and infrastructure innovations. *Ecofeminism and Climate Change*, *5*(1), 18–23. DOI: 10.26480/efcc.01.2024.18.23

Janshevaski, V. R. M. (2025). Eco-innovations in urban development: Bridging civil engineering and environmental sustainability. *Innovative Research in Civil and Environmental Engineering*, *2*(2), 22–28.

Juarez-Quispe, J., Rojas-Chura, E., Espinoza Vigil, A. J., Guillén Málaga, M. S., Yabar-Ardiles, O., Anco-Valdivia, J., & Valencia-Félix, S. (2025). Advancing sustainable infrastructure management: Insights from system dynamics. *Buildings (Basel, Switzerland)*, *15*(2), 210. DOI: 10.3390/buildings15020210

Nyokum, T., & Tamut, Y. (2025). Sustainable urban infrastructure development: Integrating smart technologies for resilient and green cities. *SSRG International Journal of Civil Engineering*, *12*(4), 18–36. DOI: 10.14445/23488352/IJCE-V12I4P103

Okem, E. S., Nwokediegwu, Z. Q. S., Umoh, A. A., Biu, P. W., Obaedo, B. O., & Sibanda, M. (2024). Civil engineering and disaster resilience: A review of innovations in building safe and sustainable communities. *International Journal of Science and Research Archive*, *11*(1), 639–650. DOI: 10.30574/ijsra.2024.11.1.0107

Ramirez Lopez, L. J., & Grijalba Castro, A. I. (2020). Sustainability and resilience in smart city planning: A review. *Sustainability (Basel)*, *13*(1), 181. DOI: 10.3390/su13010181

Reda Taha, M., Ayyub, B. M., Soga, K., Daghash, S., Heras Murcia, D., Moreu, F., & Soliman, E. (2021). Emerging technologies for resilient infrastructure: Conspectus and roadmap. *ASCE-ASME Journal of Risk and Uncertainty in Engineering Systems. Part A, Civil Engineering*, *7*(2), 03121002. DOI: 10.1061/AJRUA6.0001134

Rezvani, S. M., Silva, M. J. F., & de Almeida, N. M. (2024). Urban resilience index for critical infrastructure: A scenario-based approach to disaster risk reduction in road networks. *Sustainability (Basel)*, *16*(10), 4143. DOI: 10.3390/su16104143

Sertyesilisik, B. (2019). Circular, smart, and connected cities: A key for enhancing sustainability and resilience of cities. In *Handbook of research on implementation and deployment of IoT projects in smart cities* (pp. 19–32). IGI Global. DOI: 10.4018/978-1-5225-9199-3.ch002

Shanthi, H. J., Gokulakrishnan, A., Sharma, S., Deepika, R., & Swadhi, R. (2025). Leveraging artificial intelligence for enhancing urban health: Applications, challenges, and innovations. In *Nexus of AI, climatology, and urbanism for smart cities* (pp. 275–306). IGI Global.

Sutriadi, R. (2023). Soft infrastructure in smart sustainable cities: A literature review. *Journal of Indonesia Sustainable Development Planning*, *4*(2), 195–215. DOI: 10.46456/jisdep.v4i2.428

Swadhi, R. (2025). Innovative strategies for widespread adoption in a climate-smart future: Scaling up agroforestry. In *Agroforestry for a climate-smart future* (pp. 473–496). IGI Global. DOI: 10.4018/979-8-3693-8282-0.ch016

Velasquez-Mendez, A., Lozoya-Santos, J. D. J., & Jiménez-Vargas, J. F. (2025). Validating a sustainable, smart, and circular city architecture through urban living lab experiments. *Urban Science (Basel, Switzerland)*, *9*(9), 377. DOI: 10.3390/urbansci9090377

Wimalasena, S., Weerakoon, T. G., & Wimalasena, J. (2025). Integrated approaches to sustainable urban development: the synergy between landscape architecture and civil engineering. *Mokslas – Lietuvos ateitis / Science – Future of Lithuania, 17*. DOI: 10.3846/mla.2025.22836

Compilation of References

Abbass, M., Abbas, U., Jafri, R., Arif, S. M., & Akhai, S. (2025). AI and Machine Learning Applications in Sustainable Smart Cities. In Sustainable Smart Cities and the Future of Urban Development (pp. 1-32). IGI Global.

Abbass, M., Akhai, S., Chouksey, A., Pathak, S., Abbas, U., & Abass, S. (2025). Disaster Risk Reduction and Management With Emerging Technologies: Applications of IoT, AI, and Data Analytics for Resilient Urban Infrastructure. In Revolutionizing Urban Development and Governance With Emerging Technologies (pp. 71-110). IGI Global.

Abbass, M., Abbas, U., Jafri, R., Arif, S. M., & Akhai, S. (2025). AI and Machine Learning Applications in Sustainable Smart Cities. In Al-Humairi, S., Hajamydeen, A., & Mahfoudh, A. (Eds.), *Sustainable Smart Cities and the Future of Urban Development* (pp. 1–32). IGI Global Scientific Publishing., DOI: 10.4018/979-8-3693-6740-7.ch001

Abbass, M., Akhai, S., Abbas, U., Jafri, R., & Arif, S. M. (2025). AI-enabled sustainable urban planning and management. In *Real-world applications of ai innovation* (pp. 233–260). IGI Global.

Abdel Aal, S. I. (2023). Neutrosophic Framework for Assessment Challenges in Smart Sustainable Cities based on IoT to Better Manage Energy Resources and Decrease the Urban Environment's Ecological Impact. *Neutrosophic Systems with Applications*, *6*, 9–16. DOI: 10.61356/j.nswa.2023.33

Abdellatif, S., Raza, A., Alsulamy, S., & Khadimallah, M. A. (2024). Axial load-carrying capacity of concrete-filled steel tube columns: A comparative analysis of various modeling techniques. *Mechanics of Advanced Materials and Structures*, *31*(17), 3980–4002. DOI: 10.1080/15376494.2023.2188325

Abduljabbar, R., Dia, H., Liyanage, S., & Bagloee, S. A. (2019). Applications of Artificial Intelligence in Transport: An Overview. *Sustainability (Basel)*, *11*(1), 189. DOI: 10.3390/su11010189

Adaramola, T., Nwariaku, H., Adigun, O., Arowolo, M., Omole, O., & Wada, I. (2024). Internet of thing integration in green fintech for enhanced resource management in smart cities. *World Journal of Advanced Research and Reviews*, *23*(2), 1317–1327. DOI: 10.30574/wjarr.2024.23.2.2448

Addas, A., Khan, M. N., & Naseer, F. (2024). Waste management 2.0 leveraging internet of things for an efficient and eco-friendly smart city solution. *PLoS One*, *19*(7), e0307608. DOI: 10.1371/journal.pone.0307608 PMID: 39083562

Adefemi, A., Adekoya, O., Adegbite, A., Abatan, A., & Ukpoju, E. (2023). Artificial intelligence in environmental health and public safety: A comprehensive review of USA strategies. *World Journal of Advanced Research and Reviews*, *20*(3), 1420–1434. DOI: 10.30574/wjarr.2023.20.3.2591

Ahern, J. (2011). From fail-safe to safe-to-fail: Sustainability and resilience in landscape planning. *Landscape and Urban Planning*, *100*(4), 341–343. DOI: 10.1016/j.landurbplan.2011.02.021

Ajirotutu, R. O., Adeyemi, A. B., Ifechukwu, G.-O., Iwuanyanwu, O., Ohakawa, T. C., & Garba, B. M. P. (2024). Future cities and sustainable development: Integrating renewable energy, advanced materials, and civil engineering for urban resilience. *Magna Scientia Advanced Research and Reviews*, *12*(2), 235–250. DOI: 10.30574/msarr.2024.12.2.0204

Ajzen, I. (1991). The theory of planned behavior. *Organizational Behavior and Human Decision Processes*, *50*(2), 179–211. DOI: 10.1016/0749-5978(91)90020-T

Akhai, S., & Abbass, M. (2025). Toward Resilient Futures: The Role of AI-Driven Strategies in Climate Adaptation. In Nexus of AI, Climatology, and Urbanism for Smart Cities (pp. 325-340). IGI Global.

Akhai, S., Abbass, M., Kaur, P., & Kaur, T. (2025). Digital Transformation Across Generations: Robotics and AI in Action. In Impacts of Digital Technologies Across Generations (pp. 23-40). IGI Global.

Aktan, A. E., Brownjohn, J. M. W., Moon, F. L., Sjoblom, K. J., Bartoli, I., & Karaman, S. G. (2022). Civil engineer for urban livability, sustainability and resilience. *Sustainable and Resilient Infrastructure*, *7*(5), 480–491. DOI: 10.1080/23789689.2021.1937776

Akter, S., Dwivedi, Y. K., Sajib, S., Biswas, K., Bandara, R. J., & Michael, K. (2022). Algorithmic bias in machine learning-based marketing models. *Journal of Business Research, 144*, 201–216. DOI: 10.1016/j.jbusres.2022.01.083

Alaya, I., Solnon, C., & Ghedira, K. (2007, October). Ant colony optimization for multi-objective optimization problems. In *19th IEEE international conference on tools with artificial intelligence (ICTAI 2007)* (Vol. 1, pp. 450-457). IEEE. DOI: 10.1109/ICTAI.2007.108

AlBahsh, E. R., & Hosseinian-Far, A. (2021). *16 - The implication of big data analytics on competitive intelligence: A qualitative case study of a real estate developer in the UAE*. Strategy, Leadership, and AI in the Cyber Ecosystem. The Role of Digital Societies in Information Governance and Decision Making.

Alehashemi, A., Mansouri, S. A., & Barati, N. (2017). Urban infrastructures and the necessity of changing their definition and planning Landscape infrastructure; a new concept for urban infrastructures in 21st centur. *The Monthly Scientific Journal of Bagh-e Nazar, 13*(43), 5–18.

Ali, H., & Anwaar, F. U. (2024). *The future of civil engineering: Innovations in infrastructure and urban development*. Research Corridor Journal of Engineering Science, 1(2). https://www.researchcorridor.org/index.php/RCJES/article/view/3

Aljohani, B.M., Fadul, A., Asiri, M.S., Alkhathami, A.D., & Hasan, F. (2024). Volatility transmission in the property market during two inflationary periods: The 2008–2009 global financial crisis and the COVID-19 crisis. Research in International Business and Finance, Volume 70, Part B, 102413.

Allam, Z. (2024). *The metaverse and real estate development*. The Metaverse and Smart Cities Urban Environments in the Age of Digital Connectivity Smart Cities. DOI: 10.1016/B978-0-443-22351-8.00010-6

Almulhim, A. I. (2025). Building Urban Resilience Through Smart City Planning: A Systematic Literature Review. *Smart Cities, 8*(1), 22. DOI: 10.3390/smartcities8010022

Alotaibi, E., & Nassif, N. (2024). Artificial intelligence in environmental monitoring: In-depth analysis. *Discover Artificial Intelligence, 4*(1), 84. Advance online publication. DOI: 10.1007/s44163-024-00198-1

Alqahtani, F., Al-Makhadmeh, Z., Tolba, A., & Said, W. (2020). Internet of things-based urban waste management system for smart cities using a Cuckoo Search Algorithm. *Cluster Computing, 23*(3), 1769–1780. DOI: 10.1007/s10586-020-03126-x

Alqasi, M. A. Y., Alkelanie, Y. A. M., & Alnagrat, A. J. A. (2024). Intelligent Infrastructure for Urban Transportation: The Role of Artificial Intelligence in Predictive Maintenance. *Brilliance: Research of Artificial Intelligence*, *4*(2), 625–637. DOI: 10.47709/brilliance.v4i2.4889

Alsamhi, S. H., Afghah, F., Sahal, R., Hawbani, A., Al-Qaness, M. A. A., Lee, B., & Guizani, M. (2021). Green internet of things using UAVs in B5G networks: A review of applications and strategies. *Ad Hoc Networks*, *117*, 102505. DOI: 10.1016/j.adhoc.2021.102505

Alshahrani, S. M., Alotaibi, H. F., & Alqarni, M. (2024). Modeling and validation of drug release kinetics using hybrid method for prediction of drug efficiency and novel formulations. Frontiers in Chemistry, 12, 1395359.

Alves, R. M. C., Lousada, S. A. N., Cabezas, J., & Gómez, J. M. N. (2023b). Modelling of Flood Risk, Vulnerability and Adaptation of Machico City Center. In *Sustainable Development of Urban Tourism Systems* (pp. 19-41). IGI Global. DOI: DOI: 10.4018/978-1-6684-6449-6.ch002

Alves, R. M. C., Lousada, S. A. N., Cabezas, J., & Gómez, J. M. N. (2023c). Port of Machico, Madeira: Assessment of the Consequences of Overtopping Using AHP Methodology. In *Sustainable Development of Urban Tourism Systems* (pp. 61-80). IGI Global. DOI: DOI: 10.4018/978-1-6684-6449-6.ch004

Alves, R. M. C., Lousada, S. A. N., Cabezas, J., & Gómez, J. M. N. (2023d). Flood Risk and Vulnerability: An Approach to Urban Floods in Ribeira de Machico. In *Sustainable Development of Urban Tourism Systems* (pp. 81-103). IGI Global. DOI: DOI: 10.4018/978-1-6684-6449-6.ch005

Alves, R. M. C., Lousada, S. A. N., Cabezas, J., & Gómez, J. M. N. (2023e). Urban Flooding Risk in Machico and Planning Its Territory as a Form of Prevention. In *Sustainable Development of Urban Tourism Systems* (pp. 42–60). IGI Global., DOI: 10.4018/978-1-6684-6449-6.ch003

Alves, R., Lousada, S., & Cabezas, J. (2025). Understanding Mobility Patterns in Machico's Low-Density Rural Areas Through Stated Preference Surveys. In *Sustainable Civil Engineering for a More Resilient Society* (pp. 235–257). IGI Global., DOI: 10.4018/979-8-3693-2448-6.ch012

Alves, R., Lousada, S., Cabezas, J., & Gómez, J. M. N. (2022). The Importance of the Strategic Urban Rehabilitation Plan in the Sustainable Development of the Municipality of Machico. *Sustainability (Basel)*, *14*(24), 16816. DOI: 10.3390/su142416816

Alves, R., Lousada, S., Cabezas, J., & Gómez, J. M. N. (2023a). Local Housing Strategy: Analysis of Importance and Implementation in Machico Municipality, Madeira. *Land (Basel)*, *12*(9), 1778. DOI: 10.3390/land12091778

Alves, R., Lousada, S., Gómez, J. M. N., & Cabezas, J. (2024). Maintenance Challenges in Maritime Environments and the Impact on Urban Mobility: Machico Stayed Bridge. *Infrastructures*, *9*(10), 180. DOI: 10.3390/infrastructures9100180

Anderies, J. M., Janssen, M. A., & Ostrom, E. (2016). A framework to analyze the robustness of social-ecological systems from an institutional perspective. *Ecology and Society*, *9*(1), 18. DOI: 10.5751/ES-00610-090118

Anderson, S., & Adesina, A. (2026). Trends in smart, sustainable construction materials for disaster-resilient urban infrastructure. In *Recent trends in the sustainability improvement and use of wastes in cementitious materials* (pp. 587–614). Woodhead Publishing. DOI: 10.1016/B978-0-443-15670-0.00002-8

Anthopoulos, L., Janssen, M., & Weerakkody, V. (2023). Introduction to the Special Issue on Citizen Centricity in Smart Cities. *Digital Government: Research and Practice*, *4*(3), 1–4. DOI: 10.1145/3624556

Anwar, G. A., Dong, Y., & Li, Y. (2020). Performance-based decision-making of buildings under seismic hazard considering long-term loss, sustainability, and resilience. *Structure and Infrastructure Engineering*, *17*(4), 454–470. DOI: 10.1080/15732479.2020.1845751

Aparicio, A. C., & Ruiz-Teran, A. M. (2007). Tradition and innovation in teaching structural design in civil engineering. *Journal of Professional Issues in Engineering Education and Practice*, *133*(4), 340–349. DOI: 10.1061/(ASCE)1052-3928(2007)133:4(340)

ASCE. (2020). *Climate-resilient infrastructure: Adaptive design and risk management*. American Society of Civil Engineers.

Asghari, F., Piadeh, F., Egyir, D., Yousefi, H., Rizzuto, J. P., Campos, L. C., & Behzadian, K. (2023). Resilience assessment in urban water infrastructure: A critical review of approaches, strategies, and applications. *Sustainability (Basel)*, *15*(14), 11151. DOI: 10.3390/su151411151

Atanane, O., Mourhir, A., Benamar, N., & Zennaro, M. (2023). Smart buildings: Water leakage detection using tinyml. *Sensors (Basel)*, *23*(22), 9210. DOI: 10.3390/s23229210 PMID: 38005596

Atıcı, E., & Elen, A. (2024). Optimization of Feed Ration Cost in Dairy Cattle by Genetic Algorithm. *Mühendislik Bilimleri ve Araştırmaları Dergisi*, *6*(1), 65–76. DOI: 10.46387/bjesr.1435749

Attaran, M., & Celik, B. G. (2023). Digital Twin: Benefits, use cases, challenges, and opportunities. *Decision Analytics Journal*, *6*, 100165. DOI: 10.1016/j.dajour.2023.100165

Auliya, Y. A., Fadah, I., Zarkasi, M., Milniadi, A. D., & Baihaqi, Y. (2021). Implementation of a decision support system for optimizing feed rations in the cattle fattening industry. *International Journal of Science and Development Research*, *6*(1), 56–63.

Aungkulanon, P., Hirunwat, A., Atthirawong, W., Phimsing, K., Chanhom, S., & Luangpaiboon, P. (2024). Optimizing maintenance responsibility distribution in real estate management: A complexity-driven approach for sustainable efficiency. *Journal of Open Innovation*, *10*(1), 100239. DOI: 10.1016/j.joitmc.2024.100239

Ayasrah, F. T., N, K., Alsharafa, N. S., S, S., Sengan, S., & B, S. (2024). Strategizing Low Carbon Urban Planning Through Environmental Impact Assessment by Artificial Intelligence Driven Carbon Foot-print Forecasting. *Journal of Machine and Computing*, 1140–1151. DOI: 10.53759/7669/jmc202404105

Ayyub, B. M. (2014). *Risk analysis in engineering and economics* (2nd ed.). CRC Press. DOI: 10.1201/b16663

Babb, C. (2016). The End of Automobile Dependence: How Cities are Moving Beyond Car-based Planning. *Urban Policy and Research*, *34*(1), 90–92. DOI: 10.1080/08111146.2016.1155194

Baldin, M., Breunig, T., Cue, R., De Vries, A., Doornink, M., Drevenak, J., Fourdraine, R., George, R., Goodling, R., Greenfield, R., Jorgensen, M. W., Lenkaitis, A., Reinemann, D., Saha, A., Sankaraiah, C., Shahinfar, S., Siberski, C., Wade, K. M., Zhang, F., & Cabrera, V. E. (2021). Integrated Decision Support Systems (IDSS) for Dairy Farming: A Discussion on How to Improve Their Sustained Adoption. *Animals (Basel)*, *11*(7), 2025. DOI: 10.3390/ani11072025 PMID: 34359153

Balta, D., Kaç, S. B., Balta, M., & Eken, S. (2023). Water quality estimation and anomaly detection: A review. *EAI Endorsed Transactions on Internet of Things*, *9*(4), e2. DOI: 10.4108/eetiot.v9i4.3660

Balthazar, B. S. (2022). *Gentrificação, Turismo e Conflitos: Análise comparada entre o Bonfim (Porto, Portugal) e Koukaki (Atenas, Grécia)* [Gentrification, Tourism and Conflicts: A comparative analysis between Bonfim (Porto, Portugal) and Koukaki (Athens, Greece)]. Master's Thesis, University of Minho.

Bandura, A. (1986). *Social foundations of thought and action: A social cognitive theory*. Prentice-Hall.

Bao, W., Xu, K., & Leng, Q. (2024). Research on the Financial Credit Risk Management Model of Real Estate Supply Chain Based on GA-SVM Algorithm: A Comprehensive Evaluation of AI Model and Traditional Model. *Procedia Computer Science*, *243*, 900–909. DOI: 10.1016/j.procs.2024.09.108

Bardhan, A., Biswas, R., Kardani, N., Iqbal, M., Samui, P., Singh, M. P., & Asteris, P. G. (2022). A novel integrated approach of augmented grey wolf optimizer and ANN for estimating axial load carrying-capacity of concrete-filled steel tube columns. *Construction & Building Materials*, *337*, 127454. DOI: 10.1016/j.conbuildmat.2022.127454

Barney, J. (1991). Firm resources and sustained competitive advantage. *Journal of Management*, *17*(1), 99–120. DOI: 10.1177/014920639101700108

Batty, M. (2018). *Inventing future cities*. MIT Press. DOI: 10.7551/mitpress/11923.001.0001

Bauer, M., Cirillo, F., Fürst, J., Solmaz, G., & Kovacs, E. (2021). Urban digital twins–a fiware-based model. *Automatisierungstechnik*, *69*(12), 1106–1115. DOI: 10.1515/auto-2021-0083

Bayeroju, O. F., Sanusi, A. N., & Nwokediegwu, Z. Q. S. (2021). Review of circular economy strategies for sustainable urban infrastructure development and policy planning. *Journal of Sustainable Infrastructure*, *9*(2), 101–118.

Bayeroju, O. F., Sanusi, A. N., & Nwokediegwu, Z. Q. S. (2023). Framework for resilient construction materials to support climate-adapted infrastructure development. *Shodhshauryam: International Scientific Refereed Research Journal*, *6*(5), 403–428.

Bayya, A. K. (2022). Advocating Ethical Data Management and Security. *International Journal of Scientific Research in Computer Science* [IJSRCSEIT]. *Engineering and Information Technology*, *8*, 396–417.

Bellingeri, A., Gallo, A., Liang, D., Masoero, F., & Cabrera, V. E. (2020). Development of a linear programming model for the optimal allocation of nutritional resources in a dairy herd. *Journal of Dairy Science*, *103*(11), 10898–10916. DOI: 10.3168/jds.2020-18157 PMID: 32952013

Berardi, U., GhaffarianHoseini, A. H., & GhaffarianHoseini, A. (2014). State-of-the-art analysis of the environmental benefits of green roofs. *Applied Energy*, *115*, 411–428. DOI: 10.1016/j.apenergy.2013.10.047

Bergmeister, K., Böhm, M., & Brunner, G. (2020). Artificial intelligence in civil engineering. *CEAS Aeronautical Journal*, *11*(2), 401–413.

Bezrukov, O. (2025, May 29). AI in Transportation: Ethical Balance between Safety and Privacy. Blog. https://tech-stack.com/blog/ai-in-transportation/

Bibri, S. E., Sharifi, A., Alexandre, A., & Krogstie, J. (2023). Environmentally sustainable smart cities and their converging AI, IoT, and big data technologies and solutions: An integrated approach to an extensive literature review. *Energy Informatics*, *6*(1), 9. Advance online publication. DOI: 10.1186/s42162-023-00259-2 PMID: 37032812

Boğa, M., & Çevik, K.K. (2012). Ruminant hayvanlar için karma yem hazırlama programı. XIV. Akademik Bilişim Konferansı (Akademik Bilişim'12), Uşak Üniversitesi, 249-256.

Boopathi, S. (2024). *Advancements in Machine Learning and AI for Intelligent Systems in Drone Applications for Smart City Developments* (pp. 15–45). igi global. DOI: 10.4018/978-1-6684-9596-4.ch002

Boschert, S., & Rosen, R. (2016). Digital twin—the simulation aspect. *Mechatronic futures: Challenges and solutions for mechatronic systems and their designers*, 59-74.

Boudhaouia, A., & Wira, P. (2018). *Water consumption analysis for real-time leakage detection in a smart tertiary building*. In ICASS 2018. DOI: 10.1109/ICASS.2018.8651976

Bruneau, M., Chang, S. E., Eguchi, R. T., Lee, G. C., O'Rourke, T. D., Reinhorn, A. M., Shinozuka, M., Tierney, K., Wallace, W. A., & von Winterfeldt, D. (2003). A framework to quantitatively assess and enhance the seismic resilience of communities. *Earthquake Spectra*, *19*(4), 733–752. DOI: 10.1193/1.1623497

Budak, F., & Budak, F. (2014). Yem bitkilerinde kalite ve yem bitkileri kalitesini etkileyen faktörler. *Türk Bilimsel Derlemeler Dergisi*, *7*(1), 1–6.

Bühler, M., Hollenbach, P., Köhler, L., & Armstrong, R. (2024). Unlocking resilience and sustainability with earth-based materials: A principled framework for urban transformation. *Frontiers in Built Environment*, *10*, 1385116. Advance online publication. DOI: 10.3389/fbuil.2024.1385116

Cabot, J. H., & Ross, E. G. (2023). Evaluating prediction model performance. *Surgery, 174*(3), 723–726. DOI: 10.1016/j.surg.2023.05.023 PMID: 37419761

Cakiroglu, C., Islam, K., Bekdaş, G., Isikdag, U., & Mangalathu, S. (2022). Explainable machine learning models for predicting the axial compression capacity of concrete filled steel tubular columns. Construction & Building Materials, 356, 129227.

Camacho, J. D. J., Anthony, B., Aguirre, B., Ponce, P., & Molina, A. (2024). Leveraging Artificial Intelligence to Bolster the Energy Sector in Smart Cities: A Literature Review. *Energies, 17*(2), 353. DOI: 10.3390/en17020353

Câmara Municipal de Lisboa. (2018). *Estudo Urbanístico do Turismo em Lisboa* [Urban Study of Tourism in Lisbon].

Cancela-Outeda, C. (2024). The EU's AI act: A framework for collaborative governance. *Internet of Things : Engineering Cyber Physical Human Systems, 27*, 101291. DOI: 10.1016/j.iot.2024.101291

Canga, D., Boğa, M., Kılıç, H. N., & Bulut, M. (2022). *Information and communication technologies (ICT) and precision livestock (PLF) applications in farm conditions*. Livre de Lyon.

Cao, B., Lv, Z., Fan, S., Chen, X., & Li, R. (2020). Optimization of Classified Municipal Waste Collection Based on the Internet of Connected Vehicles. *IEEE Transactions on Intelligent Transportation Systems, 22*(8), 5364–5373. DOI: 10.1109/TITS.2020.2981564

Cardone, B., Di Martino, F., & Senatore, S. (2024). Real estate price estimation through a fuzzy partition-driven genetic algorithm. *Information Sciences, 667*, 120442. DOI: 10.1016/j.ins.2024.120442

Castro, M. J., & Lopez, R. A. (2021). Green finance growth in Mexican banks: Problems and chances. *Latin American Business Review, 22*(3), 245–267. DOI: 10.1080/10978526.2021.1895432

Cervero, R. (2014). Transport Infrastructure and the Environment in the Global South: Sustainable Mobility and Urbanism. *Jurnal Perencanaan Wilayah Dan Kota, 25*(3), 174–191. DOI: 10.5614/jpwk.2015.25.3.1

Cesario, E. (2023). Big data analytics and smart cities: Applications, challenges, and opportunities. *Frontiers in Big Data, 6*, 1149402. Advance online publication. DOI: 10.3389/fdata.2023.1149402 PMID: 37252127

Chandanshive, V. B., & Kazi, A. M. (2017, March). Application of Internet of Things in Civil Engineering construction projects-A State of the Art. In *Proceedings of the 11th INDIACom, 4th International Conference on Computing for sustainable global development* (Vol. 4, pp. 1836-1839).

Chang, F.-J., Chen, J.-F., & Chang, L.-C. (2023). Artificial Intelligence Techniques in Hydrology and Water Resources Management. *Water (Basel), 15*(10), 1846. DOI: 10.3390/w15101846

Charitonidou, M. (2022). Urban scale digital twins in data-driven society: Challenging digital universalism in urban planning decision-making. *International Journal of Architectural Computing, 20*(2), 238–253. DOI: 10.1177/14780771211070005

Cheng, J., Jiang, Z., Wu, H., & Zhang, X. (2025). Water Pipeline Leak Detection Method Based on Transfer Learning. *Water (Basel), 17*(3), 368. DOI: 10.3390/w17030368

Chen, R. S., Zhang, H. Y., Hao, X. K., Yu, H. X., Shi, T., Zhou, H. S., & Wang, P. (2024, October). Experimental study on ultimate bearing capacity of short thin-walled steel tubes reinforced with high-ductility concrete. In *Structures* (Vol. 68, p. 107109). Elsevier.

Chen, W., Chen, L., Xie, Y., Cao, W., Gao, Y., & Feng, X. (2020). Multi-Range attentive bicomponent Graph convolutional network for traffic forecasting. *Proceedings of the AAAI Conference on Artificial Intelligence, 34*(04), 3529–3536. DOI: 10.1609/aaai.v34i04.5758

Chisom, O., Obaedo, B., Biu, P., Umoh, A., Abatan, A., & Adegbite, A. (2024). Reviewing the role of AI in environmental monitoring and conservation: A data-driven revolution for our planet. *World Journal of Advanced Research and Reviews, 21*(1), 161–171. DOI: 10.30574/wjarr.2024.21.1.2720

Cicceri, G., Tricomi, G., D'Agati, L., Longo, F., Merlino, G., & Puliafito, A. (2023). A Deep Learning-Driven Self-Conscious Distributed Cyber-Physical System for Renewable Energy Communities. *Sensors (Basel), 23*(9), 4549. DOI: 10.3390/s23094549 PMID: 37177753

Cina, E., Elbasi, E., Elmazi, G., & AlArnaout, Z. (2025). The role of AI in Predictive Modelling for Sustainable Urban Development: Challenges and opportunities. *Sustainability (Basel), 17*(11), 5148. DOI: 10.3390/su17115148

Cohen, A. P. (2024). *Artificial Intelligence in transportation: Use cases and potential.* Mineta Transportation Institute. https://transweb.sjsu.edu/sites/default/files/2430-Cohen-Artificial-Intelligence-Transportation-Use-Cases.pdf

Cohen-Shacham, E., Walters, G., Janzen, C., & Maginnis, S. (Eds.). (2016). *Nature-based solutions to address global societal challenges*. IUCN. DOI: 10.2305/IUCN.CH.2016.13.en

Cordeiro, I. D., Korossy, N., & Selva, V. (2016). Determinação da capacidade de carga turística a partir do método Cifuentes et al. (1992): Aplicação à Praia dos Carneiros (Tamandaré/PE) [Determination of tourism carrying capacity using the Cifuentes et al. (1992) method: Application to Carneiros Beach (Tamandaré/PE)]. *Revista Brasileira de Ecoturismo*, *9*(5), 991-1011.

Costa, C. (2018). A Saturação do Turismo e a Governança: Desafios para as Comunidades Locais. O caso do Alojamento Local em Portugal [Tourism Saturation and Governance: Challenges for Local Communities. The case of Short-Term Rentals in Portugal]. *Proceedings of the XI Portuguese Congress of Sociology*, Lisbon, (p. 132).

Cruz, C. O., & Sarmento, J. M. (2020). "Mobility as a Service" Platforms: A Critical Path towards Increasing the Sustainability of Transportation Systems. *Sustainability (Basel)*, *12*(16), 6368. DOI: 10.3390/su12166368

Dani, A. A. H., Supangkat, S. H., Lubis, F. F., Nugraha, I. G. B. B., Kinanda, R., & Rizkia, I. (2023). Development of a smart city platform based on digital twin technology for monitoring and supporting decision-making. *Sustainability (Basel)*, *15*(18), 14002. DOI: 10.3390/su151814002

Das, D. K. (2025). Resilient and sustainable urban infrastructure in the Global South: Strategies to address climate change-linked vulnerabilities. *Urban Planning*, *10*, 10037. Advance online publication. DOI: 10.17645/up.10037

Deb, K., & Jain, H. (2014). An evolutionary many-objective optimization algorithm using reference-point-based nondominated sorting approach, part I: Solving problems with box constraints. *IEEE Transactions on Evolutionary Computation*, *18*(4), 577–601. DOI: 10.1109/TEVC.2013.2281535

Dener, M. (2017). WiSeN: A new sensor node for smart applications with wireless sensor networks. *Computers & Electrical Engineering*, *64*, 380–394. DOI: 10.1016/j.compeleceng.2017.06.027

Depietri, Y., & McPhearson, T. (2017). Integrating the grey, green, and blue in cities: Nature-based solutions for climate change adaptation and risk reduction. In N. Kabisch, H. Korn, J. Stadler, & A. Bonn (Eds.), *Nature-based solutions to climate change adaptation in urban areas* (pp. 29-49). Springer.

Desai, D., & Dawda, N. H. (2025). New approaches for integrated multimodal transport systems. SSRN Electronic Journal. DOI: 10.2139/ssrn.5080740

Diário da República. (2014). *Decreto-Lei n.º 128/2014, de 29 de agosto* [Decree-Law No. 128/2014, of August 29]. Diário da República, 1st series, no. 166. Consolidated Version.

Dietz, M. E. (2007). Low impact development practices: A review of current research and recommendations for future directions. *Water, Air, and Soil Pollution, 186*(1-4), 351–363. DOI: 10.1007/s11270-007-9484-z

Dikshit, S., Shahid, M., Thusu, A., Dwivedi, V., & Atiq, A. (2023). The Use of Artificial Intelligence to Optimize the Routing of Vehicles and Reduce Traffic Congestion in Urban Areas. *EAI Endorsed Transactions on Energy Web, 10*. Advance online publication. DOI: 10.4108/ew.4613

Direção Regional de Estatística da Madeira (DREM). (2023). *Censos 2021: Resultados Definitivos - Região Autónoma da Madeira* [2021 Census: Definitive Results - Autonomous Region of Madeira].

Direção Regional de Estatística da Madeira (DREM). (2024). *Estatísticas da Construção e Habitação da Região Autónoma da Madeira - 2023* [Construction and Housing Statistics of the Autonomous Region of Madeira - 2023].

Djenna, A., Harous, S., & Saidouni, D. E. (2021). Internet of things meet internet of threats: New concern cyber security issues of critical cyber infrastructure. *Applied Sciences (Basel, Switzerland), 11*(10), 4580. DOI: 10.3390/app11104580

Doan, X. H., & Nguyen, T. P. L. (2025). Moderated mediation model of relationship between artificial intelligence awareness and counterproductive work behavior, turnover intention. *Journal of Organizational Change Management, 38*(2), 501–519. DOI: 10.1108/JOCM-07-2024-0398

Dorigo, M., & Gambardella, L. M. (1997). Ant colonies for the travelling salesman problem. *Bio Systems, 43*(2), 73–81. DOI: 10.1016/S0303-2647(97)01708-5 PMID: 9231906

Dritsas, E., & Trigka, M. (2024). Machine Learning for Blockchain and IoT Systems in Smart Cities: A Survey. *Future Internet, 16*(9), 324. DOI: 10.3390/fi16090324

Duong, T. H., Le, T. T., & Le, M. V. (2023). Practical Machine Learning Application for Predicting Axial Capacity of Composite Concrete-Filled Steel Tube Columns Considering Effect of Cross-Sectional Shapes. *International Journal of Steel Structures, 23*(1), 263–278. DOI: 10.1007/s13296-022-00693-0

Eccles, R. G., Ioannou, I., & Serafeim, G. (2014). The impact of corporate sustainability on organizational processes and performance. *Management Science, 60*(11), 2835–2857. DOI: 10.1287/mnsc.2014.1984

El Bilali, H., Strassner, C., & Ben Hassen, T. (2021). Sustainable Agri-Food Systems: Environment, Economy, Society, and Policy. *Sustainability (Basel), 13*(11), 6260. DOI: 10.3390/su13116260

El-baky, M. A. A., Allah, M. M. A., Kamel, M., & Abd-Elaziem, W. (2022). Lightweight cost-effective hybrid materials for energy absorption applications. *Scientific Reports, 12*(1), 21101. DOI: 10.1038/s41598-022-25533-3 PMID: 36473932

Elvas, L. B., & Ferreira, J. C. (2021). Intelligent transportation systems for electric vehicles. *Energies, 14*(17), 5550. DOI: 10.3390/en14175550

Esri. (n.d.). *Transportation Equity | Improving Mobility for Underserved Communities.* Retrieved from https://www.esri.com/en-us/industries/roads-highways/transportation-equity

Eze, C. C. (2025). Integration of AI-Driven multimodal transport systems for optimizing Real-Time urban and intercity mobility solutions. *International Journal of Research Publication and Reviews, 6*(4), 9288–9304. DOI: 10.55248/gengpi.6.0425.1574

Farahani, M. S., & Ghasemi, G. (2024). How artificial intelligence plays a role in achieving sustainable development goals? *Sustainable Economies, 2*(3), 66. DOI: 10.62617/se.v2i3.66

Farrokhi, Z., Hassanpouri Baesmat, K., & Regentova, E. E. (2024). Enhancing urban intelligence energy management: Innovative load-forecasting techniques for electrical networks. *Journal of Power and Energy Engineering, 12*(11), 72–88. DOI: 10.4236/jpee.2024.1211005

Fenner, R. A., Ainger, C. M., Cruickshank, H. J., & Guthrie, P. M. (2020). Embedding sustainable development in the civil engineering curriculum. *Proceedings of the Institution of Civil Engineers. Engineering Sustainability, 173*(2), 52–64.

Firouzi, F., Farahani, B., Weinberger, M., DePace, G., & Aliee, F. S. (2020). Iot fundamentals: Definitions, architectures, challenges, and promises. In *Intelligent internet of things: from device to fog and cloud* (pp. 3–50). Springer International Publishing. DOI: 10.1007/978-3-030-30367-9_1

Freeman, R. E. (1984). *Strategic management: A stakeholder approach.* Pitman.

Gao, Y., Wang, B., Liu, C., Hui, D., Xu, Q., Zhao, Q., & Hong, X. (2022). Experimental study on basic mechanical properties of recycled steel fiber reinforced concrete. *Reviews on Advanced Materials Science, 61*(1), 417–429. DOI: 10.1515/rams-2022-0041

Garcia-Meca, E., Pucheta-Martinez, M. C., & Cuadrado-Ballesteros, B. (2015). The influence of ownership structure on the costs of corporate debt in the context of intellectual capital. *Journal of Business Research*, *68*(4), 860–869. DOI: 10.1016/j.jbusres.2014.11.040

Gardes, N. (2025). Ally or adversary? AI and the perceived threat to autonomy in services – The case of real estate professions. *Journal of Strategy & Innovation*, *36*(1), 200533. DOI: 10.1016/j.jsinno.2025.200533

Garroussi, Z., Legrain, A., Gambs, S., Gautrais, V., & Sansò, B. (2025). A systematic review of data privacy in Mobility as a Service (MaaS). *Transportation Research Interdisciplinary Perspectives*, *31*, 101254. DOI: 10.1016/j.trip.2024.101254

Gebresselassie, M., & Sanchez, T. W. (2019). Banister: Inequality in transport. *Journal of the American Planning Association*, *85*(4), 593–594. DOI: 10.1080/01944363.2019.1641385

Ghahramani, M., Zhou, M., Molter, A., & Pilla, F. (2022). IoT-Based Route Recommendation for an Intelligent Waste Management System. *IEEE Internet of Things Journal*, *9*(14), 11883–11892. DOI: 10.1109/JIOT.2021.3132126

GIH. (2023). *Global infrastructure outlook*. Global Infrastructure Hub.

Gkontzis, A. F., Verykios, V. S., Feretzakis, G., & Kotsiantis, S. (2024). Enhancing Urban Resilience: Smart City Data Analyses, Forecasts, and Digital Twin Techniques at the Neighborhood Level. *Future Internet*, *16*(2), 47. DOI: 10.3390/fi16020047

Gómez-Villarino, M. T., Ruiz-Garcia, L., & Gómez Villarino, M. (2020). Implementation of Urban Green Infrastructures in Peri-Urban Areas: A Case Study of Climate Change Mitigation in Madrid. *Agronomy (Basel)*, *11*(1), 31. DOI: 10.3390/agronomy11010031

Gonzalez, A., Martinez, L., & Sanchez, P. (2020). The transformation of the energy market and corporate adaptation in Mexico: Evidence from the industrial sector. *Energy Policy*, *145*, 111750. DOI: 10.1016/j.enpol.2020.111750

Gonzalez, T. A., Romero, L. C., Karmaziene, E., & Yuan, X. (2025). Green gains: The impact of REITs' environmental performance on sustainability-linked loan interest rates. *Finance Research Letters*, *71*, 106415. DOI: 10.1016/j.frl.2024.106415

Gümüştaş, E., & Pehlivanlı, A. Ç. (2021). In-Silico Mutajenisite Tahmininde İstatistiksel Öğrenme Modeli. *Süleyman Demirel Üniversitesi Fen Bilimleri Enstitüsü Dergisi*, *25*(2), 365–370. DOI: 10.19113/sdufenbed.867067

Guo, Y. T., Chen, J., Nie, X., Tao, M. X., Wang, J. J., & Fan, J. S. (2020). Investigation of the shear resistances of steel–concrete–steel composite structures with bidirectional webs. *Journal of Constructional Steel Research*, *164*, 105846. DOI: 10.1016/j.jcsr.2019.105846

Gupta, M., Kumari, M., & Ved, C. (2021). Digital Twin in Agriculture Sector: Detection of Disease Using Deep Learning. In *Digital Twin Technology* (pp. 47-60). CRC Press.

Gusti, M. L. H., & Aryanny, E. (2023). Determining The Optimal Milk Distribution Route Using The Ant Colony Optimization Method on Milk Industry. *Spektrum Industri*, *21*(2), 141–148. DOI: 10.12928/si.v21i2.92

Haghiabi, A. H., Nasrolahi, A. H., & Parsaie, A. (2018). Water quality prediction using machine learning methods. *Water Quality Research Journal of Canada*, *53*(1), 3–13. DOI: 10.2166/wqrj.2018.025

Hajihosseinlou, M., Maghsoudi, A., & Ghezelbash, R. (2024). Regularization in machine learning models for MVT Pb-Zn prospectivity mapping: Applying lasso and elastic-net algorithms. *Earth Science Informatics*, *17*(5), 4859–4873. DOI: 10.1007/s12145-024-01404-5

Haleem, A., Javaid, M., & Vaishya, R. (2021). Industry 4.0 and its applications in orthopaedics. *Journal of Clinical Orthopaedics and Trauma*, *12*, 51–55.

Hardhienata, M. K. D., Priandana, K., Putra, D. R., Sriatun, M., Wulandari, A. B., & Mohamed, R. (2024). Modification of the Ant Colony Optimization Algorithm for Solving Multi-Agent Task Allocation Problem in Agricultural Application. *Optimization (ACO). Journal of Advanced Research in Applied Sciences and Engineering Technology*, *34*(1), 90–105. DOI: 10.37934/araset.34.1.90105

Hassanpouri Baesmat, K., Farrokhi, Z., Chmaj, G., & Regentova, E. E. (2025). Parallel multi-model energy-demand forecasting with cloud redundancy: Leveraging trend correction, feature selection, and machine learning. *Forecasting*, *7*(2), 1–18. DOI: 10.3390/forecast7020025

Hassanpouri Baesmat, K., & Shiri, A. (2019). A new combined method for future energy forecasting in electrical networks. *International Transactions on Electrical Energy Systems*, *29*(3), e2749. DOI: 10.1002/etep.2749

Hassanpouri Baesmat, K., Shokoohi, F., & Farrokhi, Z. (2025). *SP-RF-ARIMA: A sparse random forest and ARIMA hybrid model for electric-load forecasting*. Global Energy Interconnection.

Haurum, K. R., Mia, R., & Long, W. (2024). Real Estate with AI: An agent based on LangChain. *Procedia Computer Science*, *242*, 1082–1088. DOI: 10.1016/j.procs.2024.08.199

Hernandez, J. C., & Rodriguez, M. A. (2021). Financial literacy, renewable energy certification, and business sustainability in Mexico. *Renewable Energy*, *165*, 789–801. DOI: 10.1016/j.renene.2020.11.089

Hoang, T. V. (2024). Impact of Integrated Artificial Intelligence and Internet of Things Technologies on Smart City Transformation. *Journal of Technical Education Science*, *19*(1), 64–73. DOI: 10.54644/jte.2024.1532

Hochreiter, S., & Schmidhuber, J. (1997). Long short-term memory. *Neural Computation*, *9*(8), 1735–1780. DOI: 10.1162/neco.1997.9.8.1735

Holling, C. S. (1973). Resilience and stability of ecological systems. *Annual Review of Ecology and Systematics*, *4*(1), 1–23. DOI: 10.1146/annurev.es.04.110173.000245

Hsieh, F. (2025). Emerging Research Issues and Directions on MaaS, Sustainability and Shared Mobility in Smart Cities with Multi-Modal Transport Systems. *Applied Sciences (Basel, Switzerland)*, *15*(10), 5709. DOI: 10.3390/app15105709

Huang, Y.-P., & Khabusi, S. P. (2025). Artificial Intelligence of Things (AIoT) Advances in Aquaculture: A Review. *Processes (Basel, Switzerland)*, *13*(1), 73. DOI: 10.3390/pr13010073

Hub, G. I. (2020, November 4). Mobility as a service. https://www.gihub.org/infrastructure-technology-use-cases/case-studies/mobility-as-a-service/#:~:text=Cost:%20Whim%20offers%20three%20tiers,Vienna%2C%20Antwerp%20and%20the%20US.&text=Implementation:%20Created%20by%20the%20Transport,Timeframe:%20Launched%20in%20September%202018.&text=Implementation:%20Pilot%20in%20Scotland's%20Dundee,month%20pilot%20held%202017%2D2018

Huiling, E., & Goh, B. (2017, December 31). AI, Robotics and Mobility as a Service: the Case of Singapore. https://journals.openedition.org/factsreports/4411

Hussain, D. I., Elomri, D. A., Kerbache, D. L., & Omri, D. A. E. (2024). Smart city solutions: Comparative analysis of waste management models in IoT-enabled environments using multiagent simulation. *Sustainable Cities and Society*, *103*, 105247. DOI: 10.1016/j.scs.2024.105247

Ibrahim, F. J. M. (2023). Sustainable urban infrastructure: A comprehensive review of environmental safety practices in civil engineering. *i-Manager's Journal on Civil Engineering*, *13*(4), 25–34. DOI: 10.26634/jce.13.4.20316

Idealista. (2025). *Analysis of the property sales market in the municipality of Machico*. [Data collected in July 2025].

Ighalo, J. O., Marques, G., & Adeniyi, A. G. (2020). *Internet of Things for Water Quality Monitoring and Assessment: A Comprehensive Review* (pp. 245–259). springer. DOI: 10.1007/978-3-030-51920-9_13

Ilugbusi, B. S., Adisa, O., Obi, O., Awonuga, K. F., Adelekan, O. A., Asuzu, O. F., & Ndubuisi, N. L. (2024). Urban resilience to climate change: A review of adaptation strategies and infrastructure innovations. *Ecofeminism and Climate Change*, *5*(1), 18–23. DOI: 10.26480/efcc.01.2024.18.23

India, T. O. (2025, August 4). AI-driven adaptive signals to cut wait times at 165 junctions. The Times of India. https://timesofindia.indiatimes.com/city/chennai/ai-driven-adaptive-signals-to-cut-wait-times-at-165-junctions/articleshow/123103169.cms

Inter-American Development Bank. (2018). *Financing renewable energy in Latin America: Challenges and opportunities*. IDB Publications.

International Energy Agency. (2022). *CO_2 Emissions from Fuel Combustion*. IEA Publications. https://www.iea.org/reports/co2-emissions-in-2022

International Transport Forum (ITF). (2021). *Reforming Urban Transport: Lessons from Global Practices*. OECD Publishing. https://www.oecd.org/en/publications/2021/05/itf-transport-outlook-2021_fff2ec9f.html

IPCC. (2022). *Climate Change 2022: Impacts, Adaptation and Vulnerability*. Cambridge University Press.

Ismail, M. (2020). New approach to seismic-resistant design and structural torsion mitigation. *Engineering Structures*, *207*, 110092. DOI: 10.1016/j.engstruct.2019.110092

Iyer, L. S. (2021). AI enabled applications towards intelligent transportation. *Transportation Engineering*, *5*, 100083. DOI: 10.1016/j.treng.2021.100083

Jain, R., Upreti, K., Ekatpure, S. R., Gupta, V. K., Rai, A. K., & Gothania, J. (2025). The convergence of IoT, ML, and big data in Self-Service innovation. In IGI Global eBooks (pp. 241–270). https://doi.org/DOI: 10.4018/979-8-3373-1295-8.ch009

Jain, R., Singh, S. K., Palaniappan, D., Parmar, K., & T, P. (2025). Data-Driven Civil Engineering: Applications of artificial intelligence, machine learning, and deep learning. *Turkish Journal of Engineering*, *9*(2), 354–377. DOI: 10.31127/tuje.1581564

Jamali, H., Dascalu, S. M., & Harris, F. C. (2024). AI-driven analysis and prediction of energy consumption in NYC's municipal buildings. In *2024 IEEE/ACIS 22nd International Conference on Software Engineering Research, Management and Applications (SERA)* (pp. 277–283). IEEE. DOI: 10.1109/SERA61261.2024.10685594

Jamali, H., Watson, J., Dascalu, S. M., & Harris, F. C. (2025). Harmonized data drive: Standardizing and unifying smart-car information storage for enhanced forensics and interoperability. In *2025 13th International Symposium on Digital Forensics and Security (ISDFS)* (pp. 1–6). IEEE.

Jamali, H., Karimi, A., & Haghighizadeh, M. (2018). A new method of cloud-based computation model for mobile devices: Energy-consumption optimization in mobile-to-mobile computation offloading. In *Proceedings of the 6th International Conference on Communications and Broadband Networking* (pp. 32–37). ACM. DOI: 10.1145/3193092.3193103

Jamil, H., Noorosmawie, A. B., Rabu, H. W., & Razak, L. A. (2025). From archives to AI: Residential property data across three decades in Brunei Darussalam. *Data in Brief*, *60*, 111505. DOI: 10.1016/j.dib.2025.111505 PMID: 40235696

Janshevaski, V. R. M. (2025). Eco-innovations in urban development: Bridging civil engineering and environmental sustainability. *Innovative Research in Civil and Environmental Engineering*, *2*(2), 22–28.

Jha, M. K., Yogeshwari, A., Rubini, P., & Singh, M. (2023, August). Converge of iot and ai in metaverse: challenges and opportunities. In *2023 5th International Conference on Inventive Research in Computing Applications (ICIRCA)* (pp. 1462-1467). IEEE. DOI: 10.1109/ICIRCA57980.2023.10220628

Jia, L., Hu, Z., & Li, Z. (2024). Applications of the Internet of Things in Renewable Power Systems: A Survey. *Energies*, *17*(16), 4160. DOI: 10.3390/en17164160

Jin, J. G., Tang, L. C., Sun, L., & Lee, D. (2014). Enhancing metro network resilience via localized integration with bus services. *Transportation Research Part E, Logistics and Transportation Review*, *63*, 17–30. DOI: 10.1016/j.tre.2014.01.002

Juarez-Quispe, J., Rojas-Chura, E., Espinoza Vigil, A. J., Guillén Málaga, M. S., Yabar-Ardiles, O., Anco-Valdivia, J., & Valencia-Félix, S. (2025). Advancing sustainable infrastructure management: Insights from system dynamics. *Buildings (Basel, Switzerland)*, *15*(2), 210. DOI: 10.3390/buildings15020210

Jun, C., Sun-Zheng, Y. U., & Jing-Li, L. (2009). The design of a wireless data acquisition and transmission system. *Journal of Networks*, *4*(10), 1042–1049.

Kabisch, N., Korn, H., Stadler, J., & Bonn, A. (2022). *Nature-based solutions and climate change: The state of the art*. Springer.

Kapoor, S., Cantrell, E. M., Peng, K., Pham, T. H., Bail, C. A., Gundersen, O. E., Hofman, J. M., Hullman, J., Lones, M. A., Malik, M. M., Nanayakkara, P., Poldrack, R. A., Raji, I. D., Roberts, M., Salganik, M. J., Serra-Garcia, M., Stewart, B. M., Vandewiele, G., & Narayanan, A. (2024). REFORMS: Consensus-based Recommendations for Machine-learning-based Science. *Science Advances*, *10*(18), eadk3452. DOI: 10.1126/sciadv.adk3452 PMID: 38691601

Karabulut, A., Ergül, M., Ak, İ., Kutlu, H.R., & Alçiçek, A. (2000). Karma yem endüstrisi. V. Türkiye Ziraat Mühendisliği Teknik Kongresi, TMMOB Ziraat Mühendisleri Odası, 985-1008.

Karpatne, A., Atluri, G., Faghmous, J. H., Steinbach, M., Banerjee, A., Ganguly, A., Shekhar, S., Samatova, N., & Kumar, V. (2017). Theory-guided data science: A new paradigm for scientific discovery from data. *IEEE Transactions on Knowledge and Data Engineering*, *29*(10), 2318–2331. DOI: 10.1109/TKDE.2017.2720168

Kavzoglu, T., & Teke, A. (2022). Predictive Performances of ensemble machine learning algorithms in landslide susceptibility mapping using random forest, extreme gradient boosting (XGBoost) and natural gradient boosting (NGBoost). *Arabian Journal for Science and Engineering*, *47*(6), 7367–7385. DOI: 10.1007/s13369-022-06560-8

Kenea, L. M., Kumar, L., Singh, S. K., Chala, G., & Chimdi, J. (2021). Assessment of asphalt road pavement rehabilitation and maintenance practices: The case of West Showa Zone in Oromia, Ethiopia. *IOP Conference Series. Materials Science and Engineering*, *1187*(1), 012033. DOI: 10.1088/1757-899X/1187/1/012033

Kennedy, J., & Eberhart, R. (1995). Particle swarm optimization. In Proceedings of ICNN'95 - International Conference on Neural Networks (Vol. 4, pp. 1942–1948). IEEE. DOI: 10.1109/ICNN.1995.488968

Keskintürk, T., & Söyler, H. (2006). Global Karınca Kolonisi Optimizasyonu. *Gazi Üniversitesi Mimarlık ve Mühendislik Dergisi*, *21*(4), 689–698.

Kessler, R. (2011). Stormwater strategies: cities prepare aging infrastructure for climate change.

Kiasari, M., Ghaffari, M., & Aly, H. H. (2024). A Comprehensive Review of the Current Status of Smart Grid Technologies for Renewable Energies Integration and Future Trends: The Role of Machine Learning and Energy Storage Systems. *Energies*, *17*(16), 4128. DOI: 10.3390/en17164128

Kılıç, H. N., Boğa, M., & Çanga, D. (2023). Sensitive Livestock and Information Communication Technology Applications to Prevent the Spread of Covid-19. *Black Sea Journal of Agriculture*, *6*(1), 104–107. DOI: 10.47115/bsagriculture.1194817

Kimery, A. (2025, August 8). Kansas City equips buses with AI-powered FRT ahead of World Cup. Biometric Update | Biometrics News, Companies and Explainers. https://www.biometricupdate.com/202508/kansas-city-equips-buses-with-ai-powered-frt-ahead-of-world-cup

Kitchin, R. (2014). The data revolution: Big data, open data, data infrastructures and their consequences. *Sage (Atlanta, Ga.)*. Advance online publication. DOI: 10.4135/9781473909472

Koç, H., Dünder, E., Gümüştekin, S., Koç, T., & Cengiz, M. A. (2018). Particle Swarm Optimization-based Variable Selection in Poisson Regression Analysis via Information Complexity-type Criteria. *Communications in Statistics. Theory and Methods*, *47*(21), 5298–5306. DOI: 10.1080/03610926.2017.1390129

Konstantinov, A. V., & Utkin, L. V. (2021). Interpretable machine learning with an ensemble of gradient boosting machines. *Knowledge-Based Systems*, *222*, 106993. DOI: 10.1016/j.knosys.2021.106993

Krupnova, T. G., Bondarenko, K. A., Tretyakova, V. D., & Rakova, O. V. (2022). Environmental Justice and the Use of Artificial Intelligence in Urban Air Pollution Monitoring. *Big Data and Cognitive Computing*, *6*(3), 75. DOI: 10.3390/bdcc6030075

Kumar, L. R., Kondaiah, T., Goel, L., Sahil Singh, R. K., Dixit, S., Keerthana, S., Sai Akanksha, S., Satyanarayana, K., Bobba, P. B., Perveen, A., & Debnath, S. (2024). Smart Air Quality Monitoring Using IoT. *MATEC Web of Conferences*, *392*, 01084. DOI: 10.1051/matecconf/202439201084

Kushwaha, A. (2024). Application of AI and IoT in Traffic Management of Large Metropolitan Cities. *International Journal of Innovative Science and Research Technology (IJISRT)*, 3426–3431. DOI: 10.38124/ijisrt/IJISRT24APR2069

Lempert, R. J., Groves, D. G., Popper, S. W., & Bankes, S. C. (2018). A general, analytic method for generating robust strategies and narrative scenarios. *Management Science*, *52*(4), 514–528. DOI: 10.1287/mnsc.1050.0472

Lenton, T. M., Rockström, J., Gaffney, O., Rahmstorf, S., Richardson, K., Steffen, W., & Schellnhuber, H. J. (2019). Climate tipping points—Too risky to bet against. *Nature*, *575*(7784), 592–595. DOI: 10.1038/d41586-019-03595-0

Leonardi, S., Distefano, N., & Gruden, C. (2025). Advancing Sustainable Mobility: Artificial intelligence approaches for autonomous vehicle trajectories in roundabouts. *Sustainability (Basel), 17*(7), 2988. DOI: 10.3390/su17072988

Lin, R. F.-Y., Ou, C., Tseng, K.-K., Bowen, D., Yung, K. L., & Ip, W. H. (2021). The Spatial neural network model with disruptive technology for property appraisal in real estate industry. *Technological Forecasting and Social Change, 173*, 121067. DOI: 10.1016/j.techfore.2021.121067

Li, T., Lu, C., & Chen, Z. (2023). The unintended consequence of collateral-based financing: Evidence from corporate cost behavior. *Journal of Contemporary Accounting & Economics, 19*(1), 100338. DOI: 10.1016/j.jcae.2022.100338

Liu, M., Fang, S., Dong, H., & Xu, C. (2021). Review of digital twin about concepts, technologies, and industrial applications. *Journal of Manufacturing Systems, 58*, 346–361. DOI: 10.1016/j.jmsy.2020.06.017

Liu, R., Zayed, T., & Xiao, R. (2024). Advanced acoustic leak detection in water distribution networks using integrated generative model. *Water Research, 254*, 121434. DOI: 10.1016/j.watres.2024.121434 PMID: 38484549

Li, W., Ma, Z., Li, J., Li, Q., Li, Y., & Yang, J. (2024). Digital twin smart water conservancy: Status, challenges, and prospects. *Water (Basel), 16*(14), 2038. DOI: 10.3390/w16142038

Li, X., Zhao, N., Jin, R., Liu, S., Sun, X., Wen, X., Wu, D., Zhou, Y., Guo, J., Chen, S., Xu, Z., Ma, M., Wang, T., Qu, Y., Wang, X., Wu, F., & Zhou, Y. (2019). Internet of Things to network smart devices for ecosystem monitoring. *Science Bulletin, 64*(17), 1234–1245. DOI: 10.1016/j.scib.2019.07.004 PMID: 36659604

Lopez, C., Rivera, J., & Torres, S. (2020). Renewable energy adoption in Latin America: The role of financial literacy and institutional factors. *Energy Economics, 89*, 104785. DOI: 10.1016/j.eneco.2020.104785

LTA | Intelligent Transport Systems. (n.d.). https://www.lta.gov.sg/content/ltagov/en/getting_around/driving_in_singapore/intelligent_transport_systems.html

Lucas, K. (2019). A new evolution for transport-related social exclusion research? *Journal of Transport Geography, 81*, 102529. DOI: 10.1016/j.jtrangeo.2019.102529

LUNDASSUL. (2022). *Relatório de Análise do Caniçal* [Analysis Report of Caniçal].

LUNDASSUL. (2023). *EECM – Estudo de Estacionamento da Cidade de Machico.* Relatório Final. [EECM – Parking Study for the City of Machico (Final Report)]

Lusardi, A., & Mitchell, O. S. (2014). The economic importance of financial literacy: Theory and evidence. *Journal of Economic Literature*, *52*(1), 5–44. DOI: 10.1257/jel.52.1.5 PMID: 28579637

Lv, Z., Chen, D., Feng, H., Singh, A. K., Wei, W., & Lv, H. (2022). Computational intelligence in security of digital twins big graphic data in cyber-physical systems of smart cities. [TMIS]. *ACM Transactions on Management Information Systems*, *13*(4), 1–17. DOI: 10.1145/3522760

Ly, H. B., Pham, B. T., Le, L. M., Le, T. T., Le, V. M., & Asteris, P. G. (2021). Estimation of axial load-carrying capacity of concrete-filled steel tubes using surrogate models. *Neural Computing & Applications*, *33*(8), 3437–3458. DOI: 10.1007/s00521-020-05214-w

Ma, L., & Sang, D. (2024). Would carbon tax policy promote real estate prices? International Review of Financial Analysis, Volume 96, Part A, 103668.

Ma, H., Xie, H., & Brown, D. (2017). Eco-Driving assistance system for a manual transmission bus based on machine learning. *IEEE Transactions on Intelligent Transportation Systems*, *19*(2), 572–581. DOI: 10.1109/TITS.2017.2775633

Mahdavi, Z., Samavat, T., Javanmardi, A. S. J., Dashtaki, M. A., Zand, M., Nasab, M. A., Nasab, M. A., Padmanaban, S., & Khan, B. (2024). Providing a control system for charging electric vehicles using ANFIS. *International Transactions on Electrical Energy Systems*, *2024*(1), 9921062. DOI: 10.1155/2024/9921062

Mamat, H., Ibrahim, B. H., & Sulong, M. P. (2019, November). Network topology comparison for internet communication and IoT connectivity. In *2019 IEEE Conference on Open Systems (ICOS)* (pp. 1-5). IEEE. DOI: 10.1109/ICOS47562.2019.8975702

Manahov, V., & Li, M. (2025). The digitalisation of the real estate market: New evidence from the most prominent crypto hacker attacks. International Review of Financial Analysis, Volume 103, 104166. Guarini, M.R., Roma, A., Sabatelli, E., & Segura-de-la-Cai, A. (2025). Intrinsic and extrinsic attributes in real estate pricing: Insights for sustainable urban planning strategies. *Land Use Policy*, *153*, 107543.

Mandal, S. (2024, March). A Privacy Preserving Federated Learning (PPFL) Based Cognitive Digital Twin (CDT) Framework for Smart Cities. *Proceedings of the AAAI Conference on Artificial Intelligence*, *38*(21), 23399–23400. DOI: 10.1609/aaai.v38i21.30400

Marahatta, A., Singh, A., Korba, P., Shrestha, A., Gonzalez-Longatt, F., Thapa, A., Shin, S., & Rajbhandari, Y. (2021). Evaluation of a LoRa Mesh Network for Smart Metering in Rural Locations. *Electronics (Basel), 10*(6), 751. DOI: 10.3390/electronics10060751

Marini, F., & Walczak, B. (2015). Particle swarm optimization (PSO): A tutorial. *Chemometrics and Intelligent Laboratory Systems, 149*, 153–165. DOI: 10.1016/j.chemolab.2015.08.020

Markowitz, H. (1952). Portfolio selection. *The Journal of Finance, 7*(1), 77–91. DOI: 10.1111/j.1540-6261.1952.tb01525.x

Marques, J. A., & Cariço, C. (2023). Capacidade de Carga Turística: Uma Análise sob a Ótica do Planejamento Urbano e Regional [Tourism Carrying Capacity: An Analysis from the Perspective of Urban and Regional Planning]. *Revista Brasileira de Pesquisa em Turismo.*

Martin. (2025, July 30). Cities - United Nations Sustainable Development Action 2015. United Nations Sustainable Development. https://www.un.org/sustainabledevelopment/cities/#:~:text=Goal%2011%3A%20Make%20cities%20inclusive%2C%20safe%2C%20resilient%20and%20sustainable&text=Cities%20are%20centers%20of%20innovation,%2C%20safe%2C%20and%20resilient%20cities

Martínez, I., Cambra, D., Blanco, T., Casas, R., Trillo-Lado, R., & Zalba, B. (2021). Internet of Things (IoT) as Sustainable Development Goals (SDG) Enabling Technology towards Smart Readiness Indicators (SRI) for University Buildings. *Sustainability (Basel), 13*(14), 7647. DOI: 10.3390/su13147647

Ma, S., & Chen, Z. (2022). Low-Carbon Development of Livestock in China: Evolution, Predicaments and Paths Ahead. *Low Carbon Economy, 13*(2), 53–69. DOI: 10.4236/lce.2022.132004

Meadows, D. H., Randers, J., & Meadows, D. L. (2018). *The limits to growth: The 30-year update*. Chelsea Green Publishing.

Meerow, S., Newell, J. P., & Stults, M. (2016). Defining urban resilience: A review. *Landscape and Urban Planning, 147*, 38–49. DOI: 10.1016/j.landurbplan.2015.11.011

Megahed, K., Mahmoud, N. S., & Abd-Rabou, S. E. M. (2023). Application of machine learning models in the capacity prediction of RCFST columns. Sci Rep 13, 20878.

Mehta, R., Tanwar, S., Patel, Y., Mishra, S., & Prasad, V. K. (2023). *Evolving technologies: IoT and artificial intelligence for healthcare informatics* (pp. 231–254). institution of engineering technology. DOI: 10.1049/PBHE041E_ch8

Mendes, L. (2018). *Gentrificação turística em Lisboa: Impactos do alojamento local na resiliência e sustentabilidade social do centro histórico* [Touristic gentrification in Lisbon: Impacts of short-term rentals on the resilience and social sustainability of the historic center]. Poder Local. *Revista de Administração Democrática, 155*, 58–73.

Mendoza, R., Silva, A., & Vargas, E. (2020). Financial literacy and the adoption of sustainable agriculture and renewable energy practices in Mexico. *Agricultural Systems, 185*, 102943. DOI: 10.1016/j.agsy.2020.102943

Mexican Business Council for Sustainable Development. (2019). *Green finance survey: Current status and opportunities in Mexico*. CESPEDES.

Miller, J. (2025, June 28). Barcelona implements AI technology to monitor traffic violations in bus lanes. Rentacar Trends, News and Special Offers. https://blog.getrentacar.com/sr/news/barcelona-ai-urban-transportation/

Milly, P. C. D., Betancourt, J., Falkenmark, M., Hirsch, R. M., Kundzewicz, Z. W., Lettenmaier, D. P., & Stouffer, R. J. (2008). Stationarity is dead: Whither water management? *Science, 319*(5863), 573–574. DOI: 10.1126/science.1151915

Mirindi, D. (2024). A review of the advances in artificial intelligence in transportation system development. *Journal of Civil Construction and Environmental Engineering, 9*(3), 72–83. DOI: 10.11648/j.jccee.20240903.13

Mishra, P., & Singh, G. (2023). Energy Management Systems in Sustainable Smart Cities Based on the Internet of Energy: A Technical Review. *Energies, 16*(19), 6903. DOI: 10.3390/en16196903

Mnih, V., Kavukcuoglu, K., Silver, D., Rusu, A. A., Veness, J., Bellemare, M. G., Graves, A., Riedmiller, M., Fidjeland, A. K., Ostrovski, G., Petersen, S., Beattie, C., Sadik, A., Antonoglou, I., King, H., Kumaran, D., Wierstra, D., Legg, S., & Hassabis, D. (2015). Human-level control through deep reinforcement learning. *Nature, 518*(7540), 529–533. DOI: 10.1038/nature14236

Möderl, M., Kleidorfer, M., Sitzenfrei, R., & Rauch, W. (2009). Identifying the critical infrastructure of a storm water drainage system. *Water Science and Technology, 60*(10), 2537–2544.

Mohd Yussof, N. A., & Ho, H. W. (2022). Review of water leak detection methods in smart building applications. *Buildings (Basel, Switzerland), 12*(10), 1535. DOI: 10.3390/buildings12101535

Morales, P., Gutierrez, D., & Ramos, F. (2021). Green bond markets in Mexico: Determinants of corporate participation and financing costs. *Journal of Sustainable Finance & Investment, 11*(2), 156–178. DOI: 10.1080/20430795.2020.1762911

Muddada, M. K., Vankara, J., Nandini, S. S., Karetla, G. R., & Naidu, K. S. (2023). Multi-objective Ant Colony Optimization (MOACO) approach for multi-document text summarization. *Engineering Proceedings*, *59*(1), 218. DOI: 10.3390/engproc2023059218

Mutambik, I. (2025). IoT-Enabled Adaptive Traffic Management: A multiagent framework for urban mobility optimisation. *Sensors (Basel)*, *25*(13), 4126. DOI: 10.3390/s25134126 PMID: 40648380

Narayana, S. S. (2024). Synergizing edge computing and cloud integration for enhanced efficiency in the oil and gas industry. *J. Comput. Eng. Technol*, *15*, 63–71.

Narayan, S., Beck, M. W., Reguero, B. G., Losada, I. J., van den Bergh, J., de Jager, J., Stadtmüller, T., Burks-Copes, K., Marion, C., & Infante, D. (2016). The effectiveness, costs and coastal protection benefits of natural and nature-based defences. *PLoS One*, *11*(5), e0154735. DOI: 10.1371/journal.pone.0154735

Necula, S.-C. (2023). Assessing the Potential of Artificial Intelligence in Advancing Clean Energy Technologies in Europe: A Systematic Review. *Energies*, *16*(22), 7633. DOI: 10.3390/en16227633

Nguyen, D. N., Imamura, F., & Usuda, Y. (2024). Gaps in and Opportunities for Disaster Risk Reduction in Urban Areas Through International Standardization of Smart Community Infrastructure. *Sustainability (Basel)*, *16*(21), 9586. DOI: 10.3390/su16219586

NOAA. (2023). *Sea level rise trends*. National Oceanic and Atmospheric Administration.

NRC. (2022). *Climate adaptation infrastructure: Research and implementation gaps*. National Research Council.

Nwokediegwu, Z., Dada, M., Obaigbena, A., Majemite, M., & Ugwuanyi, E. (2024). AI-DRIVEN WASTE MANAGEMENT SYSTEMS: A COMPARATIVE REVIEW OF INNOVATIONS IN THE USA AND AFRICA. *Engineering Science & Technology Journal*, *5*(2), 507–516. DOI: 10.51594/estj.v5i2.828

Nwokediegwu, Z., & Ugwuanyi, E. (2024). IMPLEMENTING AI-DRIVEN WASTE MANAGEMENT SYSTEMS IN UNDERSERVED COMMUNITIES IN THE USA. *Engineering Science & Technology Journal*, *5*(3), 794–802. DOI: 10.51594/estj.v5i3.903

Nyokum, T., & Tamut, Y. (2025). Sustainable urban infrastructure development: Integrating smart technologies for resilient and green cities. *SSRG International Journal of Civil Engineering*, *12*(4), 18–36. DOI: 10.14445/23488352/IJCE-V12I4P103

Nyoni, V., Piller, W. B., & Vigren, O. (2023). Sustainability action in the real estate sector — An organizational and institutional perspective. *Cleaner Production Letters, 5*, 100049. DOI: 10.1016/j.clpl.2023.100049

OECD. (2022). *Climate-resilient infrastructure: Getting the incentives right.* OECD Publishing.

OECD. Kamal-Chaoui, L., Matta, A., Fritz, K., Kim, B., Kim, S.-J., Akhmouch, A., Ministry of Land, Infrastructure and Transport of Korea (MOLIT), Permanent Delegation of Korea to the OECD, & Philip, P. (2020). Smart cities and inclusive growth. https://www.oecd.org/content/dam/oecd/en/publications/reports/2020/08/smart-cities-and-inclusive-growth_332850c0/8a4ce475-en.pdf

Okem, E. S., Nwokediegwu, Z. Q. S., Umoh, A. A., Biu, P. W., Obaedo, B. O., & Sibanda, M. (2024). Civil engineering and disaster resilience: A review of innovations in building safe and sustainable communities. *International Journal of Science and Research Archive, 11*(1), 639–650. DOI: 10.30574/ijsra.2024.11.1.0107

Omar, S., Ngadi, A., & Jebur, H. H. (2013). Machine learning techniques for anomaly detection: An overview. *International Journal of Computer Applications, 79*(2). Advance online publication. DOI: 10.5120/13715-1478

Önder, H., Sitskowska, B., Kurnaz, B., Piwczyński, D., Kolenda, M., Şen, U., Tırınk, C., & Çanga Boğa, D. (2023). Multi-Trait Single-Step Genomic Prediction for Milk Yield and Milk Components for Polish Holstein Population. *Animals (Basel), 13*(19), 3070. DOI: 10.3390/ani13193070 PMID: 37835676

Orissa International & Intralink Limited. (2025). Opportunities in Singapore. In Trade Report for the Ministry of Foreign Affairs of Republic of Estonia [Report]. https://www.vm.ee/sites/default/files/documents/2025-03/Singapore%20Strategy_Smart%20City.pdf

Özkan, Ş., & Hill, J. (2015). Implementing innovative farm management practices on dairy farms: A review of feeding systems. *Turkish Journal of Veterinary and Animal Sciences, 39*(1), 1–9. DOI: 10.3906/vet-1311-95

Özsağlam, M. Y., & Çunkaş, M. (2008). Optimizasyon Problemlerinin Çözümü için Parçaçık Sürü Optimizasyonu Algoritması. *Politeknik Dergisi, 11*(4), 299–305.

Padmanaban, S., Nasab, M. A., Samavat, T., Nasab, M. A., & Zand, M. (2023). Securing smart power grids against cyber-attacks. In Samavat, T., Zand, M., & Padmanaban, S. (Eds.), *IoT and Analytics in Renewable Energy Systems* (Vol. 1, pp. 17–36). CRC Press.

Pal, S., & Nag, A. (2022). Energy Harvesting in IoT-Enabled Flexible Sensors: Smart Sensing and Secure Access Control. In Nag, A., & Mukhopadhyay, S. C. (Eds.), *Flexible Sensors for Energy-Harvesting Applications. Smart Sensors, Measurement and Instrumentation* (Vol. 42). Springer., DOI: 10.1007/978-3-030-99600-0_9

Panduman, Y. Y. F., Sukaridhoto, S., Funabiki, N., Fajrianti, E. D., & Fang, S. (2024). A Survey of AI Techniques in IoT Applications with Use Case Investigations in the Smart Environmental Monitoring and Analytics in Real-Time IoT Platform. *Information (Basel)*, *15*(3), 153. DOI: 10.3390/info15030153

Parlar, T., & Koc, F. (2020). Toplam rasyon karışımı kullanılan bir süt sığırı işletmesinin besleme açısından değerlendirilmesi. *Erciyes Tarım ve Hayvan Bilimleri Dergisi*, *3*(1), 24–32.

Patil, V. S., Shinde, S. A., Sarwade, M. P., & Dhawale, N. M. (2023). Smart robotic soil pH measurement system. In *2023 National conference on smart agriculture technologies (NCSAT). Indian Soc Tech Educ J* (Vol. 46, pp. 22-30).

Perera, C., Toxopeus, H., Klein, S., & Merfeld, K. (2024). Enabling justice for nature-based solutions in real estate development. Nature-Based Solutions, Volume 6, 100148. Dell'Anna, F. (2025). Machine learning framework for evaluating energy performance certificate (EPC) effectiveness in real estate: A case study of Turin's private residential market. Energy Policy, Volume 198, 114407. Lenaers, I., & De Moor, L. (2023). Exploring XAI techniques for enhancing model transparency and interpretability in real estate rent prediction: A comparative study. Finance Research Letters, Volume 58, Part A, 104306. Elnagar, S. (2024). Using deep learning to enhance electronic service quality: Application to real estate websites. *Intelligent Systems with Applications*, *21*, 200330.

Pesari, V. S., & Porika, S. (2024). A deep learning framework for leakage diagnosis and time series water consumption prediction. *AIP Conference Proceedings*, 030002. DOI: 10.1063/5.0195057

Phyn. (n.d.). *Phyn home water monitor*. https://www.phyn.com

Plevris, V., & Papazafeiropoulos, G. (2024). AI in structural health monitoring for infrastructure maintenance and safety. *Infrastructures*, *9*(12), 225. DOI: 10.3390/infrastructures9120225

Potrawa, T., & Tetereva, A. (2022). How much is the view from the window worth? Machine learning-driven hedonic pricing model of the real estate market. *Journal of Business Research*, *144*, 50–65. DOI: 10.1016/j.jbusres.2022.01.027

Prasad, N., & Kulkarni, M. (2022). AI in water utilities: A review. *Journal of Water Resources Planning and Management, 148*(6), 04022051.

Público. (2025, July 26). *Ainda sem regulamento, Lisboa já travou 400 novos Alojamentos Locais desde Outubro* [Even without a regulation, Lisbon has already blocked 400 new Short-Term Rentals since October].

Pu, G., Zhang, Y., & Chou, L.-C. (2022). Estimating financial information asymmetry in real estate transactions in China - An application of two-tier frontier model. *Information Processing & Management, 59*(2), 102860. DOI: 10.1016/j.ipm.2021.102860

Rajan, P. (2024). Integrating IoT Analytics into Marketing Decision Making: A Smart Data-Driven Approach. *International Journal of Data Informatics and Intelligent Computing, 3*(1), 12–22. DOI: 10.59461/ijdiic.v3i1.92

Ramirez Lopez, L. J., & Grijalba Castro, A. I. (2020). Sustainability and resilience in smart city planning: A review. *Sustainability (Basel), 13*(1), 181. DOI: 10.3390/su13010181

Ramirez, S., Fernandez, M., & Castillo, J. (2020). Financial capabilities and renewable energy adoption in Mexican manufacturing: Implications for energy efficiency. *Journal of Cleaner Production, 267*, 122089. DOI: 10.1016/j.jclepro.2020.122089

Ramu, S. P., Boopalan, P., Pham, Q. V., Maddikunta, P. K. R., Huynh-The, T., Alazab, M., Nguyen, T. T., & Gadekallu, T. R. (2022). Federated learning enabled digital twins for smart cities: Concepts, recent advances, and future directions. *Sustainable Cities and Society, 79*, 103663. DOI: 10.1016/j.scs.2021.103663

Rao, A., Pang, M., Kim, J., Kamineni, M., Lie, W., Prasad, A. K., Landman, A., Dreyer, K., & Succi, M. D. (2023). Assessing the utility of ChatGPT throughout the entire clinical workflow: Development and usability study. *Journal of Medical Internet Research, 25*, e48659. DOI: 10.2196/48659 PMID: 37606976

Razmjoo, A., Mirjalili, S., Ahmadzadeh, S., Gandomi, A., Memon, S., Astiaso Garcia, D., Rezvani, A., & Mahlooji, M. (2022). An Investigation of the Policies and Crucial Sectors of Smart Cities Based on IoT Application. *Applied Sciences (Basel, Switzerland), 12*(5), 2672. DOI: 10.3390/app12052672

Reda Taha, M., Ayyub, B. M., Soga, K., Daghash, S., Heras Murcia, D., Moreu, F., & Soliman, E. (2021). Emerging technologies for resilient infrastructure: Conspectus and roadmap. *ASCE-ASME Journal of Risk and Uncertainty in Engineering Systems. Part A, Civil Engineering, 7*(2), 03121002. DOI: 10.1061/AJRUA6.0001134

Reis, M. J. C. S. (2025). Internet of Things and Artificial Intelligence for Secure and Sustainable green mobility: A multimodal data fusion approach to enhance efficiency and security. *Multimodal Technologies and Interaction*, *9*(5), 39. DOI: 10.3390/mti9050039

Renigier-Bilozar, M., Zrobek, S., Walacik, M., Borst, R., Grover, R., & d'Amato, M. (2022). International acceptance of automated modern tools use must-have for sustainable real estate market development. *Land Use Policy*, *113*, 105876. DOI: 10.1016/j.landusepol.2021.105876

Renneboog, L., Ter Horst, J., & Zhang, C. (2008). Socially responsible investments: Institutional aspects, performance, and investor behavior. *Journal of Banking & Finance*, *32*(9), 1723–1742. DOI: 10.1016/j.jbankfin.2007.12.039

Reynolds, L. P., Wulster-Radcliffe, M. C., Aaron, D. K., & Davis, T. A. (2015). Importance of animals in agricultural sustainability and food security. *The Journal of Nutrition*, *145*(7), 1377–1379. DOI: 10.3945/jn.115.212217 PMID: 25972529

Rezvani, S. M., Silva, M. J. F., & de Almeida, N. M. (2024). Urban resilience index for critical infrastructure: A scenario-based approach to disaster risk reduction in road networks. *Sustainability (Basel)*, *16*(10), 4143. DOI: 10.3390/su16104143

Ribeiro, M. T., Singh, S., & Guestrin, C. (2016). "Why should I trust you?": Explaining the predictions of any classifier. *Proceedings of the 22nd ACM SIGKDD International Conference on Knowledge Discovery and Data Mining*, 1135-1144.

Rojek, I., Mroziński, A., Mikołajewski, D., & Macko, M. (2024). Green Energy Management in Manufacturing Based on Demand Prediction by Artificial Intelligence—A Review. *Electronics (Basel)*, *13*(16), 3338. DOI: 10.3390/electronics13163338

Rong, R., Ma, S., Ren, N., Lin, Q., & Jia, N. (2025). Generative artificial intelligence in intelligent transportation systems: A systematic review of applications. *Frontiers of Engineering Management*, *12*(4), 1020–1036. Advance online publication. DOI: 10.1007/s42524-025-4241-9

Roy, P. (2025). AI-DRIVEN FLEET ANALYTICS: REVOLUTIONIZING MODERN FLEET MANAGEMENT. International Research Journal of Modernization in Engineering Technology and Science, 07. https://www.irjmets.com/uploadedfiles/paper//issue_3_march_2025/70285/final/fin_irjmets1745248405.pdf

S, B. (2024). *Advancements in Optimizing Smart Energy Systems Through Smart Grid Integration, Machine Learning, and IoT* (pp. 33–61). igi global. DOI: 10.4018/979-8-3693-0492-1.ch002

Sahbi, A., Alec, C., & Beust, P. (2024). Automatic Ontology Population from Textual Advertisements: LLM vs. Semantic Approach. *Procedia Computer Science*, *246*, 3083–3092. DOI: 10.1016/j.procs.2024.09.364

Sahbi, A., Alec, C., & Beust, P. (2025). Semantic vs. LLM-based approach: A case study of KOnPoTe vs. Claude for ontology population from French advertisements. *Data & Knowledge Engineering*, *156*, 102392. DOI: 10.1016/j.datak.2024.102392

Saki, S., & Soori, M. (2025). Artificial intelligence, machine learning and deep learning in advanced transportation systems, a review. *Multimodal Transportation*, *100242*. Advance online publication. DOI: 10.1016/j.multra.2025.100242

Sallang, N. C. A., Islam, M. T., Islam, M. S., & Arshad, H. (2021). A CNN-Based Smart Waste Management System Using TensorFlow Lite and LoRa-GPS Shield in Internet of Things Environment. *IEEE Access : Practical Innovations, Open Solutions*, *9*, 153560–153574. DOI: 10.1109/ACCESS.2021.3128314

Samavat, T., Nazari, M., Li, F., & Liu, Y. (2025). Optimizing stand-alone PV systems: A metaheuristic-enhanced fuzzy approach for adaptive MPPT. *China Communications*, *22*(1), 61–74. DOI: 10.23919/JCC.fa.2024-0224.202501

Sardinha, L. F. S., Fernandes, D. N., & Moura, J. (2018). Alojamento Local na Região Autónoma da Madeira - Tipologia e distribuição do Imóvel. O Perfil do Investidor [Short-Term Rentals in the Autonomous Region of Madeira - Typology and distribution of the Property. The Investor Profile]. *Revista de Economia. Empresas e Empreendedores na CPLP*, (3), 11–29.

Sariyer, G., Mangla, S. K., Sozen, M. E., Li, G., & Kazancoglu, Y. (2024). Leveraging explainable artificial intelligence in understanding public transportation usage rates for sustainable development. *Omega*, *127*, 103105. DOI: 10.1016/j.omega.2024.103105

Saxena, P., & Khanna, N. (2015). Optimization of dairy cattle feed by nonlinear programming. In *Proceedings of the 2015 2nd International Conference on Computing for Sustainable Global Development (INDIACom)* (pp. 139–144). IEEE.

Sertyesilisik, B. (2019). Circular, smart, and connected cities: A key for enhancing sustainability and resilience of cities. In *Handbook of research on implementation and deployment of IoT projects in smart cities* (pp. 19–32). IGI Global. DOI: 10.4018/978-1-5225-9199-3.ch002

Servou, E., Behrendt, F., & Horst, M. (2023). Data, AI and governance in MaaS – Leading to sustainable mobility? *Transportation Research Interdisciplinary Perspectives*, *19*, 100806. DOI: 10.1016/j.trip.2023.100806

Shah, H. (2023). Beyond Smart: How ICT Is Enabling Sustainable Cities of the Future. *Sustainability (Basel), 15*(16), 12381. DOI: 10.3390/su151612381

Shah, W. U., Younis, I., Missaoui, I., & Liu, X. (2025). Environmental transitions effect of renewable energy and fintech markets on Europe's real estate stock market. *Renewable Energy, 243*, 122603. DOI: 10.1016/j.renene.2025.122603

Shanthi, H. J., Gokulakrishnan, A., Sharma, S., Deepika, R., & Swadhi, R. (2025). Leveraging artificial intelligence for enhancing urban health: Applications, challenges, and innovations. In *Nexus of AI, climatology, and urbanism for smart cities* (pp. 275–306). IGI Global.

Sharma, V., Kumawat, P., Kumar, P., Yaminipriya, S., Chamarthi, S., & Kumar, J. (2024, December). Reinforcement Learning for Autonomous Vehicles. In *2024 1st International Conference on Advances in Computing, Communication and Networking (ICAC2N)* (pp. 1548-1553). IEEE. DOI: 10.1109/ICAC2N63387.2024.10895165

Sharma, A., Kosasih, E., Zhang, J., Brintrup, A., & Calinescu, A. (2022). Digital twins: State of the art theory and practice, challenges, and open research questions. *Journal of Industrial Information Integration, 30*, 100383. DOI: 10.1016/j.jii.2022.100383

Shi, J., Sun, J., & Zhang, Q. (2021). Multi-objectivization inspired metaheuristics for the sum-of-the-parts combinatorial optimization problems. *Applied Soft Computing, 103*, 107157. DOI: 10.1016/j.asoc.2021.107157

Shi, L., Chu, E., Anguelovski, I., Aylett, A., Debats, J., Goh, K., Schenk, T., Seto, K. C., Dodman, D., Gumenius, S., & Roberts, D. (2016). Roadmap towards justice in urban climate adaptation research. *Nature Climate Change, 6*(2), 131–137. DOI: 10.1038/nclimate2841

Shimamoto, H., & Kitazono, M. (2025). Feasibility assessment of ride-sharing system in mountainous area in Japan using a matching model. *Transportation Research Procedia, 82*, 11–23. DOI: 10.1016/j.trpro.2024.12.014

Shuford, J. (2024). Interdisciplinary Perspectives: Fusing Artificial Intelligence with Environmental Science for Sustainable Solutions. *Journal of Artificial Intelligence General Science (JAIGS) ISSN:3006-4023, 1*(1), 106–123. DOI: 10.60087/jaigs.v1i1.87

Shukla, D., & Verma, A. (2021). Modeling temporal patterns in urban water use. *Urban Infrastructure Journal, 12*(4), 81–90.

Sietchiping, R., Permezel, M. J., & Ngomsi, C. (2012). Transport and mobility in sub-Saharan African cities: An overview of practices, lessons and options for improvements. *Cities (London, England), 29*(3), 183–189. DOI: 10.1016/j.cities.2011.11.005

Singh, S. K. Dr. Nikhil Agrawal, & Dr. Surendra Yadav. (2017). Review on key elements of traffic & transportation planning approaches for a mid-sized city. *International Conference on Recent Development in Engineering Science, Humanities and Management*, 86–95. Singh, S. K., & Jain, R. (2025). Future of mobility: Potential challenges and ethical considerations. In *AIoT for Smart Transportation: Transforming the future of mobility* (pp. 235–253). IET. DOI: 10.1049/PBTR047E_ch12

Singh, A. (2023). The Current Scenario of Sustainable Planning and Development of the Indian Built Environment. In Cobîrzan, N., Muntean, R., & Felseghi, R. (Eds.), *Circular Economy Implementation for Sustainability in the Built Environment* (pp. 106–138). IGI Global Scientific Publishing., DOI: 10.4018/978-1-6684-8238-4.ch005

Singh, A. (2023). The Current Trends and Need of Infrastructure Projects in the Country India. In Popescu, C., Yu, P., & Wei, Y. (Eds.), *Achieving the Sustainable Development Goals Through Infrastructure Development* (pp. 235–261). IGI Global Scientific Publishing., DOI: 10.4018/979-8-3693-0794-6.ch010

Singh, A. (2023). The Present and Future Role of Artificial Intelligence in the Promotion of Sustainable Practices in the Construction Sector. In Kahraman, C., Sari, I. U., Oztaysi, B., Cebi, S., Cevik Onar, S., & Tolga, A. Ç. (Eds.), *Intelligent and Fuzzy Systems. INFUS 2023. Lecture Notes in Networks and Systems* (Vol. 759). Springer., DOI: 10.1007/978-3-031-39777-6_2

Singh, A. (2023). The Recent Trend of Artificial Neural Network in the Field of Civil Engineering. In Vasant, P. (Eds.), *Intelligent Computing and Optimization. ICO 2023. Lecture Notes in Networks and Systems* (Vol. 855). Springer., DOI: 10.1007/978-3-031-50158-6_32

Singh, A. (2023). The Significance of Digitalization of the Construction Sector. In Vasant, P., Weber, G. W., Marmolejo-Saucedo, J. A., Munapo, E., & Thomas, J. J. (Eds.), *Intelligent Computing & Optimization. ICO 2022. Lecture Notes in Networks and Systems* (Vol. 569). Springer., DOI: 10.1007/978-3-031-19958-5_100

Singh, A. (2025). The Evolution of Civil Engineering Field With the Emergence and Incorporation of Artificial Intelligence and Internet of Things. In Tariq, M. (Ed.), *Merging Artificial Intelligence With the Internet of Things* (pp. 431–486). IGI Global Scientific Publishing., DOI: 10.4018/979-8-3693-8547-0.ch014

Singh, S. K., Alasali, B. E., & Belbase, Y. (2025). Urban Transportation Planning: Artificial Neural Network Applications. In Syam Sundar, L., Balakrishnan, D., & Sousa, A. C. M. (Eds.), *Expert Artificial Neural Network Applications for Science and Engineering* (pp. 469–504). IGI Global., DOI: 10.4018/979-8-3693-7250-0.ch017

Singh, S. K., & Saraswat, A. (2019). Design service volume, capacity, level of service calculation and forecasting for a semi-urban city. *Revue d'Intelligence Artificielle, 33*(2), 139–143. DOI: 10.18280/ria.330209

Site, A., Vasudevan, S., Afolaranmi, S. O., Lastra, J. L. M., Nurmi, J., & Lohan, E. S. (2022). A Machine-Learning-Based Analysis of the Relationships between Loneliness Metrics and Mobility Patterns for Elderly. *Sensors (Basel), 22*(13), 4946. DOI: 10.3390/s22134946 PMID: 35808440

Soga, K., Ewais, A., Fern, J., & Park, J. (2019). Advances in geotechnical sensors and monitoring. In *Geotechnical fundamentals for addressing new world challenges* (pp. 29–65). Springer International Publishing. DOI: 10.1007/978-3-030-06249-1_2

Song, H. W., Jeon, T., Choi, W., & Oh, J. H. (2022). Recent Advances in Smart Organic Sensors for Environmental Monitoring Systems. *ACS Applied Electronic Materials, 5*(1), 77–99. DOI: 10.1021/acsaelm.2c01315

Spangler ML, Robbins KR, Bertrand JK, Macneil M, Rekaya R. (2009). Ant colony optimization as a method for strategic genotype sampling. Anim Genet. Jun;40(3):308-14.. Epub 2009 Feb 10. PMID: 19220227.DOI: 10.1111/j.1365-2052.2008.01835.x

Sugandha, S., Favaro, P., & Freestone, R. (2022). The social sustainability of smart cities: A conceptual framework. *City, Culture and Society, 29*, 100460. DOI: 10.1016/j.ccs.2022.100460

Sustainability Directory. (2025). *AI In Transportation Equity*. Retrieved from https://sustainability-directory.com/area/ai-in-transportation-equity/

Sutriadi, R. (2023). Soft infrastructure in smart sustainable cities: A literature review. *Journal of Indonesia Sustainable Development Planning, 4*(2), 195–215. DOI: 10.46456/jisdep.v4i2.428

Sutton, R. S., & Barto, A. G. (2018). *Reinforcement learning: An introduction* (2nd ed.). MIT Press.

Swadhi, R. (2025). Innovative strategies for widespread adoption in a climate-smart future: Scaling up agroforestry. In *Agroforestry for a climate-smart future* (pp. 473–496). IGI Global. DOI: 10.4018/979-8-3693-8282-0.ch016

Szpilko, D., Naharro, F. J., Gallegos, A. D. L. T., Lăzăroiu, G., & Nica, E. (2023). Artificial Intelligence in the Smart City — A Literature Review. *Engineering Management in Production and Services, 15*(4), 53–75. DOI: 10.2478/emj-2023-0028

Tao, F., Sui, F., Liu, A., Qi, Q., Zhang, M., Song, B., Guo, Z., Lu, S. C.-Y., & Nee, A. Y. (2019). Digital twin-driven product design framework. *International Journal of Production Research, 57*(12), 3935–3953. DOI: 10.1080/00207543.2018.1443229

Teece, D. J., Pisano, G., & Shuen, A. (1997). Dynamic capabilities and strategic management. *Strategic Management Journal, 18*(7), 509–533. DOI: 10.1002/(SICI)1097-0266(199708)18:7<509::AID-SMJ882>3.0.CO;2-Z

Thakur, R., Sajid, Z., & Khan, F. (2026). Artificial intelligence (ai) safety system for safe & trustworthy autonomy. *Digital Chemical Engineering, 19*, 100308. DOI: 10.1016/j.dche.2026.100308

Therias, A., & Rafiee, A. (2023). City digital twins for urban resilience. *International Journal of Digital Earth, 16*(2), 4164–4190. DOI: 10.1080/17538947.2023.2264827

Torres, L., & Silva, R. (2021). Financial literacy and stakeholder management as differentiators between sustainable and non-sustainable service firms in Mexico. *Business Strategy and the Environment, 30*(4), 1745–1759. DOI: 10.1002/bse.2712

TrafficInfraTech - Editor. (2025, June 26). Artificial intelligence in India's intelligent transportation systems: driving the next mobility revolution - TrafficInfraTech magazine. TrafficInfraTech Magazine. https://trafficinfratech.com/artificial-intelligence-in-indias-intelligent-transportation-systems-driving-the-next-mobility-revolution/

Tukur, M., Schneider, J., Househ, M., Dokoro, A. H., Ismail, U. I., Dawaki, M., & Agus, M. (2024). The Metaverse digital environments: A scoping review of the techniques, technologies, and applications. *Journal of King Saud University. Computer and Information Sciences, 36*(2), 101967. DOI: 10.1016/j.jksuci.2024.101967

Turismo de Portugal, I. P. (2025). *Registo Nacional de Alojamento Local.* RNAL. [National Registry of Local Accommodation (RNAL)]. [Data extracted in July 2025]

Turismo de Portugal, I. P. (2025). *Registo Nacional de Empreendimentos Turísticos.* RNET. [National Registry of Tourist Enterprises (RNET)]. [Data extracted in July 2025]

Turkish Feed Industrialists Association. (2019). *Karma Yem Sanayi Raporu 2019.* Poyraz Ofset Matbaacılık.

Turner, J. (2019, July 2). Lines of data: using IIoT and AI to improve predictive rail maintenance. Railway Technology. https://www.railway-technology.com/features/ai-and-iot-in-rail/

Tysiac, P., Janowski, A., & Walacik, M. (2024). UAV measurements and AI-driven algorithms fusion for real estate good governance principles support. *International Journal of Applied Earth Observation and Geoinformation*, *134*, 104229. DOI: 10.1016/j.jag.2024.104229

Uhlemann, T. H. J., Schock, C., Lehmann, C., Freiberger, S., & Steinhilper, R. (2017). The digital twin: Demonstrating the potential of real time data acquisition in production systems. *Procedia Manufacturing*, *9*, 113–120. DOI: 10.1016/j.promfg.2017.04.043

Ullah, F., Sepasgozar, S. M. E., Shirowzhan, S., & Davis, S. (2021). Modelling users' perception of the online real estate platforms in a digitally disruptive environment: An integrated KANO-SISQual approach. *Telematics and Informatics*, *63*, 101660. DOI: 10.1016/j.tele.2021.101660

Ullah, F., Sepasgozar, S. M. E., Thaheem, M. J., & Al-Turjman, F. (2021). Barriers to the digitalisation and innovation of Australian Smart Real Estate: A managerial perspective on the technology non-adoption. *Environmental Technology & Innovation*, *22*, 101527. DOI: 10.1016/j.eti.2021.101527

Ullah, F., Sepasgozar, S. M. E., Thaheem, M. J., Wnag, C. C., & Imran, M. (2021). It's all about perceptions: A DEMATEL approach to exploring user perceptions of real estate online platforms. *Ain Shams Engineering Journal*, *12*(4), 4297–4317. DOI: 10.1016/j.asej.2021.04.023

Ullo, S. L., & Sinha, G. R. (2020). Advances in Smart Environment Monitoring Systems Using IoT and Sensors. *Sensors (Basel)*, *20*(11), 3113. DOI: 10.3390/s20113113 PMID: 32486411

Unel, F. B., & Yalpir, S. (2023). Sustainable tax system design for use of mass real estate appraisal in land management. *Land Use Policy*, *131*, 106734. DOI: 10.1016/j.landusepol.2023.106734

United Nations. (2019). Shifting demographics | United Nations. https://www.un.org/en/un75/shifting-demographics

UNRIC. (2024). *Building an Accessible Future for All: AI and the Inclusion of Persons with Disabilities*. Retrieved from https://unric.org/en/building-an-accessible-future-for-all-ai-and-the-inclusion-of-persons-with-disabilities/

Vallim Filho, A. R. D. A., Farina Moraes, D., Bhering de Aguiar Vallim, M. V., Santos da Silva, L., & da Silva, L. A. (2022). A machine learning modeling framework for predictive maintenance based on equipment load cycle: An application in a real-world case. *Energies*, *15*(10), 3724. DOI: 10.3390/en15103724

Van Oorschot, J., Slootweg, M., Remme, R. P., Sprecher, B., & Van Der Voet, E. (2024). Optimizing green and gray infrastructure planning for sustainable urban development. *NPJ Urban Sustainability*, *4*(1), 41. Advance online publication. DOI: 10.1038/s42949-024-00178-5

Velasquez-Mendez, A., Lozoya-Santos, J. D. J., & Jiménez-Vargas, J. F. (2025). Validating a sustainable, smart, and circular city architecture through urban living lab experiments. *Urban Science (Basel, Switzerland)*, *9*(9), 377. DOI: 10.3390/urbansci9090377

Villain, G., Garnier, V., Sbartaï, Z. M., Derobert, X., & Balayssac, J. P. (2018). Development of a calibration methodology to improve the on-site non-destructive evaluation of concrete durability indicators. *Materials and Structures*, *51*(2), 40. DOI: 10.1617/s11527-018-1165-4

Villanueva, C., Moreno, J., & Perez, A. (2019). Environmental disclosure and firm performance in Latin America: The moderating role of financial literacy. *Corporate Social Responsibility and Environmental Management*, *26*(5), 1167–1179. DOI: 10.1002/csr.1746

Vishnu, S., Srinivasan, S., Anagnostopoulos, T., Senith, S., Ramson, S. R. J., Fan, X., Abu-Mahfouz, A. M., & Kirubaraj, A. A. (2021). IoT-Enabled Solid Waste Management in Smart Cities. *Smart Cities*, *4*(3), 1004–1017. DOI: 10.3390/smartcities4030053

Vujadinovic, V. L., Damnjanovic, A., Cakic, A., Petkovic, D. R., Prelevic, M., Pantovic, V., Stojanovic, M., Vidojevic, D., Vranjes, D., & Bodolo, I. (2024). AI-Driven Approach for Enhancing Sustainability in urban public Transportation. *Sustainability (Basel)*, *16*(17), 7763. DOI: 10.3390/su16177763

Wang, C., & Chan, T. M. (2023). Machine learning (ML) based models for predicting the ultimate strength of rectangular concrete-filled steel tube (CFST) columns under eccentric loading. *Engineering Structures*, *276*, 115392. DOI: 10.1016/j.engstruct.2022.115392

Wang, C., Shang, Q., Liu, K., & Zhang, W. (2025). Traffic congestion recognition based on convolutional neural networks in different scenarios. *Engineering Applications of Artificial Intelligence*, *148*, 110372. DOI: 10.1016/j.engappai.2025.110372

Wang, J., & Yang, H. (2023). Low carbon future of vehicle sharing, automation, and electrification: A review of modeling mobility behavior and demand. *Renewable & Sustainable Energy Reviews*, *177*, 113212. DOI: 10.1016/j.rser.2023.113212

Wang, M., Liu, J., & Chen, W. (2021). Multi-objective optimization for sustainable infrastructure systems: A review. *Renewable & Sustainable Energy Reviews, 135*, 110221.

Wang, Z. (2020). Digital twin technology. In *Industry 4.0-impact on intelligent logistics and manufacturing*. IntechOpen. DOI: 10.5772/intechopen.80974

Waqar, A., Othman, I., Almujibah, H., Khan, M. B., Alotaibi, S., & Elhassan, A. A. (2023). Factors influencing adoption of digital twin advanced technologies for smart city development: Evidence from Malaysia. *Buildings (Basel, Switzerland), 13*(3), 775. DOI: 10.3390/buildings13030775

Weber, O. (2017). Corporate sustainability and financial performance of Chinese banks. *Sustainability Accounting. Management and Policy Journal, 8*(3), 358–385. DOI: 10.1108/SAMPJ-09-2016-0066

White, G., Zink, A., Codecá, L., & Clarke, S. (2021). A digital twin smart city for citizen feedback. *Cities (London, England), 110*, 103064. DOI: 10.1016/j.cities.2020.103064

White, I., O'Hare, P., Wrenn, A., Connelly, A., & Aziz, A. (2018). Beyond the technical: A socio-technical systems perspective on the autonomous realization of sustainable urban drainage. *Journal of Flood Risk Management, 11*, S806–S818.

Wimalasena, S., Weerakoon, T. G., & Wimalasena, J. (2025). Integrated approaches to sustainable urban development: the synergy between landscape architecture and civil engineering. *Mokslas – Lietuvos ateitis / Science – Future of Lithuania, 17*. DOI: 10.3846/mla.2025.22836

Wolifson, P., Maalsen, S., & Rogers, D. (2024). Discrimination in the time of digital real estate: Illustrating a rental schema in the Australian setting. *Digital Geography and Society, 6*, 100088. DOI: 10.1016/j.diggeo.2024.100088

World Bank. (2020). *Mexico country climate and development report*. World Bank.

World Bank. (2021). *Climate-smart infrastructure: Building resilience for a changing world*. World Bank Publications.

Xiao, Z., & Zhang, H. (2025). Real estate market regulation and local government debt risk. *Finance Research Letters, 80*, 107362. DOI: 10.1016/j.frl.2025.107362

Xue, H., Zhao, K., Xia, F., Wang, G., & Shen, A. (2023). Numerical Investigation of the Ultimate Load-Carrying Capacity of Square Concrete-Filled Steel Tube Columns Considering Initial Stresses Generated during Construction. *Buildings (Basel, Switzerland), 13*(11), 2830. DOI: 10.3390/buildings13112830

Yanev, B. (2023). The supply and demand of infrastructure robustness, resilience and sustainability. *Građevinski materijali i konstrukcije, 66*(3), 157-168.

Yang, B., Lv, Z., & Wang, F. (2022). Digital twins for intelligent green buildings. *Buildings (Basel, Switzerland), 12*(6), 856. DOI: 10.3390/buildings12060856

Yan, J., Ma, C., Kang, B., Wu, X., & Liu, H. (2021). Extreme learning machine and AdaBoost-based localization using CSI and RSSI. *IEEE Communications Letters, 25*(6), 1906–1910. DOI: 10.1109/LCOMM.2021.3058420

Yan, M., Xu, P., Li, J., Zhang, X., Wei, L., Peng, S., Li, X., Zhou, W., Wu, X., Huang, X., & Guan, Y. (2024). A buried gas pipeline leakage model. *Journal of Pipeline Systems Engineering and Practice, 15*(4), 04024046. DOI: 10.1061/JPSEA2.PSENG-1638

Yao, J. F., Yang, Y., Wang, X. C., & Zhang, X. P. (2023). Systematic review of digital twin technology and applications. *Visual Computing for Industry, Biomedicine, and Art, 6*(1), 10. DOI: 10.1186/s42492-023-00137-4 PMID: 37249731

Yao, Y., He, C., Xiao, P., Gong, C., & Weng, J. (2024). AI-powered Strategies for Optimizing Waste Management in Smart Cities in Beijing. *World Journal of Innovation and Modern Technology, 7*(5), 22–29. DOI: 10.53469/wjimt.2024.07(05).02

Ye, X. W., Su, Y. H., & Han, J. P. (2014). Structural health monitoring of civil infrastructure using optical fiber sensing technology: A comprehensive review. *TheScientificWorldJournal, 2014*(1), 652329. DOI: 10.1155/2014/652329 PMID: 25133250

Yigitcanlar, T., Desouza, K., Butler, L., & Roozkhosh, F. (2020). Contributions and Risks of Artificial Intelligence (AI) in Building Smarter Cities: Insights from a Systematic Review of the Literature. *Energies, 13*(6), 1473. DOI: 10.3390/en13061473

Yin, A., Chen, X., Haitao, H., Morris, A., Yuan, Q., Ma, X., & Yang, Z. (2025). Shared e-bikes demand in urban mobility: Temporal heterogeneity, driving factors, and strategic implications. *Travel Behaviour & Society, 41*, 101075. DOI: 10.1016/j.tbs.2025.101075

Yukun, G., & Yan, Z. (2022). Discussion of Commercial Real Estate with A Short Remaining period of Land Use Rights on Evaluation Thoughts of Income Approach. *Procedia Computer Science, 199*, 103–109. DOI: 10.1016/j.procs.2022.01.013

Yu, Y., & Zhang, Q. (2025). The Value of Cleaner Waterways: Evidence from the Black-and-Odorous Water Program. *Journal of Environmental Economics and Management, 132*, 103159. DOI: 10.1016/j.jeem.2025.103159

Zameer, A.. (2023). Chlorine residual estimation using IoT and regression techniques. *Water Research Advances*, *12*, 1093–1102.

Zemmouchi-Ghomari, L. (2025). Artificial intelligence in intelligent transportation systems. Journal of Intelligent Manufacturing and Special Equipment. https://doi .org/DOI: 10.1108/JIMSE-11-2024-0035

Zhang, Y., Liu, A., Chen, B., Zhang, J., Pi, Y. L., & Bradford, M. A. (2020). Experimental and numerical study of shear connection in composite beams of steel and steel-fibre reinforced concrete. Engineering Structures, 215, 110707.

Zhang, C., Jin, Z., Feng, G., Tai, Y., Gao, R., & Li, C. (2024). Joint load-bearing behaviors and interface failure mechanism of encased coal samples under uniaxial compression. *Composite Structures*, *347*, 118480. DOI: 10.1016/j.compstruct.2024.118480

Zhang, X. (2025). Artificial intelligence in intelligent traffic signal control. *Applied and Computational Engineering*, *118*(1), 113–120. DOI: 10.54254/2755-2721/2025.20846

Zheng, Y. (2023). Community resilience and house prices: A machine learning approach. Finance Research Letters, Volume 58, Part B, 104400.

Zillow, https://www.zillowgroup.com/about-us/business/

Zou, Y., Wang, L., Sun, Z., Pan, J., Chen, M., & Wu, Y. (2023). Experimental and numerical studies of concrete-filled corrugated steel tubular column under axial compression. *Engineering Structures*, *276*, 114813. DOI: 10.1016/j.engstruct.2022.114813

About the Contributors

Sérgio António Neves Lousada holds an International PhD. in Civil Engineering – Hydraulics, at the University of Minho (UMinho), Portugal and an International PhD. in Sustainable Territorial Development at the University of Extremadura (UEx), Spain. He teaches at the University of Madeira (UMa), field of Hydraulics, Environment and Water Resources and Construction. Furthermore, he collaborates with the Research Group on Environment and Spatial Planning (MAOT) - University of Extremadura (UEx), Badajoz, Spain; VALORIZA - Research Center for the Enhancement of Endogenous Resources, Polytechnic Institute of Portalegre (IPP), Portugal; CITUR-Madeira - Centre for Tourism Research, Development and Innovation, Madeira, Portugal; CERIS - Civil Engineering Department of University of Aveiro, Aveiro, Portugal; OSEAN – Outermost Regions Sustainable Ecosystem for Entrepreneurship and Innovation, Madeira, Portugal. Moreover, he holds and International Interuniversity Master's Degree in Ports and Coasts Engineering at the University of Las Palmas de Gran Canaria (ULPGC) and at the University of La Laguna (ULL).

***

Shyam Agasthya is a Civil Engineering Professional with an inclination towards Business Management, Financial Management and Business Strategy. MBA in Finance from State University of New York at Oswego. Five Years of Industry Experience handling over 40 construction projects at different stages of completion. Fourteen years of Academic Teaching Experience. Currently working at National University of Science and Technology, College of Engineering, Muscat, Sultanate of Oman as a Faculty in Department of Civil and Environmental Engineering.

Aishwarya Arya is pursuing Ph.D. in Management, Accomplished her Bachelor's and Master's from Institute of Hotel Management and Catering

Technology, Lucknow in Hospitality Administration thereafter another Master's in Travel & Tourism Management. She started her academic career in the year 2019 from IHM Lucknow, previously worked with Babu Banarsi Das University, Lucknow as Assistant Professor. She presented and attended many National and International Conferences, seminars and webinars. Her area of specialization includes Room Division, Hospitality Management & Sales and Marketing. She has patents published in Under the office of the controller General of Patents, Designs and Trademarks Department of Industrial Policy and Promotions, Ministry of Commerce and Industry, Government of India.

Harpreet Singh Bedi, PhD, is an Associate Professor at Lovely Professional University (LPU), Phagwāra, India. He has 15 years of experience in both teaching and industry, specializing in areas such as Communication & Signal Processing, Information and Communication Technology, Cellular Communication, Satellite Communication, Photovoltaic Systems, and Algorithms. His contributions to research and innovation include over 30 published patents, 2 granted patents, and 2 registered copyrights under his name. He has also authored a book and published more than 50 research papers in renowned scientific and technical journals, showcasing his dedication to advancing knowledge in his field.

Mustafa Boğa completed his undergraduate education in Animal Science at Çukurova University, Faculty of Agriculture, between 1997-2001. He received his master's degree from the Department of Animal Science at the same university's Institute of Science between 2002-2004, with his thesis titled "Effects of yeast and Lactobacillus cultures-based probiotics on milk yield and composition in dairy cattle." He completed his doctoral studies between 2004-2008 at the same university's Institute of Science, with his thesis on "Effects of climate, preferential feeding, and yield level on feeding behavior and milk yield characteristics in dairy cattle." His academic career began as a Research Assistant in the Department of Animal Science at Çukurova University, Faculty of Agriculture, between 2002-2009. Between 2009-2015, Prof. Dr. Boğa served as an Assistant Professor in the Department of Plant and Animal Production at Niğde University's Bor Vocational School. He was promoted to Associate Professor at the same institution between 2015-2016. From 2016-2020, he continued his duties as an Associate Professor in the Department of Food Processing at Niğde University's Bor Vocational School. Since 2020, he has been continuing his academic work as a Professor in the Department of Food Processing at Niğde Ömer Halisdemir University's Bor Vocational School. Prof. Dr. Boğa's areas of expertise include animal nutrition, feed technology, ration preparation, and biotechnology. He has numerous articles published in national and international peer-reviewed journals, presentations at scientific meetings, and book chapters.

Demet Çanga Boğa completed her undergraduate education in Mathematics at Kahramanmaraş Sütçü İmam University between 2000-2003. She received her master's degree from the Department of Mathematics at the same university's Institute of Science between 2004-2006, with her thesis titled "Spectrum of Schrödinger differential equation." She completed her doctoral studies between 2013-2018 at Kahramanmaraş Sütçü İmam University's Department of Animal Science, with her thesis on "Contrast usage in comparing means." Her academic career began as a Lecturer in the Food Technology Program at Kahramanmaraş Sütçü İmam University's Bahçe Vocational School between 2007-2009. Since 2009, she has been serving as a Lecturer in the Department of Chemistry and Chemical Processing Technologies at Osmaniye Korkut Ata University's Bahçe Vocational School. Between 2014-2017, she served as the Head of the Food Processing Department at the same university. As of 2024, she is the Head of the Department of Quantitative Methods in the Faculty of Economics and Administrative Sciences, Business Administration at Niğde Ömer Halisdemir University. Assistant Professor Demet Çanga Boğa's research interests include statistical analysis, data mining, animal science, and food technology. Dr. Demet ÇANGA BOĞA conducts studies in data science and statistical modeling in the field of animal science biometrics within the livestock sector. She continues to integrate her expertise in data science and biometrics into sustainable agriculture and precision livestock applications. With her deep expertise in MARS algorithm and multivariate analysis techniques, she offers innovative solutions to sectoral problems. She has numerous articles published in national and international peer-reviewed journals, presentations at scientific meetings, and book chapters.

Parag Chatterjee received the B.E. degree in Computer Science & Engineering from the Karnatak University, Dharwad, India, in 1998 and received M.Tech. degree from IIEST Shibpur, Kolkata, India, in 2007. He has an academic experience of 20 years. Currently he is working as an Assistant Professor in the Department of CSE at Global Institute of Management and Technology, Krishnanagar, West Bengal, India. His current research interests include Data ware Housing, Data mining and Wireless sensor. He has 15 publication in national and international journals and 3 Patents.

Swathi Mirthika G. L is an Assistant Professor in the Department of Computing Technologies, School of Computing, Kattankulathur Campus, SRM Institute of Science and Technology. She completed her Ph.D. in 2025 and holds over 7 years of academic experience. She earned her Master's degree in Information Technology from Vel Tech Multi Tech Dr. RR Dr. SR Engineering College (2013) and her B.Tech in Information Technology from Sasurie College of Engineering (2010). Her research primarily focuses on medical recommender systems using Knowledge Graphs and Large Language Models (LLMs), along with related areas such as Graph Theory, Artificial Intelligence, and Deep Learning. She has presented her work in various national and international conferences, and has published research papers in reputed journals and book chapters. She also contributes to the scientific community by serving as a reviewer for multiple journals. She is also MIE, an active member of the IEI chapter, contributing to academic and professional engagements in the computing domain.

Abby Balkis Gorayah is a Chartered Quantity Surveyor (MRICS) and Senior Lecturer in Oman with over 30 years of industry and academic experience. Specialising in contract law, cost management, and project management, he has led major international projects, supervised research, and contributed to accreditation, consultancy, and professional development in construction education.

Jaishri Gothania is an academician, author, researcher, and data scientist with over 10 years of experience. She has published 01 book, 18 book chapters, and 12 research articles, presented 6 conference papers, and holds multiple patents. Her work bridges data science and innovation, contributing significantly to academia, research, and technological advancement.

M.K Harikeerthan has an experience in teaching for more than 22 years. Currently working as an Associate Professor at Dayanand Sagar Academy of Technology and Management, Bangalore, India. Skilled in testing of Highway Materials and Concrete, AutoCAD. Educational professional with a PhD in Highway Engineering from B. M. S. College of Engineering and focused in teaching Highway Technology and Transportation subjects in the undergraduate course of Civil Engineering.

J. Rengamani has got more than 30 years of teaching and research experience in the field of International Business, Logistics and Management Disciplines. The author has published more than 105 research articles in Scopus Indexed Journals, UGC Approved & UGC-CARE journals and other High Impact Factor journals. He has authored 6 books and guiding 7 Ph.D scholars. He has presented articles in many Conferences and Seminars. He has received 9 awards. Presently, he is working as the Professor of AMET Business School, AMET University, Chennai.

Rituraj Jain is currently working in the Department of Information Technology, Marwadi University, Rajkot, Gujarat, India. He has published many papers in

International Journals and Conferences. He has published 4 books. He is an Oracle Database SQL Certified Expert. He is also associated with professional bodies like the Indian Society of Technical Education (ISTE), Computer Science Teachers Association (CSTA), Academy & Industry Research Collaboration Center (AIRCC), International Association of Computer Science and Information Technology (IACSIT), International Association of Engineers (IAENG). His research interests include Cloud Computing, Machine Learning, Deep Learning, Fuzzy Logic, IoT, Genetic Algorithms, Big Data and Software Engineering.

Abhishek Kumar is working as a professor of electronics and electrical engineering at Lovely Professional University, India. With over 15 years of academic and deep-tech research experience. His primary research interests lie at the intersection of VLSI Design, low-power CMOS VLSI Circuits, and Hardware Security.

Bhavesh P is an undergraduate researcher and author affiliated with the Department of Computational Intelligence, SRM Institute of Science and Technology, Chennai, India. His research interests include artificial intelligence, machine learning, data analytics, intelligent sensing systems, and smart infrastructure applications. He has actively contributed to academic research through journal articles, conference papers, and book chapters, with a focus on applying AI-driven methodologies to real-world problems such as smart water management, IoT-enabled systems, and predictive analytics. His work emphasizes interdisciplinary approaches combining data-driven intelligence, optimization, and practical system design.

Deepa Rajesh, Vice President (Academics) and Executive Director of AMET Business School, AMET University, Chennai has rich academic, research and administrative experience. She specialized in Human Resources and System and completed her Doctorate in University of Madras, Chennai. She also serve as a Director, Human Resource Development Centre, AMET University and Nodal officer for AMET University Academic and Administrative Development Centre (AADC), Association of Indian University.

R. Swadhi serves as the Additional Coordinator, Academic and Examination Cell, at Saraswathi Institute of Medical Sciences, Hapur, India. She completed her PhD in Physics at the Academy of Maritime Education and Training (AMET), Chennai, specializing in nano biomaterials. Her academic career includes roles as Assistant Professor at Nirmala College, Coimbatore, and Lecturer and Examination Coordinator at DMI–St. Eugene University, Zambia. She has published 25 peer-reviewed research papers and contributed to multiple edited book chapters, demonstrating her active research engagement and expertise in emerging materials science.

R. Vettriselvan, an Associate Professor at AMET University, specializes in HRM and Marketing. He is acting as a mentor at Saraswathi Institute of Medical Sciences, Hapur. He served as a Review Board member of the National Council for Higher Education, Malawi; Head of the Department of Commerce and Management Studies at St. Eugene University, Zambia; and Director of Research and Publication at St. John the Baptist University, Malawi. He has published 30+ books and 100+ articles in peer-reviewed journals and edited book chapters. Under his guidance, three have completed and seven are pursuing their PhD.

Gowtham Rajendiran is currently working as an assistant professor in the Department of Computing Technologies, School of Computing, College of Engineering and Technology, SRM Institute of Science and Technology, Kattankulathur Campus, Chengalpattu District, Tamil Nadu, India. I have five years of research experience with good number of publications in reputed journals and conferences.

Nayantra Ramakrishnan is a B.Tech undergraduate in Computer Science and Engineering (Artificial Intelligence & Machine Learning) at SRM Institute of Science and Technology. He is deeply curious about deep learning and its real-world applications, with a strong interest in building intelligent systems that combine data, algorithms, and scalable software design. My academic focus includes machine learning, deep neural networks, computer vision, and AI-driven problem solving. He enjoys exploring how advanced models can be applied to domains such as healthcare, activity recognition, and intelligent recommendation systems. He is motivated to continuously learn, experiment, and translate theoretical concepts into practical, impactful solutions.

R. Ramya is a dedicated healthcare administrator with a profound commitment to enhancing public health systems and empowering communities through innovative outreach programs. With extensive experience and advanced academic qualifications in Public Administration and Hospital Management, she currently serves as the Vice Chairperson at the Saraswathi Group of Institutions in Hapur. In this role, she leads efforts to achieve excellence in healthcare delivery and medical education. Through her leadership and initiatives supported by charitable trusts, Ramya actively addresses healthcare disparities and champions community welfare, reflecting her unwavering dedication to social impact and public health advancement. Her strategic vision and compassionate approach are pivotal in creating a holistic healthcare environment that prioritizes patient care, community health, and educational excellence, ensuring the Saraswathi Group of Institutions remains a leader in medical education and public health initiatives.

Krishikaa Mathi Bharathi S S is currently an undergraduate student studying for a degree in Computer Science with a major in Artificial Intelligence and Machine Learning. She is interested in becoming a researcher with specific inclinations towards Deep Learning, Machine Learning, Computer Vision, and Generative Artificial Intelligence. Her research and academic interests are focused on building intelligent and explainable models attempting to solve problems in healthcare, automation, as well as user-centered applications. She is extremely interested in bringing advanced neural architecture with explainable AI approaches for enabling transparency and robustness in decision-making mechanisms. Focused on a mission-oriented path towards connecting theoretical research with real-world applications, she aims at contributing towards developing trustworthy and human-centric AI systems for enabling innovation as well as societal welfare.

Shivansu Sachan is presently pursuing a Doctor of Philosophy (Ph.D.) in Management and is recognized as an emerging academician, researcher, innovator, and educator in the domains of Hospitality, Tourism, and Management Studies. He completed his Master's degree from Integral University and earned his Bachelor's degree in Hospitality and Hotel Administration from Institute of Hotel Management Lucknow. With over six years of cumulative academic and professional experience, he has developed a strong interdisciplinary foundation integrating hospitality education, management research, innovation, and industry engagement. Currently serving as an Assistant Professor at Chhatrapati Shahu Ji Maharaj University, Mr. Sachan has consistently contributed toward academic excellence through teaching, curriculum enrichment, research supervision, scholarly publications, and institutional development activities. His pedagogical approach emphasizes experiential learning, research-oriented teaching methodologies, and the integration of contemporary technological advancements into hospitality and management education. He has actively participated in numerous Faculty Development Programs (FDPs), workshops, seminars, conferences, and professional training initiatives organized by reputed national and international institutions to strengthen his academic and research competencies. His research interests encompass hospitality management, sustainable tourism development, gastronomy tourism, artificial intelligence applications in tourism and hospitality, digital transformation, consumer behavior, smart tourism ecosystems, and innovation management. Through his scholarly endeavors, he has contributed significantly to emerging academic discussions surrounding sustainability, technology adoption, and transformative practices within the hospitality and tourism sector. Mr. Sachan possesses an extensive publication portfolio and has authored and co-authored more than 19 Scopus-indexed book chapters and research publications published in reputed international journals, edited books, and conference proceedings. His scholarly contributions primarily focus on interdisciplinary and contemporary themes associated with tourism innovation, sustainability practices, digital hospitality, and management transformation. In addition to his indexed publications, he has authored and edited more than 12 academic and reference books with recognized national and international publishers, thereby contributing to academic literature and higher education resources in the fields of hospitality, tourism, and management studies. Demonstrating a strong orientation toward innovation and intellectual property development, Mr. Sachan has also published more than 40 patents in diverse interdisciplinary domains. These patents reflect his research and innovation capabilities in areas including educational technologies, sustainable systems, artificial intelligence applications, tourism innovation, hospitality solutions, and management practices. His intellectual contributions signify a growing engagement with applied research and translational innovation aimed at bridging the gap between academia and industry. Beyond his teaching and research responsibilities, he has delivered numerous invited lectures, keynote sessions, and expert talks on hospitality management, tourism innovation, research methodology, and emerging technologies in higher education. He actively participates in national and international academic collaborations, editorial activities, conference committees, and scholarly networks, thereby strengthening academic discourse and interdisciplinary research engagement. Through his continuous pursuit of academic excellence, innovative research, and professional development, Mr. Shivansu Sachan has established himself as a committed scholar and educator dedicated to advancing knowledge creation, intellectual contribution, and sustainable development within the broader domains of hospitality, tourism, and management studies.

Aditya Singh ISDS, M.ASCE, AMIE, is an Independent Researcher, Editorial Review Board Member in 6 American International Journals (IJSSMET, IJSDS, IJPAE, IJESGT, IJDSST, IJRLEDM), Reviewer in 10 American, 6 Singaporean, 1 Indonesian, 1 Swiss International Journals (IJUPSC, IJCVIP, IRMJ, JTA, IJAIBM, JITR, IJSSSP, IJAIEF, IJEOE, IJITN, JDSIS, AAES, JCCE, SWT, JOPR, JCWR, IJEECS, Buildings) which includes some Scopus-indexed & WoS indexed Journals, Alumni Mentor registered by Amrita University Alumni (Amrita Vishwa Vidyapeetham, India), Think Tank Expert (Beijing Zhonghuan Electric Power Energy Big Data Research Institute), Ex-Research Mentor at Academy Innova World, India & Ex-Alumni Mentor (LPU) at present. He is a Multidisciplinary Expert recognized globally. He had completed his regular B. Tech degree in Civil Engineering from Lovely Professional University, India, in 2020, completed regular M.Tech

(Structural & Construction Eng.) in 2025 from Amrita Vishwa Vidyapeetham, while simultaneously maintaining an uninterrupted global professional career from 2020 onwards, which was built on his research journey as a scholar from late 2018 to mid 2020 during his B.Tech days. He is affiliated with Amrita Vishwa Vidyapeetham, India. This leads to his roughly 8 years of Research Experience & roughly 6 years of global professional experience. He has 40 publications under his name, in which he is a sole author in 29 of them, which is exceptionally high ratio especially in engineering. Most of his publications are Indexed in Scopus. He had also worked as a Reviewer/Editorial Review Board Member/Lead Editor for 240 times in the evaluation of manuscripts so far in international conferences, international books and international journals. His global presence as a professional could also be seen with 42 Conference Presentations in 37 Conferences in 16 countries (Thailand, Sri Lanka, Egypt, India, Cambodia, Bangladesh, Turkey, UK, Germany, Austria, Italy, Czech Republic, UAE, Jordan, Japan, Spain) as Paper Presenter & Invited Speaker/Eminent Speaker/Keynote Speaker; as well as his position as Program Committee Member, Scientific Committee Member, Session Chair in 5 International Conferences (India, UK, Canada, Czech Republic, Jordan), and Reviewer in total 11 International Conferences (including countries Thailand, Cambodia, USA, India, China, Turkey, Denmark), and Keynote Speaker in Faculty Development Programme, Editorial Advisory Board(Scopus indexed International Research Book), Co-Editor in International Research Book, etc. He had also won 2 International Awards((Thailand) & (Cambodia)) as a Reviewer in 2023 for his excellent performance as a Reviewer. He had worked on 81 positions so far in his professional career, including Journals, Conferences, Research Books of International stature. He is also a lifetime full member of ISDS (Japan) & International Association of Engineers (IAENG), and 14 IAENG Societies (ISAI, ISB, ISCS, ISDM, ISEE, ISIME, ISINDE, ISISE, ISICWS, ISME, ISOR, ISSC, ISSE, ISWN), AMIE, M.ASCE(Annual), SEI(Annual), ISDTDE (Founding Member)., ISET (Lifetime), Zhonghuan Research Institute/ EEDAT(Lifetime).

Sitesh Kumar Singh is an accomplished academic and Chartered Engineer with a Ph.D. in Civil Engineering and over ten years of diverse experience in education, research, consultancy, and educational administration. Currently serving as an Assistant Professor and Program Manager in the Department of Civil and Environmental Engineering at the National University of Science and Technology, Muscat, Sultanate of Oman, he specializes in Civil and Transportation Engineering, Project Management, and Data Analytics. Dr. Singh holds multiple advanced degrees including M.Tech, B.Tech, PGDIPR, PGDUPDL and PGDE. His career encompasses curriculum development, research supervision, quality assurance, and strategic academic planning. He has supervised over 40 postgraduate and 2 doctoral students, contributing to academic leadership through workshops, conferences, and editorial roles, and has published over 05 books, 27 book chapters, 43 research articles, and several patents.

B. Sivakumar is an Associate Professor at the Department of Computing Technologies, School of Computing, Kattankulathur Campus, SRM Institute of Science and Technology. He is having enormous teaching experience with computer engineering of around 19 years of experience at various reputed institutes. He has attended various reputed conferences across the globe and also having reasonable number of publications with peer reviewed quality journals. He is an active researcher in cutting edge technologies like Machine Learning, IoT, Cloud Computing, Artificial Intelligence and Deep learning. He is also serving as a reviewer of many reputed Journals and reviewed quality papers of peer researchers in unbiased manner. He has also completed one AICTE funded project and also actively working on filing new patents. He is also working on various international project funding agencies proposals to boost his research findings. Also taking active part in Industry-Academia collaborations in academics and research front.

Index

F

G

H

I

L

M

N

O

P

R

S

www.ingramcontent.com/pod-product-compliance
Lightning Source LLC
LaVergne TN
LVHW062330300926
849161LV00007B/1643

9798260030387